# Sciences of the Ancient Hindus: Unlocking Nature in the Pursuit of Salvation

Alok Kumar

ISBN-13: 978-1497374331

ISBN-10: 1497374332

Published by CreateSpace Independent Publishing Platform, Charleston, South Carolina, USA.

This book is dedicated
to my parents,
(Late) Ganga Saran Sarswat, father
and
(Late) Shanti Devi, mother
Who taught me the virtues of life.

# Contents

**Acknowledgements** i

**1 Introduction** 1
1.1 Some Neglected Chapters . . . 3
1.2 The Multicultural Nature of Science . . . 6
1.3 A Case for the Ancient Hindus . . . 12
1.4 Salient Features of This Book . . . 15

**2 The Cultural Context . . .** 23
2.1 Who is a Hindu? . . . 25
2.2 Traits of the Ancient Hindus . . . 26
2.3 Hindu Ethics . . . 29
2.4 Hindu Science: Indigenous or Implanted? . . . 31
2.5 Respect for Knowledge: The Role of a Guru . . . 35
2.6 *Śāstrārtha* (Debate) to Acquire Knowledge . . . 40
2.7 *Smṛti*, Information on the Internal Drive . . . 46
2.8 Discovery and Invention for *Mokṣa* . . . 58

**3 The Hindu Numerals** 61
3.1 The Decimal System . . . 62
3.1.1 The Word-Numerals . . . 63
3.2 The Place-value Notations . . . 65
3.3 Hindu Numerals to Arabic Numerals . . . 70
3.4 From *Śūnyatā* and *Neti-Neti* to Zero . . . 73
3.4.1 Epistemology of Zero . . . 76
3.5 The Binary Number System . . . 79
3.6 Diffusion of Hindu Numerals to Other Cultures . . . 86
3.6.1 Support of Pope Sylvester II . . . 96

**4 Mathematics** **101**
4.1 Arithmetic (*Aṅka-gaṇita*) . . . . . . . . . . . . . . . . . . . 102
4.1.1 The Fibonacci Sequence . . . . . . . . . . . . . . . . 105
4.1.2 The Square-root Operation . . . . . . . . . . . . . . . . 107
4.2 Algebra, Algorithm, and al-Khwārizmī . . . . . . . . . . . . . 112
4.2.1 Unknown in an Algebraic Equation . . . . . . . . . . . 113
4.2.2 Product of Two Numbers Using Squares . . . . . . . . 114
4.2.3 Loan with Complex Interest Scenario . . . . . . . . . . 114
4.2.4 Sum of a Series . . . . . . . . . . . . . . . . . . . . . 115
4.2.5 Sum of a Series with $\Sigma n^2$ and $\Sigma n^3$ . . . . . . . . . . . 116
4.2.6 Simultaneous Quadratic Equations . . . . . . . . . . . 116
4.2.7 System of Linear Equations . . . . . . . . . . . . . . . 117
4.3 Geometry . . . . . . . . . . . . . . . . . . . . . . . . . . . 118
4.3.1 Area of a Triangle . . . . . . . . . . . . . . . . . . . . 120
4.3.2 Transforming a Square into a Circle . . . . . . . . . . . 120
4.3.3 Height of a Tall Object . . . . . . . . . . . . . . . . . 122
4.3.4 The Value of $\pi$ . . . . . . . . . . . . . . . . . . . . . 124
4.4 The Pythagorean Theorem . . . . . . . . . . . . . . . . . . . 126
4.5 Trigonometry: From *Jyā* to Sine . . . . . . . . . . . . . . . 129
4.5.1 Spread of the Sine Function to Other Cultures . . . . . 133

**5 Astronomy** **137**
5.1 Heliocentric Solar System . . . . . . . . . . . . . . . . . . 146
5.1.1 Ujjain, Greenwich of the Ancient World . . . . . . . . 148
5.1.2 Diurnal Motion of the Earth . . . . . . . . . . . . . . . 150
5.2 Hindu Calendar . . . . . . . . . . . . . . . . . . . . . . . . 158
5.3 Constellations . . . . . . . . . . . . . . . . . . . . . . . . . 167
5.4 Hindu Cosmology . . . . . . . . . . . . . . . . . . . . . . . 169
5.5 Diffusion of Hindu Astronomy . . . . . . . . . . . . . . . . 174
5.5.1 Hindu Astronomy in the Middle East . . . . . . . . . . 174
5.5.2 Hindu Astronomy in Europe . . . . . . . . . . . . . . . 178
5.5.3 Hindu Astronomy in China . . . . . . . . . . . . . . . 181

**6 Physics** **185**
6.1 Space (*Ākāśa*) . . . . . . . . . . . . . . . . . . . . . . . . 185
6.2 Time . . . . . . . . . . . . . . . . . . . . . . . . . . . . . 189
6.3 Matter and Mass . . . . . . . . . . . . . . . . . . . . . . . 196
6.3.1 Conservation of matter . . . . . . . . . . . . . . . . . 200

6.4 Atom (*Paramāṇu*) . . . . . 201
6.5 Motion . . . . . 209
6.6 Gravitation and Ocean Tides . . . . . 210

**7 Chemistry** **215**
7.1 Sanskrit Names in Modern Chemistry . . . . . 217
7.2 Mining and Metallurgy . . . . . 218
7.2.1 Delhi's Iron Pillar . . . . . 223
7.3 *Wootz* or Damascus Steel . . . . . 227
7.4 *Agni-astra* and Gunpowder . . . . . 231
7.5 Fermentation . . . . . 239
7.6 *Rasāyan-śāstra* as Alchemy . . . . . 241

**8 Biology** **247**
8.1 Sacred Rivers and Mountains: Ecological Perspectives . . . . . 248
8.2 Sacred *Tulsī* and Sacred Cow . . . . . 251
8.3 Life in Plants . . . . . 255
8.4 Biomimicry . . . . . 261
8.5 Metempsychosis; Kinship of All Life Forms . . . . . 264
8.6 *Ahiṁsā* and Vegetarianism . . . . . 269
8.6.1 Vegetarianism and Global Warming . . . . . 275
8.6.2 Vegetarianism and World Hunger . . . . . 276

**9 Yoga** **279**
9.1 Some Basic Processes of Hatha Yoga . . . . . 284
9.1.1 Āsana (Posture) . . . . . 284
9.1.2 *Prāṇāyāma* (Breathing) . . . . . 285
9.1.3 *Dhyāna* (Meditation) . . . . . 289
9.2 Benefits of Yoga . . . . . 291
9.2.1 Cardiovascular . . . . . 294
9.2.2 Cancer . . . . . 295
9.2.3 Mental Illnesses and Addictions . . . . . 297
9.2.4 Pain . . . . . 299

**10 Medicine** **301**
10.1 The Biological Nature of Diseases . . . . . 304
10.1.1 Caraka . . . . . 305
10.1.2 Suśruta . . . . . 307

10.2 Food as Medicine: Curry for Cure . . . . . . . . . . . . . . . . . . 308
10.3 Doctors, Nurses, Pharmacies, and Hospitals . . . . . . . . . . . 317
10.4 Ayurveda . . . . . . . . . . . . . . . . . . . . . . . . . . . . . . 322
10.4.1 *Pañca-karma* . . . . . . . . . . . . . . . . . . . . . . . . 332
10.5 Modern Research on Ayurvedic Prescriptions . . . . . . . . . . 333
10.6 Surgery . . . . . . . . . . . . . . . . . . . . . . . . . . . . . . 336
10.6.1 Leech Therapy . . . . . . . . . . . . . . . . . . . . . . . 341
10.6.2 Plastic Surgery . . . . . . . . . . . . . . . . . . . . . . . 343
10.7 Hindu Medicine in Other World Cultures . . . . . . . . . . . . 347

**11 A World View** **353**
11.1 Impacts during the Ancient . . . . . . . . . . . . . . . . . . . . 357
11.1.1 Plotinus . . . . . . . . . . . . . . . . . . . . . . . . . . . 362
11.2 Impacts During the Modern Period . . . . . . . . . . . . . . . . 364
11.2.1 Erwin Schrödinger: the Guru of Dual Manifestations . 369
11.2.2 Emerson and Thoreau – Two Celebrated American Scholars . . . . . . . . . . . . . . . . . . . . . . . . . 372

**A The Travels of Pythagoras** **379**
A.0.3 About Pythagoras . . . . . . . . . . . . . . . . . . . . . . 383

**B Timeline of the Hindu Manuscripts** **391**

**A Glossary of Sanskrit Terms** **395**

**Bibliography** **399**

# Acknowledgements

It has been a long arduous journey for me to complete this book. The number of wordprocessors I have gone through in writing this book can easily define this challenging path: MultiMate, WordStar (2 versions), WordPerfect (4 versions), Microsoft Word (two versions), and Latex 2e. However, it has been an emotionally satisfying journey. My work on this book has made me aware of my roots, it has provided me a unique perspective of my own culture, and has allowed me to know the gaps that exist in the popular texts of the history of science. I have a mixed feeling in completing this book. I am relieved and exhilarated about completing the book. I also sense a vacuum in my heart that is similar to what parents feel when their children first leave home for higher studies.

This journey was not taken alone. I had plenty of help from people around me. First I would like to mention my parents, (Late) Ganga Saran Sarswat and (Late) Shanti Devi, who taught me that giving back to community and humanity are noble causes. My father often told me to do things that are close to my heart. He did not want me to be much concerned about the worldly successes. He strongly believed that successes of all kinds chase after the people who follow their own heart. I did face challenges in my profession as a physics teacher since ancient world history, that too non-Western history, is not a valued area of research in most physics departments. However, it was the encouragement of my parents that allowed me to keep working on this book. For this reason, I have dedicated this book to them.

Mr. Toshimitsu Saito from Japan is perhaps the most important person for this project during the eighties. I often walked with him on the beaches of Southern California and discussed a variety of ideas that are included in this book. He always encouraged me to complete this project.

Eighties was the period before the Internet revolution when information was not so easy to access. Sidney Sims, Librarian, at California State University, Long Beach was of great help to me. I also ordered a large number of books and articles from the Interlibrary Loan at California State University, Long Beach. I found the staff there quite helpful and professional.

I thought that my book was complete in 1989 and shared it with (Late) Professor D. P. Khandelwal, Founder-President, Indian Association of Physics Teachers. He read the manuscript, made his suggestions, and wanted me to publish this book. He strongly believed that this book will provide impetus to other scholars to work in this field. He even offered his services in finding a publisher. However, after reading his comments, I decided to delay its publication and kept on improving the text.

I never claimed my work on this book as a part of my scholarly activities in my annual reports as a professor. I did it to avoid the ridicules of several colleagues who do not much value such scholarship. I had to work on this project during my off hours. This meant that I had to cut my hours at home with my wife, Kiran, and my daughter, Aarti. Kiran, despite hardships, always encouraged me to complete this book. Kiran and Aarti are the silent heroes in this project.

I was also helped by Mr. Rajiv Malhotra, President, Infinity Foundation. The Infinity Foundation gave me a small grant to design a course that is partly based on the contents of this book. This grant, though quite small, legitimized my activities as a scholar in this field at a crucial stage.

Ronald A. Brown, my colleague in the physics department at Oswego, often suggested me to write what I had in my mind. These gaps between the thoughts and the written words distinguish good writers from the average ones. Whenever I needed to evolve my thoughts, I could always bounce my ideas on Ron. Even during busy hours in his teaching schedule, he always found some time for me to discuss topics contained in this book. Ron read this book carefully and gave his honest opinion. I am grateful to him for his support.

I am also thankful to Professor John Kares Smith, Professor Emeritus

Carlton Salvagin, Professor Emeritus Douglas Deal, and Michele Reed, all from SUNY Oswego, for reading this book. Their comments were quite valuable in improving the style and language used in this book.

I have received helps from a large number of friends and acquaints who helped me in a variety of ways in this project. It is difficult to name them all. However, I would like to mention John Boresma (Rochester), Anne Caraley (Oswego), Andy Groves (Oswego), Patrick Kenealy (Long Beach), (Late) Gaunce Lewis (Syracuse), Patrick Murphy (Oswego), Peter Oppewall (Grand Rapids), (Late) S. I. Salem (Long Beach), Ajeet Sethi (India), and Bruce Zellar (Oswego) for their helps.

I was advised by two international publishers to avoid the term Hindus from the title and replace it with Indians. One reviewer warned me of "the deeply contested nature of the adjective Hindu and its association with a particular kind of nationalist politics." I have ignored such comments since I want to be truthful in this book. I do not have any involvement with the national politics of India, nor do I want to push any political agenda through this book.

I have tried hard to avoid printing and scholarly mistakes. However, if some left, I would appreciate if you could bring them to my notice (alok.kumar@oswego.edu). If you like the book, the credit goes to the people mentioned above. I am responsible for the errors.

Alok Kumar

2014

# Chapter 1

# Introduction

"India has contributed much to science and technology in history. Can you give me a few examples of her contributions?" I decided to commence my lecture with this question to find out the prior knowledge of my audience in a junior-level science lecture for honors students at the State University of New York, Oswego. I was in this classroom as a guest speaker to share India's contributions to science and technology. Instead of an answer, some students perplexedly started looking at each other while the others started whispering among themselves. A pin-drop silence soon prevailed in the class. Not one person answered my question. I decided to allow my audience some extra time to think about this question. Still, no response. I felt that either the students did not know the scientific contributions of India, or they were reluctant to interact with me. I decided to ask another question: "What are some of the contributions of China to science and technology?" Three students, in a class of 24, offered answers to this question, mentioning gunpowder and paper. To allow the students to think more on the issue, I broadened my question and asked: "What are the non-Western contributions to science and technology in our history?" Only one more semi-accurate contribution surfaced after several replies: the so-called Arabic numerals from Arabia. I immediately realized that I should discard my prepared lecture notes and teach in an interactive style about India's contributions to science and technology – contributions that included Hindu-Arabic numerals, Damascus steel, Kaṇāda's atomic theory,[1] trigonometric functions, yoga, ayurveda, and the so-called plastic surgery, to name a few.

[1] popularly known as Kanada

Most of these topics were new to my audience. All the students had excelled in their studies and, yet, their education left them ignorant about the true history of science and its multicultural contributions. At the conclusion of my lecture, some students left the room while the others stayed with me to continue a conversation about multiculturalism in science. Several students shared their frustrations with me on the existing educational system which ignores the issues of human diversity and multiculturalism. I was delighted to observe that just one lecture could make substantial changes in the minds of several students. However, when some students asked me for an introductory book on this topic, I could not suggest one that was up-to-date and relevant to this group. Since then, I have worked tirelessly to fill this gap. I only hope that this book serves as an introduction for the future generations of students who want to learn more about the contributions of the Indus Valley civilization to science.

With the revolutions in transportation and telecommunications technologies along with the free market global economy, a large number of metropolitan cities around the world are becoming meeting grounds for different races and cultures. This new reality brings a new set of problems as racial or cultural relations turn sour between different communities. To resolve these issues, politicians as well as religious and social leaders suggest that we exercise "restrain" and abandon our hostile emotions toward others. Governing authorities legislate new laws to safeguard minorities or underprivileged populations. However, the idea of "tolerance of diversity" has met with little success. The relations between Afro-American and Anglo-Saxon populations in the United States of America, neo-Nazis and the immigrant population in Germany, Hindus and Muslims in India, colored and white populations in South Africa, Arabs and Israelis in the Middle East have not improved to meet our expectations. People simply do not tolerate others when they do not "value" them. Tolerance interpreted as "enduring differences" is inherently weak and fragile. How many people would be happy and stay in a marriage where their spouse simply tolerated them? Probably, not too many.

A more solid foundation for the stability of a diverse society demands a general respect and appreciation across the constituent boundaries of race, culture, and gender. Most problems related to human diversity and cultural diversity primarily stem from misinformation, misunderstanding, and ignorance. Many of these problems can be avoided with proper education and communication. For this reason, knowledge is considered as the soul of a democracy.

## 1.1 Some Neglected Chapters in the History of Science

Many conventional science textbooks consider that scientific rational thinking originated with the Greeks around the seventh century B.C., and flourished there for about 800 years. Greek philosopher-scientists such as Thales (636 - 546 B.C.), Pythagoras (562 - 500 B.C.), Democritus (460 - 370 B.C.), Hippocrates (460 - 360 B.C.), Plato (427 - 347 B.C.), Aristotle (386 - 322 B.C.), and Archimedes (287 - 212 B.C.) are considered as the main originators of most basic ideas in science. According to these texts, Europeans, after the fourteenth century, reacquainted themselves with the scientific tradition of the Greeks which led to the European Renaissance. Although some textbooks mention the contributions of the Arabs, due to their presence in Medieval Europe, the overwhelming majority of conventional science texts leave their readers with the impression that the Greeks laid the foundations of science and that experimental scientific and technological developments began in Europe about four centuries ago.

The intermediate period between these two disjointed stages is generally called *the Dark Ages* (475 - 1000 A.D.) or *the Middle Ages* (475 A.D. to the Renaissance). The term, *Dark Ages*, defines the lack of intellectual and scientific activities among the Europeans. The contributions of other non-European cultures have generally been ignored and science is popularly perceived as the exclusive product of European civilization.[2]

Greece is located in southern Europe and was certainly a center of learning. The Greeks did make significant contributions to natural philosophy from the seventh century B.C. until the second century A.D. The land area of Greece (50,944 square miles or 131,944 square kilometers) today is about the size of the states of Florida or Wisconsin in the United States, with a population of about 10 million. Ancient Greece included settlements in Asia

[2]Kumar, 1994; Kumar and Brown, 1999; and McDougall, 1999.

Minor and Italy. If we include the nearby regions and exclude the ancient civilizations of Egypt and Babylon in this count, the total area of ancient Greece would still not be more than severalfold of its present size. Ancient Greece was often politically divided among its islands and suffered from foreign invasions. Yet, it is impressive to notice that such a small nation has received so much glory in human history, unparalleled to any other nation.

Greece played a crucial role in the survival and propagation of Christianity. The first popular version of the New Testament in the *Bible* was written by the Greeks, or Greek-educated apostles, in the Greek language. The first popular version of the Old Testament of the *Bible* was also written in the Greek language by 72 Jewish rabbis who were called to Alexandria to translate an earlier Hebrew Version of the Old Testament.[3] This does make Greece a champion in the eyes of the Christian world.

If the conventional science texts are correct, what prevented the Chinese, Indians, Mesoamericans, and Egyptians from contributing to science and technology? Also, why were the intellectual endeavors of humans centered in two nearby geographical locations and not elsewhere?

The conventional science textbooks are mostly silent about the interactions and cultural exchanges of the ancient Greeks with the ancient civilizations of Egypt, Persia, Babylon, and India. These textbooks are also mostly silent on the interactions and exchanges of the Europeans with other civilizations that led to the European Renaissance. The visits of noted Greek philosophers–Thales, Pythagoras, Democritus, and Plato–to Egypt, Babylon, or Persia are clearly documented in the Greek literature. However, these travels are paid minimal attention by scholars and are rarely discussed in most books. Why did these scholars visit foreign lands? Did they go there to teach or to learn? At what age did they travel to other countries?[4] Were they established as natural philosophers in Greece before they left? What

[3] Anderson, 1972, p. 108.

[4] This information can allow us to know the purpose of the trip. If young, these scholars may have gone for the purpose of learning. Indeed, this was the case. Most Greek philosophers went abroad for learning. Pythagoras, Plato, Democritus, and Thales became famous in Greece only after their visits to foreign lands. Though this topic is not central to this book, travels of Pythagoras are included in Appendix A to demonstrate this point.

kind of reception did they receive from other scholars and the general population in Greece after they came back? The answers to such questions are important in determining the inheritance we owe to Greece.

Nations reap material benefits from their scientific and technological growth. Due to the role of science and technology in economic and military successes, several nations distorted facts to create a false image in the eyes of their citizens and other nations. Prime examples include the propagandist books written during the Cold War period and the Nazi period. This intellectual piracy is not a modern phenomenon; even during the Renaissance, deliberate efforts were made by scholars to give no credit to the original authors in their translations and to portray themselves as the original authors.[5]

England, France, Spain, Portugal, and the Netherlands controlled most of the world during the eighteenth century. It took them several centuries of effort to reach this stage. Portugal, England, France, and Spain set their sights on India as early as the sixteenth or seventeenth century. England, France, and Portugal succeeded eventually in their intentions by capturing various parts of India while Spain failed due to the "lost trip" of Columbus. This lost trip to India is also well known in history as it led to the discovery of the Americas by European nations.[6] The domination of England, France and Portugal over India was an outcome of well-organized socio-political policies and military invasions. Popular literature was produced by European nations to support their dream of domination. On the one hand, these works educated Europeans about these ethnic cultures,[7] while, on the other hand, they were also used to demonize and subjugate these civilizations. In this literature, Egypt, India, Persia, and China entered the scientific age only through their interactions with the Europeans. Thus, the histories of Asians, Africans, and other indigenous peoples often appear only after their encounter with the Europeans.[8]

---

[5]Høyrup, 1999; Sardar, 1989, p. 10.

[6]Christopher Columbus planned a trip to India. However, he landed in America because he used an incorrect, smaller, value for the circumference of the earth. Columbus assumed the natives of Americas to be Indians. For this reason, the Native Americans were called Indians till the mid-twentieth century.

[7]For example, Roger Bacon (c. 1214 - 1294) wrote his book, *The Opus Majus*, under the instruction of Pope Clement IV. The main purpose of the book was to improve the training of missionaries to Christianize these ethnic lands.

[8]This issue was researched extensively during the end of the twentieth century. How-

The omissions of contributions of the non-Western culture in the history of science are "deeply unjust to other civilizations. And unjust here means both untrue and unfriendly, two cardinal sins which mankind cannot commit with impunity," wrote Joseph Needham.[9] He noticed such omissions in his studies of the history of Chinese science and tried to correct the situation by compiling a monumental book, *Science and Civilization in China.*[10] Needham's book is a milestone in our understanding of the contributions of China to the modern world, and sets an example of how the efforts of a single person can drastically change our knowledge of the history of science.

## 1.2 The Multicultural Nature of Science

The regular appearance of the sun, the moon, and the stars awakened a natural desire for understanding in most ancient cultures. They speculated on what these objects were, how and why they followed their regular motions. They wondered about the phases of the moon and the relation of oceanic tides of the sun and moon, and pondered over the influence of heavenly objects on human life. The relation of the yearly seasons to the sun's apparent northward or southward motions and occurrences of solar and lunar eclipses needed satisfactory explanations. Most cultures recognized that some astronomical objects (now recognized as planets) systematically changed their positions with respect to the fixed background of stars, which provoked yet further interest in heavenly phenomena. These are some of the reasons why nearly all cultures developed their own calendars that were based upon astronomical observations. The Islamic calendar is based on the moon, the Greek, Roman, and Mayan calendars are designed on the apparent motion on the sun, the Hindu, Babylonian and Jewish calendars consider the motion of the moon as well as of the sun. Not all calendars focused on the moon and the sun. Mayans had a calendar that relied on the motion of Venus while the Egyptians had their calendar based on the appearance of Sirius, a star. These calendars served a variety of social needs and functions: administrative func-

ever, considerably more research is needed to better understand the contributions of ethnic civilizations. For more information, consult Baber, 1996; Goonatilake, 1984, 1992; Headrick, 1981, Perrett, 1999; Petitjean, 1992; and Said, 1978 and 1993.

[9]Needham in Nakayama and Sivin, 1973, p. 1.

[10]Needham, 1954 - 99.

tions such as tax collections; agricultural functions such as deciding the times for planting, harvesting, and potential floods; and religious functions such as determining the proper times for fasting and prayer. The ancient people in most cultures formulated stories to preserve their knowledge of these natural phenomena. Some cultures preserved these stories in the form of mythology. For example, in India, the moon became *Candā-māmā* (or Chanda-mama, meaning uncle moon) and the big dipper became *sapt-ṛṣi* (seven sages) for children.

Humans have wondered about heavenly phenomena since the dawn of creation. However, human curiosity is not restricted to heavenly objects. People have also contemplated the limits of the world (space), the growth and decay of all living organisms (biology and medicine), the alternation of day and night (time), the properties of biological and physical materials, and the connections and purpose of all life forms (biology).

Although science has flourished in the West and this fact is universally accepted by most historians, the West is not the only fertile ground for science. Science prospered at various times in other civilizations too. While feudalism was forcing Europe into the *Dark Ages* (about 476 - 1000 A.D.), scholars in the Middle East were involved in vigorous intellectual activities which brought together the learning of ancient Greece, China, India, and the Middle East.[11]

The pioneering work of Joseph Needham in the 1960s pointed out the gaps in global science history in relation to the contributions from China. In response, some historians have tried to incorporate Chinese contributions to science and technology in their science textbooks.[12] Citations about Arab contributions, though scanty, were already present in the science texts. A significant number of articles and books have been published in the last 25 to 30 years, including an *Encyclopedia of the History of Arabic Science*.[13] A recent surge in scholarship on Islamic sciences, particularly in the wake of 9/11, has added much new knowledge. The scholarship on the Indian, Egyptian, and Mayan civilizations is still incomplete. We need another Joseph

[11] O'Leary, 1949; Rashed, 1996; Saggs, 1989; and Selin, 1997.

[12] Cobb and Goldwhite, 1995; Katz, 1993; and Ronan, 1982.

[13] Hogendijk and Sabra, 2003; Kennedy, 1970; King, 1993; Kunitzsch, 1989 and 1983; Rashed, 1996; Saliba, 1994; Samsó, 1994; Sardar, 1989; Selin, 1997 and 2000.

Needham(s) to revive the knowledge of these civilizations.

The failure to mention multicultural contributions in present-day science education has been noted by many scholars.[14] For example, the Greek philosopher Leucippus (born ca. 490 B.C.) or his disciple Democritus (born 470 B.C.) are generally credited for being the originators of atomism.[15] The understanding of the atom has played an important role in human history and much has been written on the subject. However, the contemporaneous Indian philosopher Kaṇāda, who lived sometime between the sixth to tenth centuries B.C., has not been acknowledged in even one-percent of the science books that mention the work of Democritus. Henry Margenau, a noted philosopher-physicist, pointed to this gap in the history of science, and wrote, "But the most remarkable feature . . . which I have never seen in American textbooks on the history of science is the atomic theory of philosopher Kanada [Kaṇāda]."[16] This is not an American issue, as listed by Henry Margenau; it is an academic issue that is international in scope.

It is important to understand that the inclusion of multicultural contributions is not due to a political agenda of bringing equality to all cultures or races. This inclusion is necessary to demonstrate the true nature of science. Needless to say, there still remains a minority of historians and scholars who feel that the Greeks were perhaps the only ones in antiquity to discover science and technology due to their rational thinking.[17]

Let us illustrate the multicultural nature of science with some specific examples from various cultures. Since the third millennium before Christ, the Valley of the Nile supported a flourishing civilization for many centuries. Egypt was a leading center in natural philosophy and mathematics, prompting many Greek philosophers, including Thales, Pythagoras, Plato, and Democritus to visit Egypt for learning. These philosophers have acknowledged their gratitude to Egypt, as we know from subsequent writings. The Greek physician Hippocrates (ca. 460 - 370 B.C.), who is considered to be the fa-

[14]Gill and Levidow, 1987; Harding, 1991, 1994; Needham, 1954 - 99; Ogawa, 1995; Rashed, 1996; Teresi, 2002; and Van Sertima, 1992. This knowledge is yet to be incorporated appropriately in introductory science textbooks.

[15]Asimov, 1982, p. 11 - 12; Singer, 1990, p. 20 - 21.

[16]Margenau, 1978, p. XXX.

[17]Cromer, 1993; Grombrich, 1998.

ther of modern medicine, and the famous Roman anatomist Galen (130 - 200 A.D.), acknowledged Egypt as the source for their own knowledge.

About 150 biologists, mineralogists, linguists, mathematicians, and chemists accompanied Napoleon Bonaparte (1769 - 1821 A.D.) in his conquest of Egypt. This list included mathematician Jean Baptiste Joseph Fourier, mineralogist Déodat Guy Grater de Dolomieu, and botanical artist Pierre Joseph Redouté.[18] These scientists did not visit Egypt for teaching purposes; they went there for learning. They wanted to know more about Egyptian natural resources and learn about Egyptian science and technology. Their efforts led to the discovery of the Rosetta Stone, along with the eventual deciphering of the Egyptian hieroglyphic language. Later, Napoleon went back to France and formed the Institut d'Égypte in Paris. We owe a great deal of our current knowledge about ancient Egypt to the scholarship at this institute.

The *Edwin Smith Surgical Papyrus*, the *Ebers Papyrus*, and the *Kahun Papyrus*, all from Egypt, are some of the earliest manuscripts on medicine.[19] These manuscripts provide information on the treatment of various diseases. Such remedies include surgeries and medicines prepared with complex combinations of plant, mineral, and animal products. The *Ebers Papyrus* has descriptions and remedies for about 700 diseases. The *Edwin Smith Surgical Papyrus* describes procedures for complex surgeries, including brain surgery. This papyrus deals with the perennial question for the severely ill patients: should we continue to treat them to prolong their life? This remains a quandary even for modern societies. The *Kahun Papyrus* has remedies for diseases specifically related to females and animals. This is highly unusual in western history. Similarly, the *Rhind Papyrus* and the *Moscow Papyrus* are some of the earliest manuscripts on Egyptian mathematics that describe the use of mathematical operations to deal with everyday problems.

The Abbasid dynasty in Baghdad (750 - 1258 A.D.) and the Umayyad dynasty in Damascus (650 - 750 A.D.) lured scholars from Persia, Egypt, Greece, and India to their kingdoms by providing them with necessary subsistence and respect. Scholars from nearby kingdoms went to these regions and enriched Islam and both dynasties. In addition to the Greek manuscripts,

---

[18]For more information, read Brier, 1999.

[19]Allen, 2005; Bryan, 1931; Ebeid, 1999; Estes, 1989; and Nunn, 1996.

Indian, Egyptian, and Persian manuscripts were also translated into the Arabic language. The *Bayt al-Ḥikma* (House of Wisdom) was established in 832 A.D. with the sponsorship of al-Ma'mun. Several Indian, Persian, and Greek manuscripts were translated into the Arabic language at *Bayt al-Ḥikma*.[20] This intellectual center became a model for the rulers in the Middle East and Europe, a place where scholars could pursue their academic projects without concern about their livelihood. Later, Toledo, Cordoba, and Seville, all in Spain, became centers of learning where scientific knowledge of the Arabs, Greeks, Egyptians, and Indians was preserved and nurtured. A large number of manuscripts were written in the Arabic language, the official language of Muslim Spain. Through the Latin translations of these manuscripts, the scientific knowledge of the Arabs was introduced into Europe between the 8th and 13th centuries.[21]

Muslims prostrate themselves in prayer five times a day facing Kaaba, the holiest mosque in the city of Mecca, in Saudi Arabia. *Qibla* is the term used to define the direction to the holy mosque. Muslims all over the world must know *qibla* for their prayers (*adhān*). Mosques were accordingly designed so the worshipers could go there and pray in the direction of Kaaba.[22] Knowing this direction is also important for human burials and the slaughter of animals to have *halāl* meat. Determining *qibla* on a spherical coordinate system was one of the most advanced problems in spherical astronomy faced by the medieval astronomers and mathematicians, and the trigonometric solution found was one of great sophistication.[23] This required a sound knowledge of arithmetic, algebra, spherical trigonometry, celestial mapping, latitude, and longitude. As a result of these religious necessities, Islamic scholars became proficient in astronomical and mathematical sciences.

Nicolaus Copernicus (1473 - 1543 A.D.), a Polish astronomer, in proposing his heliocentric solar system model, borrowed ideas and techniques from the developments achieved earlier in the Middle East. The Arab astronomers previously suspected that the geocentric model for our solar system, sup-

[20] Sourdel, 1960.

[21] Rashed, 1996; Salem and Kumar, 1991.

[22] Kaaba is a cubical building in Mecca containing a sacred black stone. The name is derived from the Arabic word *Ka'abah*, meaning square building.

[23] Hairetdinova, 1986; Ilyas, 1989; Kennedy, 1970; King, 1993; and Samsó, 1994.

ported by Aristotle and Ptolemy, was invalid.[24] It was this suspicion that led Copernicus to come up with a heliocentric model of our solar system (See Chapter 5).

The Islamic influence on science is evident from the Arabic terms that are commonly used. For example, the words alcohol, Aldebaran, algebra, almanac, alkali, algorithm, Altair, azimuth, Betelgeuse, calendar, Denab, magazine, monsoon, nadir, ream, and zenith are derived from the Arabic language.[25] These words have become a part of the Western heritage and are listed in most English dictionaries.

Prior to 1960s, westerners credited Roger Bacon (1214 - 1292 A.D.) with the invention of gunpowder. After the monumental work of Joseph Needham, the credit is given to the Chinese. Gunpowder was accidentally discovered in China when the Taoists (or Daoists) worked on saltpeter to prepare the elixir for life. The main ingredient of gunpowder is commonly called saltpeter (chemical name, potassium nitrate), which is found as a natural deposit in sanitation pipes under tropical climatic conditions. Using a combination of saltpeter, sulfur, and charcoal, the Chinese made their explosive devices that were originally used mostly for festivals and other rituals, and not for military purposes. When the Arabs learned about gunpowder, they called it Chinese snow, due to the bright white color of saltpeter.[26]

The *Nine Chapters on the Mathematical Art* can easily be labeled as the Chinese counterpart of Euclid's *Elements*. With 246 problems in nine chapters, some of which date back to the Qin dynasty (221 - 207 B.C.), it deals with land measurements, manufacturing and agriculture issues, square roots, cube roots, the volumes of various shapes, and solutions to linear equations.[27]

Acupuncture was used for anesthesia in surgery and in pain relieving treatments from the earliest period in China. It was also used to regulate the flow of energy, the *chi* (or *qi*), in humans and animals. There are 14 meridians (or channels) through which *chi* moves in the human body and

---

[24]Saliba, 1994; Rashed, 1996; and Teresi, 2002.

[25]Allen, 1963; Kunitzsch, 1983 and 1989; Ridpath, 1988; and Salem and Kumar, 1991.

[26]Crosby, 2002; Needham, 1981. The possibility of an alternative origin of gunpowder is also provided in Chapter 7.

[27]Crossley and Lun, 1999.

acupuncturists manipulate the flow of *chi* in these channels by inserting slender and sterile needles at specific points.[28] Using acupuncture, long and complicated heart and brain surgeries are performed using minimal anesthesia and pain management is done quite effectively.

A multicultural approach to science humanizes the discipline by demonstrating that science arises in many cultures out of human necessities, and from universal and basic urges of curiosity concerning the natural world in which we live. Science thereby becomes a normal and meaningful activity, accessible to all, and an essential part of every person's education. Science education can allow us to learn our own heritage of ideas and can serve as a glue for our diversified society.

Science does not belong to any one particular culture or gender; it belongs to all who want to unfold the mysteries of nature. Modern science certainly did not spring into a completely evolved form suddenly with the Renaissance in Europe. Influences came from various parts of the world like streams from many different sources join to form a river.

## 1.3 A Case for the Ancient Hindus

What would you say about the people or culture who gave us the place-value system of numerals, the numeral zero, the trigonometric function "sine" and several trigonometric formulae, and set standards for mass, length, and time? What about those who developed a sophisticated system of medicine with its mind-body approach known as Ayurveda, detailed anatomical and surgical knowledge of the human body, metallurgical methods of extraction and purification of metals including the so-called Damascus blade, chemical techniques to transform compounds, knowledge of various constellations and planetary motions that was good enough to assign motion to the Earth in the fifth century A.D., and the science of self improvement (yoga)? Should we credit them and provide the details about them in science texts?

The present place-value notation system (base 10) that is used to represent numbers, is Hindu in origin. The place-value notation is a system where, for example, eleven is written as 11. The one on the left, due to its second

[28] Needham, 1981; Sivin, 1987; Stux *et al*, 2003; and Tan, Tan, and Veith, 1973.

place, starting from the right, is equal to ten. The Greeks, the Romans, and the Egyptians did not use a place-value notational system. In their systems, eleven was written as ten and one (for example, XI in the Roman system). The Hindu place-value notation system simplified mathematical calculations.

Our numerals are sometimes called Arabic numerals. However, these numerals were not invented by the Arabs, they simply introduced them to the Europeans. For example, Nicolaus Copernicus (1473 - 1543 A.D.) in his book, *On the Revolutions*, advocated the heliocentric model of the solar system, and used the Arabic numerals in computation. He noted their usefulness relative to the Roman or Greek numerals. Copernicus was not the first one to use Hindu numerals. Several hundred years before him, Leonardo Fibonacci (1170 - 1250) wrote a book *Liber Abaci* and introduced the Hindu methods of numeration and computation to the Western world (See Chapter 3).

Trigonometry deals with specific functions of angles and their applications in geometry. It unites the disciplines of architecture, mathematics, physics, and astronomy. With the work of Āryabhaṭa I (ca. 500 A.D.), trigonometry began to assume its modern form. He used the half chord of an arc and the radius of a circle to define the sine of an angle.[29] The origin of the word *sine* can be traced to the Sanskrit language (See Chapter 4).

The ancient Hindus used a complex calendar that used the sun and the moon in defining day, month, and year. While days and months were defined by the moon, the year was defined by the sun. Regarding the Earth's motion, Āryabhaṭa I suggested about one millennium before Copernicus a theory in which the Earth was in axial rotation. All stars, but not planets, were at rest in this theory. Āryabhaṭa I's hypothesis of the Earth's rotational motion is clearly explained by the analogy of a boatman who observes objects on the shore moving backward. (See Chapter 5) He showed a possible explanation of the sun's apparent motion by proposing the axial motion of the Earth, incorporating a splendid use of the concept of relative motion many centuries before its more formal discussion by the noted Parisian scholar Nicholas Oresme in the fourteenth century.[30] Interestingly, Copernicus used

[29] Clark, 1930; Kumar, 1994.

[30] Kumar and Brown, 1999.

more or less the same analogy of a boatman to explain the apparent motion of the sun in his book, *On the Revolutions* (Book 1, Chapter 8).

The ancient Hindus systematically studied the physiological causes of various diseases and tried to find their cures. They became proficient in medicine and their expertise was sought after by the Arabs and the Romans during the ancient and early medieval periods. Complex surgeries related to the eyes, ears, and throat were performed by Hindu surgeons. Skin grafting (the so-called plastic surgery) from one body part to another was practiced. Suśruta, an ancient Hindu surgeon, divided surgery into incision, excision, puncturing, exploration, extraction, evacuation, and suturing. Painstaking details were given about the design of the surgical tools. The manuscripts of ayurveda, the Hindu system of medicine, are voluminous – *Caraka-Saṁhitā* alone is three times the size of the entire surviving medical literature of ancient Greece.[31] (See Chapter 10)

Yoga is a *darśana* (philosophy) to bring the body and the mind in unison. This art and science of yoga reached a milestone with the work of Patañjali. Yoga evolved with careful and systematic observations and experimentations with various life forms in nature. A large number of *āsana* (postures) were designed after careful observation of dogs, cats, peacocks, crocodiles, and other birds and animals. The ancient Hindus must have experimented with the human body, and their observation of animals, along with their knowledge of human anatomy, to design *āsana* to improve health. The discipline of yoga is an example of biomimicry, a term coined only recently. Patañjali's *Yoga-Sūtra* is the foremost book on yoga.

The ancient Hindus wondered about where the basic building blocks of the universe (earth, water, fire, space, and aether) were concocted. Where was the furnace (place) that created these building blocks? What were the raw ingredients? These investigations led them to define standards for physical measurements of space, mass, and time (See Chapter 6), and a theory of creation. This also led them to the concept of atom, as expounded by Kaṇāda in his *Vaiśeṣika-sūtra*.

Sanskrit is the language of most ancient Hindu scriptures. The *Vedas*,

[31]Chattopadhyaya, 1977, p. 20.

*Upaniṣads*, *Mahābhārata*, *Rāmāyaṇa*, and *Purāṇas* were all written in Sanskrit. This language provides the prime source of information about the ancient Hindus. It is the Sanskrit language that sits at the top of the Indo-European languages that include Latin, Greek, German and other European languages. According to linguists, Sanskrit plays an important role in the understanding of human civilization. "As soon as Sanskrit stepped into the midst of these languages [Latin and Greek], there came light and warmth and mutual recognition," writes Max Müller,[32] a linguist and Indologist of the nineteenth century, in his analysis of ancient languages. "They all [other languages] ceased to be strangers, and each fell of its own accord into its right place. Sanskrit was the eldest sister of them all, and could tell of many things which the other members of the family had quite forgotten."[33]

We have a similar situation in the sciences. The inclusion of Hindu science provides a context to Arabic and European sciences. It allows us to understand the growth of science as continuum with infinite mutations and not a sudden eruption at one time or place. A more coherent picture of the history of science emerges in this process.

## 1.4 Salient Features of This Book

This book provides information on all topics that are listed in Section 1.3. It explains the religious, social and cultural contexts that allowed the invention of key concepts in sciences. For example, the religious philosophy of the ancient Hindus may have played a crucial role in the invention of zero . (Chapter 3) The ancient Hindus tried to explain the nature of God by depriving Him of all attributes (*Nirguṇa-swarūpa*, *nirguṇa-rūpa*, or *amūrta*). This religious and philosophical approach led to the invention of zero in mathematics.

"Why do they call it by this name?" This epistemological question is often asked by students. Their curiosity needs a satisfactory answer to enhance further learning and retention. This book will explain the epistemology and origins of various words and concepts that are commonly used in science texts. The origin of the so-called Arabic numerals, zero, the trigonometric function "sine", algebra, and algorithm are explained in this book. We will

[32]Müller,1883, p. 40.
[33]Müller, 1883, p. 40.

also elaborate on gold, silver, and the so-called Damascus steel. A knowledge of these developments takes us from one culture to another and allows us to understand the multicultural nature of science.

Biomimicry is a tool of scientific investigation where nature's actions are carefully observed and reproduced in solving human problems. Plants and animals demonstrate a wide variety of behaviors to cope with their surroundings. Such studies have led scientists to find new medicines, improved energy production devices, better paints, stronger fibers, more nutritious foods, and faster computers. However, this tool is nothing new. The ancient Hindus also carefully observed the natural world around them and invented yoga as a science of self-improvement using biomimicry. (See Chapter 2) By observing animals, they came up with herbal cures for various diseases. As explained by Suśruta in *Suśruta-Saṁhitā*, tools for surgeons and physicians were designed by observing the hunting and eating behaviors of animals. (See Chapter 10)

Science evolves out of human necessities. Such necessities may be spiritual, physical, or social in nature. Religious, social and cultural traits play important roles in the growth of science. There are some common traits which may or may not simultaneously exist in all cultures: respect for intellectual activities, respect for social and personal ethics, and desire for economic prosperity. In Chapter 2, readers will learn the role of the social and cultural matrices that allowed the growth of science among the ancient Hindus. They will also learn the role of *śāstrārtha* (debate or discussion) in social and religious matters: the way spouses chose their brides (*svayaṁbara*), and the way conflicting views of scholars or social disputes were settled.

Orality was promoted among the ancient Hindus by design. It became a norm to produce literature that was easy to memorize. Hindu literature is mostly written as poetry or stories that are rich in similes and metaphors. Therefore, despite the destruction of libraries in the Indian peninsula after the Islamic invasion, their knowledge was salvaged to some extent. The absence of orality in modern culture is a major stumbling block for the growth of science. How do we recall commands of a computer program when needed? Continually looking for the program manual is not a good solution. It is time consuming to search the manual for commands and this interrupts the train of thought for scholars. Many companies like Microsoft, IBM, and other businesses deal with this issue on an ongoing basis. Science educators will find new ideas of promoting science literacy in Chapter 2.

The worldwide use of the Hindu numerals is perhaps the greatest triumph of the ancient Hindus. This did not happen in one day; political rivalries ensued with the spread of these numerals in Arabia and Europe, decrees were issued against their use, and their usefulness in mathematical calculations finally established their supremacy. The history of the Hindu numerals and their growth is described in Chapter 3.

Leonardo Fibbonacci (1170 - 1250 A.D.) is well known for Fibbonacci's Sequence. However, little is known about his gratitude to the ancient Hindus, as he acknowledged in *Liber Abaci.* The ancient Hindus played an important role in the growth of algebra. The word *algebra* is derived from the title of a book written by al-Khwārizmī (ninth-century A.D.), a scientist who was based in the House of Wisdom in Baghdad. However, his book was a compilation of algebraic techniques that are partly Hindu in origin. Many algebraic techniques that include the solution of a quadratic equation, the sum of a finite series, and the sum of a series with $\Sigma n^2$ and $\Sigma n^3$ were known to the ancient Hindus. Similarly, the growth in geometry as a religious necessity of *yajna* (group prayer with fire in the middle) led to the books of *Śulbasūtra*. (See Chapter 4)

Copernicus was not the first person to assign motion to the Earth. Āryabhaṭa I (b. 476 A.D.) assigned motion to the Earth in his book, *Āryabhaṭīya*, a thousand years before Copernicus. This was done to explain the apparent motion of the stars. Āryabhaṭa I's book and the works of several Hindu astronomers were compiled into a single book, called *Sindhind*, in Baghdad. This astronomy book became popular in Europe during the medieval period prior to Copernicus. It was translated into Latin and a copy of this modified version was on the bookshelves of Copernicus. How much of this information was used by Copernicus is a mystery of science. The possibility of a helio-centric system in *Āryabhaṭīya*, the Hindu calendar, and spread of Hindu astronomy to world cultures are presented in Chapter 5.

Kaṇāda's work is as important as Democritus' work on the concept of the atom. Kaṇāda's book, *Vaiśeṣika-Sūtra*, defines the nature of time and space, conservation of matter, gravitation, and the concept of the atom. The word "time" is commonly used in our daily conversation. However, its nature is enigmatic and subtle. Readers will learn the subtleties in the concept of time in Chapter 6. They will also learn the concept of the atom, as suggested by Kaṇāda, and its comparison with Democritus' atom.

The steel industry is indebted to an ancient process invented by the ancient Hindus to harden iron. A small circular shaped disk of hardened iron was defined as *wootz*. This invention allowed the medieval warrior to enter a battlefield without worrying about breaking or bending his sword. As the name suggests, this process became known to Europeans from Damascus, in the present Syria, and the product was defined as Damascus steel. Chapter 7, on chemistry, describes the process and the rust-proof artifacts it yielded.

The discipline of biology has united the study of animals and plants under a single banner. Plants have life; they try to protect themselves; they can be poisoned; and they need food. These little known facts of science are discussed in historical perspective in Chapter 8, which also explains why the ancient Hindus sanctified all forms of life, even the adversary's life. The role of *ahiṁsā*, vegetarianism, and its practicality in daily life are also covered in Chapter 8.

Discrimination against women has been a hot issue all over the world for more than a century – and it still is today. In some countries, females cannot drive a car and must sit only on the companion seat or back seats. In contrast, Hindus have defined gods and goddesses for different attributes, and the two most important attributes, wisdom and money, have gone to goddesses Sarasvatī and Lakṣmī, respectively. These goddesses are worshipped in most Hindu households since wisdom and prosperity is needed by all. There is no male counterpart for these two attributes. The sacred books of the Hindus instruct us to pay due respect to females and allow them to pursue all disciplines, including natural philosophy. (See Chapter 8)

Yoga is considered a mystical discipline by some. This book provides the scientific aspects of yoga and its usefulness in promoting good health. The use of proper breathing, posture, and thinking is emphasized. Chapter 9 summarizes several clinical studies on the effectiveness of yoga in various diseases.

Ayurveda, the Hindu science of healing and rejuvenation, is the subject covered in Chapter 10. The body-mind emphasis of ayurveda is becoming popular in treating people in the West. The curative and the preventive measures for optimum health as well as the role of diet, good sleep, good *dincarayā* (daily routine) are the focus of this chapter. The role of a doctor and a patient in the successful treatment of a disease, and hospital designs recommended by Caraka and Suśruta are also shared. The plastic surgery finds its roots in the surgical skills of the ancient Hindus. (See Chapter 10)

This book is written primarily for readers who are trained in the Western knowledge system and are interested in learning about Hindus' contributions to science. Many references are deliberately chosen, whenever possible, from Western sources. Such a selection is made in view of a general skepticism by some Westerners concerning Hindu accounts. For some, scholars in the East stretch their imaginations too far to establish their views and do not provide rational steps that lead to conclusions. I have tried to address such criticisms by incorporating mostly primary references and authentic secondary references in this book. The works of the Greeks, Arabs, Chinese, and Europeans, included here, support the achievements of the ancient Hindus.

This book is not comprehensive because the ancient literature is so voluminous and varied. Multiple volumes are not sufficient to document the cumulative knowledge of the ancient Hindus. This book is only the "tip of the iceberg," as the phrase goes. I have chosen the topics where my knowledge and interests lie.

Several Sanskrit and Hindi words are now used in English and have entered in English dictionaries. However, their English spellings in some cases are not in accordance with the system of transliteration. As these words are in common use in the English-speaking world, using any other spelling might create confusion for a broad range of readers. For this reason, we have kept the popular usage in some cases. For example, the spelling of the word "ayurveda," as mentioned in most dictionaries, is incorrect; the proper spelling is āyurveda. For Sanskrit terms that are not present in most English dictionaries, diacritics have been retained. To keep the book readable, simple, and enjoyable to non-scholars in the field, an amalgamated system of popular spelling as well as proper spelling is used, as is common practice among authors.

As mentioned earlier, a salient feature of this book is the use of the Chinese, Persian, Egyptian, Arabian, Greek, and Roman accounts of the ancient Hindus. These documents provide excellent cross-cultural perspectives and comparisons, and portray a coherent picture of the contributions of the ancient Hindus to the sciences.

I have used scientific norms of analysis and have sorted out the hard facts from fantasy; in other words, the analysis is rational and objective. In philosophy and mysticism, there are several areas where the ancient Hindu literature stands abreast with the later concepts in science, such as causality and duality (or dualism). But these concepts are not covered in this book, because the borderline between facts and opinion is hazy. [34]

In the Lawrence Livermore Laboratory in California, the Shiva Target Chamber is a popular facility, internationally known for laser fusion. Edward Teller, the father of the hydrogen bomb, explained nuclear fusion in the following words: "Laser light is brought in simultaneously from ten pipes on the top and ten pipes on the bottom. Compression and nuclear reaction occurs in a tiny dot at the middle of the sphere. Apparatus practically filling a whole building feeds the twenty pipes, or the arms of the god Shiva. According to Hindu Creed, Shiva [Śiva] had three eyes: two for seeing, and one (usually kept closed) to emit annihilating radiation. The Hindus obviously know about lasers." [35] Is this the case? Definitely not, in my mind. The ancient Hindus did not know about lasers. What Teller has mentioned is a mere conceptual similarity and not a historical fact.

Similarly, duality in Hindu philosophy is quite different to de-Broglie's Wave-Particle Duality. Both appear to be analogous in language in some explanations and the similarities are curious and fascinating. Yet, they are not the same and belong to different realms of knowledge. The parallelism between ancient theories and modern science is fascinating to read at times.

[34] *The Tao of Physics* by Fritzof Capra, *The Dancing Wu Li Masters* by Gary Zukav, and *Mysticism and the New Physics* by Michel Talbot are examples of such works. These books are popular in general public for their insights. These are scholarly works that brought together the disciplines of religion and science. However, the *Law of Karma* as described in the Hindu literature is not Newton's Third Law of Motion. The language used may be analogous. However, these two concepts are different

[35] Teller, 1979, p. 216.

But in many cases, this is where the connection stops. The works of noted physicists such as Brian Josephson and Erwin Schrödinger do pose an interesting dilemma of joining these two disciplines. Their worldviews played a significant role in their discoveries in science.[36]

The contents of this book draw an artificial boundary at the fifth-century when the work of Āryabhaṭa I was written. Later contributions are mentioned only to maintain continuity when desirable to understand the ancient period. Since the works of the ancient Hindus took many centuries to become known to the Europeans or to the Arabs, the later accounts from these cultures are discussed. For example, al-Bīrūnī's work during the eleventh century is discussed in relation to Āryabhaṭa I's work. Copernicus' comment about the usefulness of the place-value notation and his work in astronomy are discussed, although the works were done during the sixteenth-century. Similarly, the work of J. C. Bose on plants is discussed in relation to a quote from *Mahābhārata*. All these later contributions provide a context to ancient discoveries.

In his book, *Science in Antiquity*, Benjamin Farrington predicted that, once the contributions of India are understood after more archaeological explorations and interpretations, "the records of scientific achievement before the Greeks may need to be written again."[37] With recent archaeological excavations and new scholarship, the time is appropriate to make such corrections. "A class in arithmetic would be pleased to hear about the Hindoos [Hindus] and their invention of the 'Arabic notation'," suggested Cajori.[38] "They will marvel at the thousands of years which elapsed before people had even thought of introducing into the numeral notation that Columbus-egg – the zero."[39]

---

[36]Read Bohm, 1980; Capra, 1976; Josephson, 1987; Restivo, 1978 and 1982; Schrödinger, 1964; Talbot, 1981; and Zukav, 1979. Some these scholars are discussed in Chapter 11.

[37]Farrington, 1969, p. 15. Farrington (1891 - 1974) was trained in Dublin and later served as a professor of classics at the University of Capetown in South Africa and Swansea University in Wales. His primary training was in Greek literature.

[38]Cajori, 1980, p. 3. Florian Cajori (1859 - 1930) was a professor at the University of California at Berkeley. Many of his books on the history of mathematics are highly respected by scholars.

[39]Cajori, 1980, p. 3. Columbus's egg is a term that refers to a brilliant discovery or idea that looks simple after the fact.

# Chapter 2

# The Cultural Context of Scientific Developments

Science evolves out of human necessities and curiosities. Human beings recognize issues that are relevant to them individually or collectively and resolve them using their resources. The growth of science centers around the questions and issues that are deemed critical. What were the uses of science for the ancient Hindus? What were the geographical and social conditions that allowed the growth of science? How did the ancient Hindus preserve their scientific knowledge and transmit it to the following generations? What were the social roles of the people who contributed to Hindu science? These are important questions that need to be explored to understand the cultural contexts of the development of science among the ancient Hindus.

The lofty heights reached by the ancient Hindus in the realm of philosophy and religion are established and extensive literature exists on these topics.[1] However, not much is known in the popular literature about their contributions to the natural sciences. The sciences of the ancient Hindus are embedded in their religious books with other disciplines. In ancient India, the various domains of knowledge, including science and religion, progressed hand-in-hand and grew under the shelter of one another. Hindu scholars mathematically calculated the dates and the exact times of religious rituals that were based on careful astronomical observations of the sky. Religion flourished with the help of science and the boundaries of scientific inquiries

[1]Dasgupta, 1922 - 1955; Durant, 1954; and Radhakrishnan, 1958.

were kept in check by the religious codes.

The *Chāndogya-Upaniṣad* cites an episode in which the vagabond deity, Nārada, approaches another sage, Sanatkumāra, to achieve the ultimate knowledge. Nārada wanted to obtain the knowledge that could liberate him from the cycle of birth and death and guide him to achieve salvation. Sanatkumāra asked what knowledge Nārada already had, to have a meaningful conversation. Nārada pointed out astronomy (*Nakṣatra-vidyā*) and mathematics (*rāsi-vidyā*), along with logic, history, grammar, fine arts, and the four *Vedas* as the knowledge that he had already mastered in his efforts to achieve salvation.[2]

Nārada is considered one of highest seers (*Deva-ṛiṣi*) in the Hindu religion. The above episode reveals that acquiring knowledge of the natural sciences, along with religion and fine arts, was considered relevant to achieve salvation. In Hindu tradition, secular knowledge (*aparā-vidyā*) is considered to be helpful in achieving salvation, along with spiritual knowledge (*parā-vidyā*), as advised in *Muṇḍaka-Upaniṣad.*[3]

Similar to Nārada, Āryabhaṭa I, in his book *Āryabhaṭīya*, suggested that one can achieve salvation by becoming well versed in astronomy and mathematics.[4] Also, Kaṇāda, in his book *Vaiśeṣika-sūtra*, suggested that people learn about the physical properties of matter to achieve salvation.[5]

Science was institutionalized among the ancient Hindus. It was considered sacred and as good as their moral codes for society. Scientific activities had important functions that were valued in society. The role of astronomers to fix the calendar, to set dates of religious festivals, and to predict eclipses or other astronomical events became as important as their moral codes. All forms of knowledge, including science, music, and art, were considered sacred by the Hindus. This is the reason the famous sitarist Ravi Shanker usually suggested that his audience not talk or walk while he played music at a concert. Music was sacred and divine for him. Thus, in the Hindu religion, a person could achieve salvation by mastering physics, chemistry, astronomy,

[2] *Chāndogya-Upaniṣad*, 7: 1: 2 - 4.
[3] *Muṇḍaka-Upaniṣad*, 1: 1: 3 - 5.
[4] *Āryabhaṭīya*, Daśgītika, 13.
[5] *Vaiśeṣika-Sūtra*, 1: 1: 4.

medicine, or music. Sciences and the fine arts were bound to prosper among the Hindus.

## 2.1 Who is a Hindu?

Herodotus (c.484 - 425 B.C.), a Greek traveler and geographer, visited the Indus-Sarasvatī region and called it "Indoi" in his travel accounts. In due time, this word metamorphosed into India in the English language. Thus, the name India has been derived from the name assigned by Herodotus to this region more than two millennia ago. The name India became popular in use only after the British occupation of the region, some 250 years ago. In the process of colonization, foreign names became more popular than the native names for several other countries. For example, Misr is now popularly known as Egypt.

The Indus-Sarasvatī geographical region was also called Sapta-sindhavah[6], Āryavarta[7], or Bhāratvarṣa[8] in ancient or medieval Hindu literature. During the colonial period, this region was subjugated by the foreign powers. After the region became free, it was divided into several countries. Before the past century, the name India also described the geographical regions of Pakistan, Bangladesh, and part of Afghanistan.

The word "Hindu" is probably derived from the word "Sind," the focal region of the inhabitants of the Indus-Sarasvatī Civilization. The Sindhu river (derived from Sind; also known as Indus river) was central in supporting the needs of the region, especially after the Sarasvatī river dried up around 1900 B.C. The Iranian or the Middle Eastern textual or epigraphic references to "Hindu" were not to any religious faith but to the region. The land was called Hind by its neighbors on the northeast, and the inhabitants were called Hindus. Their social, cultural, and spiritual doctrines formed the basis of the Hindu religion.

Hindustān (place of the Hindus) is the Persian name of India which is commonly used to define the country in the Hindī language, the most popu-

[6] *Ṛgveda*, 7: 24: 27.
[7] *Manu-Saṁhitā*, 2: 22.
[8] *Viṣṇu-Purāṇa*, 2: 3: 1.

lar language of India. The Persians and the Arabs defined the geographical region as Hind, and its inhabitants were called the Hindus, prior to the British conquest. Presently, *Hind*, *Hindustan*, *Bhārat*, and India are the names synonymously used to describe India as a nation. One of the most popular daily newspapers in India is named *The Hindustan Times*, indicating Hindu as a region or country. It is not a newspaper for the Hindus; it is for all Indians. Similarly, the most popular nationalistic song in India is as follows: *Sāre Jahān se acchā Hindustān hamārā* (My Hindustan (the land of the Hindus) is better than any other country in the world.) This song was written by a famous Muslim poet Muhammad Iqbal,[9] defining Hindu as a region. This song is full of patriotic fervor and pride for India. Thus, the word Hindu defines the geographical region to some and religion to another. For this book, the term Hindu is used to define the Indus-Sarasvatī region during the period when the inhabitants composed *Vedas*, *Upaniṣads*, and *Purāṇas.*

To meet their social, religious, and personal needs, the ancient Hindus compiled their knowledge in the form of oral books that are called *Vedas*, *Upaniṣads*, and *Purāṇas.* While the words Hindu or Hindu religion were used to describe people and their religion in the Indus-Sarasvatī region, today the word "Hindu" is commonly used to describe people who follow the doctrines of the "Hindu corpus," *Vedas* being central to all of them. This allows the people living in Bali, West Indies, America, or elsewhere to be called Hindu as long as they believe in and practice the doctrines of the *Vedas*, *Upaniṣads*, or *Purāṇas.*

Today, Hinduism is practiced by about one billion people. It is the third largest religious community in the world, after Christianity and Islam. Hindus are located primarily in India, Mauritius, Nepal, Sri Lanka, and Bali; about 2% of them live outside India. The United States has about 1.5 million Hindus.

## 2.2 Traits of the Ancient Hindus

The geography of India, with its vast mineral resources, diverse plant and animal life, favorable climate, and sound social ethics provided material pros-

[9] After the partition of India, Iqbal moved to Pakistan. This song is still quite popular in India among the people of all religions.

perity to the region and fostered the intellectual endeavors of the Hindus. As Ralph Waldo Emerson (1803 - 1882 A.D.), an eminent American philosopher and poet, wrote: "The favor of the climate, making subsistence easy and encouraging an outdoor life, allows to the Eastern nations a highly intellectual organization, - leaving out of view, at present, the genius of Hindoos [Hindus] (more Orient in every sense), whom no people have surpassed in the grandeur of their ethical statement."[10] Emerson carefully studied the Hindu literature, including *Vedas* and *Upaniṣads* and the sensual aspect of Kālidāsa's poetry. The Hindu philosophy reaffirmed Emerson's transcendentalism and helped him in his quest to define a truly representative man.

India was visited by travelers from Greece, Rome, China, and Arabia during the ancient and early medieval periods. Many of these travelers provided accounts of the prosperity of India. An early account is from Megasthenes (350 - 290 B.C.), who was an ambassador of Seleucus I, a Greek king, who ruled India for a short period. He wrote that the Indians, "having abundant means of subsistence, exceed in consequence the ordinary stature, and are distinguished by their proud bearing. They are also found to be well skilled in the arts, as might be expected of men who inhale a pure air and drink the very finest water."[11] Megasthenes stated that India was a prosperous country and never suffered from "famine" and "general scarcity in the supply of nourishing food."[12] Megasthenes' statements were written after Alexander's invasion of India when the intellectual life in Athens was already vigorous. Strabo (63 B.C. - 21 A.D.), a Greek geographer and traveler, visited India and found that the country was "abounding in herbs and roots,"[13] indicating prosperity.

The Chinese traveler Yijing (or I-tsing; 643 - 713 A.D.), wrote that, in India, "ghee [clarified butter], oil, milk, and cream are found everywhere. Such things as cakes and fruit are so abundant that it is difficult to enumerate them here."[14] Yijing was from Canton, China. He learned the Sanskrit language during a six-month stay in Java, Indonesia. After becoming proficient in Sanskrit, he lived in India about 22 years, mostly in the region presently

[10]Emerson, 1904, vol. 8, p. 239. For more information, read Acharya, 2001.
[11]McCrindle, 1926, p. 30.
[12]McCrindle, 1926, p. 31.
[13]Strabo, 15: 1: 22; Strabo, 1961, vol. 7, p. 35.
[14]Takakusu, 1966, p. 44.

known as Bihar, which was a center of learning for the Buddhists. He carried some 400 Sanskrit texts back to China with him.[15]

Most Chinese, Greek, Persian, Arabian, and European documents written in different periods testify to the prosperity of the Indus-Sarsvatī Valley region. It is only after the later part of the British occupation, in the late nineteenth century, the region suffered due to colonial exploitation.

Hindu philosophy is a dynamic system in which different systems of knowledge flourish and coexist in harmony. "Indians recognize that only the greatest and simplest religious questions can be asked now in the same words that came to the lips more than two thousand years ago and even if the questions are the same, the answers of the thoughts are still as widely divergent as the pronouncements of the Buddha and the Brahmans . . . Indian religion eschews creeds and will not die with the spread of knowledge. It will merely change and enter a new phase of life in which much that is now believed and practised [practiced] will be regarded as the gods and rites of the *Veda* are regarded now," explains Eliot while outlining the essence of Hindu philosophy.[16]

Like science, the ancient Hindu literature was a dynamic system of knowledge that collected the wisdom of a wide spectrum of men and women over time. New ideas were often proposed and their validity was tested. If valid, they became part of the Hindu corpus. Logic, intuition, and personal experiences along with debate or discussion (*śāstrārtha*) evolved as a system of knowing. Hindu literature has become a blend of traditional and progressive thought. The dynamic nature of Hindu literature is the main reason it has withstood the stress and strain of more than two millennia.

Only a dynamic system can accommodate and survive the infinite mutations of time. Any literature that cannot meet the new demands of time cannot survive for long. The new extensions in the Hindu literature are not to accommodate any new foolish outcry of the community; they simply accommodated new needs of Hindu society.

---

[15]Takakusu, 1966, p. xvii.

[16]Eliot, 1954, volume 1, p. XCVII.

Hindu literature can be symbolically described as a Banyan tree. The seed sprouts and germinates and slowly emerges as a small bush. This bush then takes the form of a tree with leaves and flowers. When the Banyan tree grows old, some branches of the tree proceed toward the soil. After touching the soil, these branches penetrate the soil and feed the tree with nutrients. They serve as roots. These branches gradually evolve into new trunks, which produce new trees.

## 2.3 Hindu Ethics

Science is a knowledge that evolves from observation, experimentation, organization, and interpretation of facts. It is a search for truth. The practice of science thus should be limited to people of integrity.[17] As we try to understand the mysteries of nature to predict natural phenomena accurately, and devise new applications of science (meaning technology), ethical principles play an important role in these endeavors.

In response to concerns about ethical improprieties and issues in science, various scientific institutions and societies, such as the National Science Foundation (NSF), the National Institutes of Health (NIH), the American Association for the Advancement of Science (AAAS), the National Academy of Sciences (NAS), and Sigma Xi have commissioned committees to study ethical issues and improprieties in science and make policy recommendations.

Scientists should not fabricate, falsify, or misrepresent data or results. They should be objective, unbiased, and truthful in all aspects of the research process. Honesty promotes the cooperation and trust necessary for scientific research. The practice of science has always encompassed values like honesty, objectivity, and openness. Scientists' interests follow from the collective wisdom of the scientific community which draws upon an extremely wide range, not only of "natural" experiences, but also of contrived events and phenomena (e.g. nuclear accelerators, spaceships, and chemical laboratories).

The fundamental operative term for ethics is *dharma* in the Sanskrit language. The ancient Hindus were obsessed with ethics and laid down a

[17]Macrina, 1995, p. 1.

philosophy which was not restricted to mere morality; this philosophy was a code for life that allowed individuals and society to evolve together. When emotions are in conflict with duty, it is the duty that has to be followed. It is the famous doctrine of *Niṣkāma-karma* (selfless action) or *Niṣkāma-yoga* of *Bhagavad-Gītā*[18] in which Lord Kṛṣṇa asked Arjuna to perform his duties in all situations without worrying about the consequences.

For the Hindus, a person has to perform his or her duties not for any ulterior motive but for the sake of duty itself. It was *dharma* that led Lord Rāma to live in a forest for fourteen years with his brother, Lakṣmaṇa, and his wife, Sītā. It was done to honor the words of his father Daśaratha, as the story is shared in *Rāmāyaṇa*. Similarly, it was *dharma* that led King Yudhiṣṭhira to live in a forest in discomfort for thirteen year to bear the consequences of his acts, as written in *Mahābhārata*. The Law of *Karma* demands truthfulness or right action as a duty and not a mere expectation of people.[19]

Truthfulness, goodness, and beauty, as marked in *satyaṁ*, *śivaṁ*, and *sundaraṁ* [Only truth is beneficial and beautiful], became the guiding principles for the ancient Hindus. Truth is the ultimate good for an individual and a society and leads to ultimate beauty in life.

Honesty is more than a moral obligation to avoid deceit; it is revered as a duty which a person must extend to another person – a path to salvation. *Muṇḍaka-Upaniṣad* tells us that "truth alone conquers, not falsehood. By truth is laid out the path leading to the gods."[20] A teacher provides the following instruction to a student in *Taittirīya-Upaniṣad*: "Speak the truth. Practice virtue . . . One should not be negligent of truth."[21]

Ṣā'id al-Andalusī a well-known natural philosopher and historian from the eleventh century Muslim Spain, described the Hindus as wise, fair, and objective.[22] Henry David Thoreau, (1817 - 1862 A.D.), an American philosopher, naturalist, and author, praised the Hindus for their openness to new

---

[18] *Bhagavad-Gītā*, 2: 38 - 41.

[19] Hopkins, 1968; Rao, 1926; Tattvabhushan, 1912; and Widgery, 1930.

[20] *Muṇḍaka-Upaniṣad*, 3: 1: 6.

[21] *Taittirīya-Upaniṣad*, 1: 11: 1.

[22] Salem and Kumar, 1991, p. 11.

ideas. "The calmness and gentleness with which the Hindoo [Hindu] philosophers approach and discourse on forbidden themes is admirable."[23] Thoreau wrote this statement after his studies of the *Vedas* and *Upaniṣads*.

Hindu ethics allowed people to be liberated from the cycle of birth and death. This allowed the philosophical basis for the Hindus' veneration of the natural world.[24] These ethical values are "integrated and progressive, culminating in the *summum bonum* of liberation."[25] And this also allowed the ancient Hindus to become excellent scientists.

## 2.4 Hindu Science: Indigenous or Implanted?

The originality of Hindu literature is by and large accepted by most Indologists.[26] Some indologists assert that the Hindus (more appropriately Aryans) came from Central Asia and settled in the Indus-Sarsvatī Valley while others consider them to be indigenous. It is not the intent of this book to set this record straight one way or the other. However, a few points are to be noted.

1. It is certain that the holy books of the Hindus were compiled in the Indus-Sarasvatī region by the people who lived there for an extended period. The *Ṛgveda* mentions tigers, lions, elephants, rhinoceros, water buffalo, horses, and cows. All these animals are present together in India.

2. There is no scholarly tradition in India to visit foreign lands for learning. On the contrary, India attracted scholars from China, the Middle East, and even from the Mediterranean. This is in contrast to the Greek tradition in which many young Greek scholars undertook arduous journeys to Egypt, Babylon, Persia, or even to India for learning. The possible journey of Pythagoras to India is discussed in Appendix A. Similarly, the visits of several Chinese travelers such as Faxian, Xuan Zang, and Yijing to India are well documented. Indians mostly traveled to foreign lands to teach, as we know from the Chinese and Arabic

[23]Thoreau, 1906, vol. 2, p. 3.
[24]Crawford, 1995, p. 232.
[25]Crawford, 1995, p. 8.
[26]Macdonell, 1968, p. 7; Vartak, 1995, p. 1.

sources. For example, Kanaka went to Baghdad to help scholars in *Bayt al-Ḥikma*. Similarly, during the Tang Dynasty, Indian astronomical schools guided the emperors in China. There is not even single case from the ancient and medieval periods when a Hindu went abroad for learning and became famous upon his/her return. Indians started visiting foreign lands for learning only after the late nineteenth century under the British occupation.

3. There was no organized effort among the Hindus to learn foreign languages. Only, a few manuscripts with foreign origins have appeared from the early medieval period. For example, *Romaka Siddhānta* of Varāhamihira (505 - 587 A.D.) appeared in the Hindu literature, indicating influences of Rome. These names themselves indicate their foreign origins. Hindu literature provides ample examples where native scholars were considered better than their foreign counterparts. Even some foreign scholars attested to this. Al-Bīrūnī (973 - 1048 A.D.), an Islamic natural philosopher, labeled the Arabs as "illiterate people" when he compared Arabic science with Hindu science.[27]

4. On the question of the originality of ancient Hindu literature, it is important to understand the chronology of various civilizations. Science is an essential and integral part of the Hindu religion, with roots anterior to those of the Greek civilization.

Some scholars suggest that Hindu literature is based on Greek literature, rationalizing that Alexander the Great captured a part of northern India in the third century B.C. and it was then that the region erupted with scientific activities. This is questionable, as Hindu traditions have their roots in the religious scripture of *Vedas* and *Upaniṣads*, which were written before Alexander's invasion. The hymns of *Ṛgveda* were composed some 3,000 to 800 years before Homer (eighth century B.C.) composed his poetry in Greece.

[27] Sachau, 1964, vol. II, p. 81. This remark should not be taken literally. Many scholars from the ancient and medieval period have used language that can be considered "uncivil" in today's world. Al-Bīrūnī considered Hindus to be proficient in sciences. This is typical of many Islamic scholars contemporary with al-Bīrūnī. A large number of Arabic manuscripts of noted scientists such as al-Fazārī (d. 777 A.D.), al-Khwārizmī (fl. ninth-century), Ṣā'id al-Andalusī (1029 - 1070 A.D.), al-Uqlidīsī (920 - 980 A.D.) attest to the achievements of the Hindus. Incidentally, this comment of al-Bīrūnī came after the glorious period of Baghdad.

The prime philosophical contributions of the Greeks started with the works of philosopher-scientists such as Thales and Pythagoras during the seventh and sixth centuries B.C., long after the birth of the philosophical traditions of the Hindus. Keith appropriately wrote that it is not to be thought that the early philosophy of Greece exercised any influence on the philosophy of India: "Apart from every other consideration, it is clear that the rise of philosophy in Greece was long subsequent to the beginnings of Indian philosophy in the hymns of Rigveda [*Ṛgveda*], and from those hymns the history of that philosophy presents itself in the light of an ordered development . . . The only question, therefore, which can arise is whether the early schools of Greek philosophy were affected by the tenets of the schools of the Brahmans." [28]

Indeed, there are holy books of the Hindus that were written after the Greek invasion by Alexander, like the *purāṇic* literature. This allows for the possibility of a Greek influence on Hindu literature. However, most Hindu holy books that were written later are in accordance with the hymns of *Vedas*, and deviate little from the Hindu philosophy established in the earliest ancient works. Therefore, such assertions of the Greek influence come into question.

A well known Spanish philosopher Ṣā'id al-Andalusī (1029 - 1070 A.D.) labeled India as the leader in science and technology: "The first nation (to have cultivated science) is India [al-Hind, in the original manuscript]. This is a powerful nation having a large population, and a rich kingdom (possessions). India is known for the wisdom of its people. Over many centuries, all the kings of the past have recognized the ability of the Indians in all branches of knowledge."[29]

Ṣā'id al-Andalusī was one of the leading natural philosophers of eleventh century Muslim Spain. He compiled a book on the history of science, *Ṭabaqāt al-'Umam*, in 1068 A.D., that classified the contributions of various nations in the sciences. This was perhaps the first organized effort to compile a

[28]Keith, 1971, p. 601. Arthur Berriedale Keith (1879 - 1944) was a professor of Sanskrit and Indology at the University of Edinburgh, England.

[29]Salem and Kumar, 1991, p. 11. Though Ṣā'id mentioned India to be the first nation in science and technology, this description is not significant for the purposes of this book. A more important fact is that Ṣā'id considered India's contribution good enough to be considered as the leading (or first) nation in science and technology.

global history of science. Ṣā'id's book covers the earliest period of the history around the world. It also includes the later developments in Europe until the eleventh century. His work included the civilizations of Egypt, Greece, Persia, India, Chaldea, the Middle East, and Spain.

Al-Jāḥiz, Abū Uthmān 'Amr ibn Bahr (ca.776 - 868 A.D.; the name implies goggle-eyed), an Arabic philosopher-scientist from Baghdad, wrote: "When Adam descended from Paradise, it was to their land [Hind] that he made his way."[30] This statement defines the Hindu culture to be the earliest one. Al-Jāḥiz was one of the leading Islamic scientists from Basra, Iraq, and was a tutor of Caliph al-Mutawakkil of Baghdad, and was well versed in the Greek literature. Al-Jāḥiz wrote a book on animals, *Kitāb al-Hayawan*, which remained a popular book in Islamic Spain and Latin Europe.

Al-Mas'ūdī, Abu 'l-Ḥasan 'Alī ibn al-Ḥusayn ibn 'Ali (died ca. 957 A.D.), a tenth-century Islamic philosopher from Arabia, considered India to be the country where science and technology have originated.[31] This statement is similar to the statements made by al-Jāḥiz and Ṣā'id al-Andalusī. Since the Indus-Sarasvatī region contributed much to sciences, owing to the ancient origin of the culture, al-Mas'ūdī considered the region to be the first in science and technology.

Al-Mas'ūdī was born in Baghdad and lived in Egypt for most of his life. He became well versed with Greek literature there. He also lived in India for a short period. Al-Mas'udi had a sound knowledge of the Hindu as well as Greek literature.[32] He was a prolific writer, having written some 37 works on history, geography, jurisprudence, theology, genealogy, and administration. It is interesting to note that al-Jāḥiz and al-Mas'ūdī came from Iraq – a place where some people believed that the first civilization started.

It would be of little value to attempt to prove that the Hindus were the first, second, or third in science and technology. As mentioned in the previous chapter, people all over the world have always pursued science and technology to meet their personal and social needs. The above quotations

[30]Pellat, 1969, p.198.

[31]Khalidi, 1975, p. 104; al-Mas'ūdī, *Murūj.*, sec. 1371, taken from Khalidi.

[32]For more information, see Khalidi,1975; Shoul Ahmad, 1979.

simply indicate that the ancient Hindus must have been among the leaders in science and technology. Also, the Hindus must have made considerable contributions to science and technology to earn such global respect of leaders in science throughout history. Their scientific knowledge must have been of the highest quality and ingenious to earn such a reputation without borrowing much from other civilizations.

## 2.5 Respect for Knowledge: The Role of a Guru

Hinduism does not enforce any undue restraint upon the freedom of human reasoning, the freedom of thought, or the will of an individual. Hindus' respect for knowledge is inherent in the core values of the religion. "Let noble thoughts come to us from all sides," suggests the *Ṛgveda*.[33] As a mother, in nourishing her children, gives *capāti* (or *Rotī*, a flat bread) and *dal* (lentils) to one, and *khicaḍī* (a rice and lentil preparation) and yogurt to another, according to their needs, the Hindu religion offers choices to human rational minds, and allows individuals to choose their own path to salvation – an ultimate goal for all Hindus. Four methods (paths) are prescribed to achieve salvation:

1. *Karma-yoga* (the Path of Action)
2. *Bhakti-yoga* (the Path of Devotion or Love)
3. *Jñāna-yoga* (the Path of Knowledge)
4. *Rāja-yoga* (the Mystical Path)

These paths are not listed in the order of effectiveness; all paths are equally effective and important. It is up to an individual to select the appropriate path for himself or herself. Also, all four paths allow a person to achieve the same end goal – salvation. The *Ṛgveda* tells us that the God is one and we call Him by different names. Thus, the multiple gods are the multiple manifestations of just one God. This concept of multifarious gods is confusing to some Western scholars. For the Hindus, in many temples, the idols of these gods sit side by side. This is similar to the ancient Egyptian religion

[33] *Ṛgveda*, 1: 89: 1.

where multiple gods were derived from just one God, Amon Ra. For many people, Hindus and Egyptians have many gods with no connections to each other. However, for the Hindus and Egyptians, these gods are all one. This multiplicity unified by the single reality is a salient feature of the Hindu philosophy and religion.

*Karma-yoga*, the method of action, tells us that selfless action is a way to attain salvation. The scriptures preach *niṣkāma-karma* (*niṣkāma* = selfless, *karma* = action). For example, one should help another person as a *duty* and should not expect anything in return from the person.

*Niṣkāma-karma* is explained in the *Bhagavad-Gītā* during a dialog between Lord Kṛṣṇa and Arjuna at a battle site in the Mahābhārta war. Lord Kṛṣṇa suggests to Arjuna: "Your duty is to work, not to reap the fruits of work. Do not go for the rewards of what you do. Neither be fond of laziness. Be steady in Yoga, do whatever you must do. Give up attachment, be indifferent to failure and success. This equilibrium is yoga."[34] *Karma-yoga* allowed a scientist to practice astronomy, mathematics and medicine for salvation; musicians concentrated on musical instruments and *rāgās*, and dancers practiced various styles of dances: *bharat-nātayam*, *odissi*, and *kathakali*.

*Bhakti-yoga* is the yoga of love and devotion. In *Bhakti-yoga*, all actions are attributed to the God, and the divine being becomes the source of all strengths, both spiritual and physical. The person completely surrenders to his deity and finds his ultimate refuge in doing so. The Hindu religion has multiple gods. The Hindus choose their own God and develop a personal relationship – *e.g.*, that of a son, a mother, a friend, a lover, or beloved – with the divine being. For example, Rāmakṛṣṇa Parmhaṁsa, a saint who lived during the nineteenth century, had a mother-son relationship with the Goddess Kālī; Mīrā Bāi, a queen in Marātha-dynasty, had a husband-wife relationship with Lord Kṛṣṇa.

*Jñāna-yoga* (*Jñāna* means intellect) encourages the individual to understand the *ultimate truth* by raising such questions as, "Who am I?", "Why am I here?", or "What is the purpose of my life?" This mode of Socratic questioning allows self-introspection: the person discovers the ultimate truth

[34] *Bhagavad-Gītā*, 2: 41 - 42.

from within. "If speaking is through speech, if breathing is through breath, if seeing is through eyes, if hearing is through the ears, if touching is through the skin, if meditation is through the mind, if breathing out is through the out breath, if emission is through the generative organ, then who am I?"[35] Like Socrates suggested in Greece, the *Bhagavad-Gītā* tells us that one can find knowledge within and perfect it by yoga: "For there is no better means of achieving knowledge; in time one will find that knowledge within oneself, when one is oneself perfected by yoga."[36] Worshipping deities to achieve wisdom is common in the *Ṛgveda*: "All pervading Deity, sharpen the intelligence, wisdom and the insight of him who is striving for enlightenment."[37] "We contemplate that adorable glory of the deity, that is in the earth, the sky, the heaven! May He stimulate our mental power."[38] This hymn is the popular *Gāyatri-mantra* that is chanted over and over by millions of Hindus as prayer every day.

The mind is the source of our knowledge. Our five senses do not function without the mind. It is the mind that unravels most of the obstacles that we face from day to day. One can even achieve salvation through the mind. For this reason, knowledge has always remained so central to the Hindus.

If knowledge is the key to salvation, then the issue of ownership of knowledge and defining spatio-temporal attributes associated with knowledge become trivial issues. Therefore, the ancient Hindus did not care much to define the period or author of their knowledge. This practice contrasts greatly with the Western tradition. As Charles Eliot elucidates: "They [Hindus] simply ask, is it true, what can I get from it? The European critic, who expects nothing of the sort from the work, racks his brain to know who wrote it and when, who touched it up and why?"[39] Truth becomes more important than knowing who originated the idea. While this avoids the cult of personality, it also avoids the chronological records of discoveries and inventions. This explains why these chronological records, so valued by Western historians,

---

[35] *Āitareya-Upaniṣad*, 1: 3: 11.
[36] *Bhagavad-Gītā*, 4: 38.
[37] *Ṛgveda*, 8: 42: 3.
[38] *Ṛgveda*, 3: 62: 10.
[39] Eliot, 1954, vol. 1, p. LXVII. Charles Norton Edgecumbe Eliot (1862 - 1931) was a botanist, linguist, and diplomat. He was the British ambassador to Japan and was fluent in 16 language and could converse in another 20 language, including Sanskrit.

are lacking in Hinduism.

*Rāja-yoga* provides tools to unify body and mind. The union of body and mind has miraculous effects enabling people to achieve extraordinary goals. *Āsana* (postures) are prescribed for physical body-toning and meditation for mind-toning. In yoga, these two are combined. Explaining the importance of meditation for scientific research, Fritzof Capra a nuclear physicist, wrote: "Scientists are familiar with the direct intuitive modes from their research . . . But these are extremely short moments . . . In meditation, on the other hand, the mind is emptied of all thoughts and concepts and thus prepared to function for long periods through its intuitive mode."[40] Here intuition is not a spiritual revelation; it is a perception, imagination, reason, and valuation. It is an intensive state of mind that allows the person to unfold the mystery at hand.

Respect for intellectual activities is obvious from the respect that was bestowed to gurus by the Hindus. In the *gurukula* system of education, children lived with their *guru* during the *brahmacarya* stage (learning stage) of life. Young children not only gained content-based knowledge from their guru, they also learned a virtuous life style and ethics. This system worked quite well for the ancient Hindus and they excelled in the sciences, along with other disciplines.[41] A festival, *guru-pūrṇimā*, is celebrated every year and people offer their thanks, love, and devotion to the guru. On this day, Hindus renew their faith and show appreciation to the guru for the learning. This falls on the full moon (*pūrṇimā*) in the month of *Āṣāḍha* (June - July).

In Hindu tradition, gurus[42] have a very high status that is comparable to parents and God. For example, Kabir, a famous medieval period poet, shared a situation where his guru and God came together to him. In the Hindu tradition, usually dignitaries are greeted and honored by touching their feet in the order of their comparative stature. Kabir's dilemma was who should

---

[40]Capra, 1980, p. 21. I have used Capra's work since it splendidly explains the role of meditation in scientific research. Capra's work is excellent on the issue of mind-body connections. My main reservation about his book is on the analogies drawn between scientific principles and religious statements. This causes confusion for some readers who may assume both to be the same, when they are inherently different.

[41]Joshi, 1972.

[42]*Śikṣak*, *ācārya*, *śrotriya Upādhyāya*, *purohit* are other words for guru.

he first greet, his guru or God? Kabir resolved this by suggesting that the guru should be the first one since he allowed the person to know God.[43] Gurus shared their own personal experiences with disciples, guided disciples in their interactions with the community, taught social rules, taught various intellectual disciplines, and most important of all, became the spiritual guide of disciples. It is the guru who assigned a caste (*jāti* or *varṇa*) to a disciple after the completion of his or her education.

Hindus divided the life of a person into four stages (*āśrama*): *brahmacarya* (learning stage), *gṛhastha* (family stage), *vānaprastha* (stage of contemplation) and *sanyasa* (stage of renunciation). *Brahmacarya* is the first stage of life where a child, after the infancy stage, joins *gurūkula* (*gurū* = teacher, *kula* = home), a place where children leave their parents and live with the family of their guru and not with their parents. Irrespective of the social and economic status of the parents, each child had to live at the same standard of living as the guru's family. This allowed disciples to live near the guru, who could observe firsthand aptitudes and inclinations of disciples. The disciples understood the complexities of life quickly because of their continued association with the guru.

Gurus were considered to be authority figures, but were not considered infallible. It was considered a healthy practice for the disciples to raise questions about gurus' teachings to understand reality in their own way. The *Prasna-Upaniṣad*, one of the principal *Upaniṣads*, is entirely based on the questioning between disciples and their teacher.

Respect for the guru has caused confusion among the Western scholars, who feel that the spirit of questioning could not have originated in India. This incorrect inference is based on the prominence given to gurus in Hindu society. In the minds of some, the distinguished status of gurus in India hampered the questioning by students that is crucial for learning. As a result, they assume science to have mainly originated in Greece, and to have spread later to Europe.[44] This is simply false. The entire *Bhagavad-Gītā*, perhaps the most popular religious book among the Hindus, is based on the questions and

[43] *Guru Govind dāū khare, kāke lāgu pāya, balihārī guru āpnu jin Govind diyo batāya.* My teacher and God are in front of me. Whom should I prostrate first? It has to be the guru who taught me to recognize God.

[44] Cromer, 1993.

answers between Lord Kṛṣṇa and Arjuna. Maitreyī learned from her husband Yājñavalkya by questioning him as shown in *The Bṛhadāraṇyaka-Upaniṣad.* And as mentioned above, in the *Prasna-Upaniṣad*, a group of students went to Pippalāda, a teacher, to ask six questions. The whole *upaniṣad* is about the questions and the answers.

The role of the guru was and still is paramount to most Hindus. It does not stop after the *brahmacarya* stage; it continues for one's whole life. Carl Jung, a noted psychoanalyst, emphasized the importance of the guru in the following words: ". . . practically everybody of a certain education, at least, has a guru, a spiritual leader who teaches you and you alone what you ought to know. Not everybody needs to know the same thing and this kind of knowledge can never be taught in the same way."[45] For their role, teachers earned enormous respect in society. Some of the famous teachers in Indian history are Kauṭilya (fl. 300 B.C.), teacher of Candragupta Maurya (reigned c. 321 - 297 B.C.) and Vasiṣṭha (pre-historic), teacher of Lord Rāma. Kauṭilya laid the rules of administration for Candragupta, especially for psychological warfare, political philosophy, and economics, which are compiled in his book, *Arthaśāstra*. It was Kauṭilya's strategies that ended the Greek rule in India.[46] Similarly, it was Vasiṣṭha who convinced *Rāja* Daśaratha to allow Lord Rāma to relinquish the worldly comforts of his palace and live in a forest to protect Viśvāmitra from *rākṣasa* (unlawful people).

## 2.6 *Śāstrārtha* (Debate) to Acquire Knowledge

Questioning is a powerful tool to discover unchartered territory of knowledge. As a child, I was told by my teachers that asking the right kind of questions is essential to the growth of science. Questioning is a powerful tool of investigation. Usually multiple possibilities are suggested to resolve a question. How do you select one possible solution over another? How do you decide, when different people support different solutions? A debate is a possible

[45] Jung, Carl, *CW Letters*, 1973, vol. 1, p. 237. Jung (1875 - 1961) was an influential psychiatrist and thinker of the twentieth century. Hindu philosophy played an important role in his theories on symbolism and the unconscious mind.

[46] A brief account of Kautilya's diplomacy is provided by Boesche, 2003. For Kautilya's economics, read Jha and Jha, 1998.

means in which scholars can present their cases; such debates can easily filter novices from scholars. The practice of debate was so ingrained and valued among the Hindus that it was one of the eight ways to select the groom for a bride. Prospective grooms debated with established scholars or among a group of prospective grooms in public on various issues to win the bride. It was this practice that led Gautama (who later became Lord Buddha) to debate Arjuna, as mentioned in *Lalitvistara*, an ancient book (See Chapter 3).[47]

The *Bṛhadāraṇyaka-Upaniṣad* narrates an episode in which King Janaka decided to donate one thousand cows to the best Brahmin. Yājñavalkya, a revered saint, took all the cows and thus infuriated several other holy men who also needed the gift for their livelihood. To resolve the matter, a debate ensued and Yājñavalkya had to demonstrate his superior intellectual abilities by answering questions posed by the holy men.[48]

A healthy *śāstrārtha* (debate or discussion) is an essential element and practice in the Hindu religion. It has led to the idea of "monism" – there is only one existence (God) – as well as the idea of "dualism" – there are two realities, Him (God) and me. The quest to know the ultimate truth led to the evolution of various systems of knowledge which outwardly seem to be divergent. The Hindu tradition allowed divergent opinions to coexist and established *śāstrārtha* as a mean to resolve scholarly differences. Each person was allowed to test and discover the truth in his/her own way. The "direct perception of truth" was both the means and end of education.[49]

The Sanskrit term *Ānuvīkṣikī*, defined as investigation through reasoning, has a long tradition in India. It is a tool of investigation that is applicable in all aspects of learning: scientific, religious, and social. The *Arthaśāstra* of Kauṭilaya lists[50] it as one of the four cognitive (*vidyā*) disciplines, along with *trayī* (vedic learning), *daṇḍanīti* (jurisprudence), *vārttā* (economics). *Ānuvīkṣikī* is considered as a "source of all knowledge" or a "means for all activities," and a "foundation for all social and religious duties." "When seen in the light of these sciences, the science of *ānuvīkṣikī* is most beneficial to the world, keeps the mind steady and firm in weal and woe alike,

[47]Bays, 1983, p. 224.
[48]*The Bṛhadāraṇyaka-Upaniṣad*, 3: 1.
[49]Mookerji, 1989, p. 317.
[50]King, Richard, 1999, p. 34; *Arthaśāstra*, 1: 2: 6 - 7.

and bestows excellence of foresight, speech, and action. Light to all kinds of knowledge, easy means to accomplish all kinds of acts and receptacle of all kinds of virtues, is the science of *ānuvīkṣikī* ever held to be," suggests Kauṭilaya.[51]

Interrogation, cross-examination, debate, symposia, and discussion were practiced among the Hindus from the ancient period. In the Hindu tradition, scholastic debate (*vāda*) is practiced not only for the disciplines of philosophy; it is also for sciences and religion. The *Ṛgveda*,[52] *Bṛhadāraṇyaka-Upaniṣad*[53] and *Chāndogya*-Upaniṣad[54] refer to debates by councils on philosophical matters. *Tantra-Yukti* is one of the earliest documents, written around 600 B.C.,[55] providing clear evidence of *śāstrārtha* for learning. The *Suśruta-Saṁhitā* (Uttara-tantra, Chapter lxv) mentions that *Tantra-Yukti* has techniques to allow a person to establish his own position and overthrow that of his opponents who are unfair in *śāstrārtha*. As shown in Table 2.1, the *Tantra-Yukti* mentions a code of 32 practices that should be followed in a scientific argument.[56]

The medical treatise *Caraka-Saṁhitā* emphasizes the importance of debate and discussion in the learning process. "Discussion with a person of the same branch of science increases knowledge and brings happiness. It contributes towards the clarity of understanding, increases dialectical skill, broadcasts reputation, dispels doubts regarding things heard. . . Hence it is the discussion with men of the same branch of science, that is applauded by the wise."[57]

According to Caraka, these discussions can be friendly or hostile. In the Hindu tradition, even hostile discussions were an organized tradition in which scholars with differing opinions shared their point of view, and debated with

[51] *Arthaśāstra*, 1: 2: 6 - 7.

[52] I ask, unknowing, those who know, the sages, as one all ignorant for sake of knowledge What was that One who in the Unborn's image hath established and fixed firm these worlds' six regions." *Ṛgveda*, 1: 164: 6.

[53] *Bṛhadāraṇyaka-Upaniṣad*, 6: 2: 1.

[54] *Chāndogya*-Upaniṣad, 5: 31.

[55] Mookerji, 1989, p. 318

[56] Mookerji, 1989, p. 318.

[57] *Caraka-Saṁhitā*, *Vimānasthāna*, 8: 15.

Table 2.1: The Code of Practices in *śāstrārtha*, *Tantra-Yukti*

| | |
|---|---|
| *Adhikaraṇa*, subject | *Vidhāna*, arrangement |
| *yoga*, union of words | *Padārtha*, category |
| *Hetvārtha*, implication | *Uddeśa*, enunciation |
| *Nirdeśa*, declaration | *Updeśa*, instruction |
| *Apadeśa*, specification | *Atideśa*, extended application |
| *Pradeśa*, determination from a statement to be made | *Upamāna*, analogy |
| *Arthāpatti*, presumption | *saṁśaya*, doubt |
| *prasaṅga*, a connected argument | *viparyaya*, reversion |
| *vākya-śesha*, context | *anumata*, assent |
| *Vyākhyāna*, description | *Nirvachana*, etymological explanation |
| *Nidarśana*, example | *apavarga*, exception |
| *Sva-saṁjñā*, a special term | *Pūrva-paksha*, question |
| *Uttara-paksha*, reply | *Ekānta*, certain |
| *Anāgatāvedshaṇa*, anticipation | *Atikrāntāvekshaṇa*, retrospection |
| *niyoga*, injunction | *Vikalpa*, alternative |
| *Samuchchaya*, aggregation | *Ūhya*, ellipsis |

each other.[58] The rule was to avoid any "celebration for the victor" or "any insult to the loser."[59] Since knowing truth was the purpose of the debate, it was the knowledge that became central and not the person. This discouraged a feeling of triumph or defeat for the participants. It was suggested that all assertions in a debate should be made in a polite manner. A person in anger can do anything to win, even inappropriate actions. Therefore, wise people debate in a polite manner.[60] This chapter of *Caraka-Saṁhitā* reminds us of the Robert's Rules of Order that were established in the West centuries later, in the nineteenth century.[61]

Debate was considered a good way to effectively formulate the thought process, a good way to understand the subject matter, and to become established among scholars. Forty-four terms of debates or discussions are described in *Caraka-Saṁhitā*, demonstrating an evolved system of debate in a learning process.[62] It was advised to decide the purpose of debate in advance: Is it for curiosity or to subjugate the opponent? The later purpose of subjugation was practiced in the special situation of *śāstrārtha* for marriage. It was a standard practice among the Hindu to invite prospective grooms for *śāstrārtha* in which they had to answer questions posed by the bride and her family.

The *Caraka-Saṁhitā* suggests *yukti* or heuristic reasoning as a valid and independent means of knowledge. Medical practitioners were advised to free themselves from bias and search for the truth dispassionately.[63]

As the universe is varied and complex, it is impossible to know everything. The *Kena-Upaniṣad* states, "If you think you know it well, you know it lit-

---

[58] *Caraka-Saṁhitā, Vimānasthāna*, 8: 16.

[59] *Caraka-Saṁhitā, Vimānasthāna*, 8: 17.

[60] *Caraka-Saṁhitā, Vimānasthāna*, 8: 22 - 23.

[61] Henry Martyn Robert (1837 - 1923) was an engineering officer in the U.S. Army. In 1875, he wrote a book, *The Pocket Manual of Rules of Order for Deliberative Assemblies*, in two parts. The book immediately gained popularity because it provided rules for group interactions. In 1876, the book was again published by S. C. Griggs and Company with a new title, *Robert's Rules of Order*. The book is still a classic and frequently consulted by groups for smooth interactions and discussion.

[62] *Caraka-Saṁhitā, Vimānasthāna*, 8: 27.

[63] *Caraka-Saṁhitā, Sūtrasthānam*, 25: 32.

tle indeed."[64] Confronted with a world of multifarious diversity, the ancient Hindus realized that wisdom cannot be attained by collecting, analyzing, and mastering all of the manifestations of nature. Instead, they looked for the unified reality producing and permeating all the apparent multiplicity that surrounds us. Hindus looked for the answers to the epistemological questions: "What is that, being known, all else becomes known?"[65] This is analogous to the unified field theory that modern-day physicists are trying to build in explaining the nature of forces. Multiplicity unified by a single reality is a part of Hindu philosophy and has played an important role in Hindus' quest for salvation. The *Ṛgveda* tells us that God is one. We call Him by different names.[66]

The practice of unification of manifold realities into a single reality in the Hindu religion has attracted several modern scientists. Erwin Schrödinger, a Nobel laureate in physics, writes to emphasize that all events in nature are connected with each other: "Looking and thinking in that manner you suddenly come to see, in a flash, the profound rightness of the basic conviction in Vedanta [Hindu philosophy] . . . Hence the life of yours is not merely a piece of the entire existence, but in a certain sense the whole."[67]

"Knowledge is the *kāma-dhenu* [the celestial cow that yields all desires, and came from the churning (*samundra-mathana*) of the ocean] from which all boons in life may be milched," proclaims *Agni-Purāṇa*.[68] This is similar to the "knowledge is power" slogan commonly used these days to emphasize the importance of education. The ancient Hindus viewed knowledge as a liberator from all weaknesses.

As knowledge was considered divine by the ancient Hindus and its myriad representations were acceptable, no atrocities were ever committed against the proponents of a particular philosophy. No censorship was ever imposed; no book was ever burned in ancient India by the Hindus. The capacity to disagree without rancor was the norm; differences were sorted out only

[64] *Kena-Upaniṣad*, 2: 1.

[65] *Muṇḍaka-Upaniṣad*, 1: 1: 3.

[66] *Ṛgveda*, 1: 164: 46. Again, the unified field theory and the quest defined in *Muṇḍaka-Upaniṣad* are inherently different.

[67] Schrödinger, 1964, p. 21.

[68] *Agni-Purāṇa*, 211: 57.

by the established channel of *śāstrārtha*. This is in contrast to what early medieval India faced by the Islamic conquerors, what the Muslims faced from the Christian rulers in Spain during the Inquisition, or what the Chinese intellectuals faced from the emperor during the Qin dynasty. No Socrates, Bruno, Galileo, or Hypatia ever suffered or was persecuted in ancient India for his or her beliefs.

## 2.7 *Smṛti*, Information on the Internal Drive

Imagine that all hard drives, floppy disks, magnetic tapes, CDs, and other storage devices get destroyed in a fire. Imagine that all books are burned as a result of natural calamities or human actions. What would happen to our knowledge? Can we recover what was lost? If so, how fast can we retrieve it? Do we have to reinvent our knowledge? Will it hamper the growth of science? These questions are not merely hypothetical. The histories of the Qin dynasty in China, the Alexandrian Library in Egypt, and the libraries in India, Mesoamerica, and Europe make such questions quite relevant. The lessons of September 11 in America and the destruction of documents in the World Trade Center make such possibilities too real to ignore.

On another level, it is common to erase computer files by mistake. It is possible to lose your laptop, smart phone, or tablet. You can imagine the difficulties such action or loss would create. In our day-to-day lives, we rely heavily on written texts and other external media. We consult our contact list to call our friends, a spreadsheet to balance our checking account, or a shopping list for food. Such lists or books are our extended memories.

Our memories play a pivotal role in defining who we are as persons, our thought processes, our emotions, our actions, and even our outlooks on life in general. Our personal identity, our behavior, our health, and our well-being largely depend on our memory. What happens when we lose our memories? It is the memory that plays "a pivotal role in our thoughts, emotions, choices, actions, and even our personalities."[69]

The burning of the Alexandrian library was not an isolated incident. In 213 B.C., the Chinese emperor Si Huang Ti held a great feast in Hien Yang

[69] McDonald, Devan, and Hong, 2004.

city. Many noblemen came to the feast and gave speeches in honor of the emperor. However, a few scholars criticized the emperor for proclaiming that he was the "first great emperor." Si Huang Ti felt insulted and became angry. He sought the advice of his trusted ministers. One minister told the emperor: "These scholars read their books so much they think of nothing but the olden days. I propose that all books be burned and that anyone who talks about old times and criticizes the present be put to death."[70] The emperor accepted his minister's advice and ordered all books to be brought to the capitol and burned. To counter similar problems, Taoists implemented strict rules to not share information with the outsiders (non-initiated). It worked well for them. Pythagoreans, in the Greek tradition, had to restrict sharing of information with the non-initiated people, so did the ancient Hindus.

Book burning is not just an ancient phenomenon; books are burned and scientists are persecuted every time fundamentalist zealots decide to impose their values on others. In 1499 A.D., in Spain, Ferdinand II and Isabella gave only three days to the Jews and Muslims to convert back to Catholicism. The whole of Spain was illuminated by bonfires of Arabic manuscripts after the third day. Five centuries have passed since that period and Spain has yet to reemerge from the shadows to recapture its former dominance in science and technology.

In contrast, libraries were intentionally burned in India by foreign invaders to destroy Hindu heritage and culture. For example, the libraries in Nalanda and Vikramsila were destroyed in about 1200 A. D. by Bakhtiyar Khilji. It hurt the scientific progress in the region since even the people were killed. However, not all was lost. Some of the manuscripts were reproduced and the intellectual heritage was kept intact despite foreign domination and suppression for several centuries.

The limitation of space to write on, the perishable nature of the medium used to write on, and the absence of a quick reproduction (printing) system were great challenges to the ancients. They dealt with these challenges by relying on their memories. Sophisticated techniques were developed to recall facts; a class of people (Brahmin) were entrusted to preserve the knowledge.

---

[70]Seeger, 1962, p. 118.

How do we remember facts? How does our memory function? How do we store information and retrieve it when needed? These are important questions at an individual level; they are equally crucial at the societal level. How do we preserve our culture, our customs, and our traditions? Different civilizations came up with different answers in different periods.

We remember information by association. Sentences are easier to remember than words in random order. This explains why a story is easier to remember than scattered facts.[71] "Both long- and short-term memories arise from the connection between neurons, at points of contact called synapses, where one neuron's signal-emitting extension, called an axon, meets any of an adjacent neurons dozens of signal-receiving fingers, called dendrites."[72] In a short-term memory, synapses are strengthened only for a short period while they become permanently strengthened for a long-term memory.

We remember in chunks. For example, it difficult to remember 3123745897. However, it is much easier to remember 312-374-5897. No wonder this is how phone numbers are listed in telephone directories. It is for this reason, Social Security numbers are also listed in groups of three, two, and four (123-45-6789).

Stories or poems are easier to remember than abstract texts. Stories have multiple bits of information that are stored in a large number of neurons. These neurons are linked with each other and this link of neurons allows recollection. It is this link of neurons that separates good writers from the average ones, or a good teacher from an average one. Good writers provide connections and relevance of the information in such a way that allow the information to be stored for long periods.

Sensory memories use different senses to tell us how something looks, how it smells, how it feels to touch, how it sounds, etc. "Most memories do not appear as full-blown, exact duplicates of the information learned, but are generated through a process of reconstruction."[73]

---

[71] Fields, 2005., Higbee, 1993, p. 30; Underwood, 1964.
[72] Fields, 2005.
[73] Higbee, 1993, p. 30.

Retention of knowledge is a serious issue to most science educators and students. Yet, the issue of retention of information is not much discussed in science. On the lack of emphasis of retention issues in science teaching, Roald Hoffman, a Nobel Laureate in chemistry, laments: "Science seems to be afraid of storytelling, perhaps because it associates narrative with long, untestable yarns. Stories are perceived as 'just' literature. Worse, stories are not reducible to mathematics, so they are unlikely to impress."[74] However, Hoffman rejects the premise and suggests that "[a]ll theories tell a story. They have a beginning, in which people and ideas, models, molecules and governing equations take the stage. Their roles are defined; there is a puzzle to solve."[75]

Mnemonic devices are used to enhance memories. These devices can be visual or verbal. Researches indicate that "when you are trying to recall a word, other words that rhyme with it may be effective cues to help recall it . . . even when they were not together in the original list."[76]

In India, the preservation of their intellectual and cultural heritage was made possible by writing the scientific literature in poetic verses or in stories. This is similar to what Pythagoreans practiced in Greece. The native people of the Americas also kept their heritage alive through songs and stories for several hundred years despite experiencing domination by the European settlers.

The hymns of *Vedas* were recited according to certain modulations in an oral tradition. People learned these verses by heart. Only a small group of people also understood the meaning. The intellectual wisdom of the ancient Hindus was not to be shared with the uninitiated. This was probably intended to avoid distortions or misinterpretations of their knowledge. Disciples were selected with great care by teachers for the preservation of knowledge. It created a class of people (brahmin) that was entrusted with this responsibility of preserving the knowledge and culture. Oral transmission of knowledge became a practice in ancient India and continues in some respects

---

[74]Hoffman, 2005

[75]Hoffman, 2005. My interactions with Roald Hoffman strengthened my resolve to work on topics that are not in the mainstream of physics, including this book. I, along with Ivan Brady, once hosted Hoffman to push for a project on conversations across disciplines.

[76]Higbee, 1993, p 48.

even today.

The ancient Hindus memorized their literature verbatim. The spoken words, not the written words, have been the basis of literary and scientific traditions of the Hindus. The people who memorized the texts were highly respected as they became the tools that could keep the tradition alive. This situation has not changed much even today. People who memorized *Vedas* or *Upaniṣads* are well respected in today's Hindu society.[77] The first written accounts in India are from the period of Aśoka (r. 269 - 232 B.C.), the third emperor in line of the Mayura dynasty. He erected rock edicts all over India, and these edicts are a useful guide to life in ancient India. However, such inscriptions or manuscripts are limited in number.

Yijing (also I-tsing, 635 - 713), a Chinese traveler who visited India, was impressed when he met people who could recite hundreds of thousands of verses of *Vedas*. "The *Vedas* have been handed down from mouth to mouth, not transcribed on paper or leaves. In every generation there exist some intelligent Brahmans who can recite 100,000 verses . . . This is far from being a myth, for I myself have met such men."[78]

The ancient Hindus used symbols and inflections in their books that made sense only to the initiated. This encryption of ideas was frequently used in

[77] Vedi (or Bedi), Dvivedi, Trivedi, and Caturvedi are common last names (surnames) among brahmins, the people who were entrusted to preserve *Vedas*. These names literally symbolize the number of *Vedas* memorized by the person. *Dvi* means two and Dvivedi is the surname of a person who has memorized two *Vedas*. Similarly, *tri* and *catur* mean three and four, respectively. Thus, Trivedi and Caturvedi are the people who have memorized three or four *Vedas*, respectively. Of course, today these surnames are inherited from father to children without any connection to memorization.

[78] Takakusu, 1966, p. 182. I have thought of this statement along with the people who memorized *Vedas* in India. Suppose, it takes 10 seconds to narrate one verse. To narrate 100,000 verses, this equals to one million seconds. One million seconds equal a period of non-stop chanting for more than 11 days. How can one remember voluminous texts that take more than 11 days to read? Obviously, the person had to breathe in between, take rest, eat food, and sleep. This means that, in a rough estimate, the person memorized a text that took about 25-30 days to narrate. I find it difficult to comprehend. Is it possible? Similarly, I find it intriguing to think that there were people who memorized all four *Vedas*. Yet, it is a common belief among the Hindus and the surnames that are mentioned above are quite common among Brahmins. Historical documents support such memorization.

the Hindu literature. For example, in the *Arthaśāstra*, a book written by Kauṭilya, it was advocated that kings should be trained in cryptography in order to communicate secret messages without detection. Similar advice was given by Vātsyāyana to the common people in his book, *Kāmasūtra*, known especially for the art of sensuality. This method of knowledge preservation using the oral tradition later degenerated into privileges of the upper social classes in Hindu society.

The ancient Hindus understood that human minds recognize symmetry and rhythm; it is much easier to memorize poetry than prose. The ancient Hindus followed the practice of putting their literature into poetic verses. This was true not only for the religious literature, but for the sciences too. The resources of poetry were applied to all branches of knowledge – systems of philosophy, medicine and mathematics, grammar, lexicon and other technical subjects along with arts and crafts. This practice played an important role in the history of the Hindus and allowed the preservation of their culture.

The poetic verses of Hindu literature are written in an obscure and mystical language for a variety of reasons. Although this technique is well adapted to aid the memory of those who already understand the subject, it is often unintelligible to the uninitiated. One can take a crash program in German, Russian, Chinese, Arabic, or French in order to understand modern scientific literature, but this would not be sufficient for an understanding of German, Russian, Chinese, Arabic, or French poetry. It can be difficult to understand subtle meanings in a poem even after long familiarity with the language, since most poetry contains inflections, metaphors, and other hidden meanings especially to maintain rhythm.

"The Rg Veda [*Ṛgveda*] is a poetic text, and its language is metrically organized in a certain way, so that it differs, for instance, from colloquial speech. . . the solemn manner of recitation, and the musical part of the delivery of the hymn, taken as a whole, rendered the language truly divine in the *Ṛṣis'* mind," writes Elizarenkova in her book, *Language and Style of the Vedic Ṛṣis.*[79]

The *Ṛgveda* distinguishes between ordinary speech and poetic speech.

[79]Elizarenkova, 1995, p. 107.

"When men of wisdom create by their intellect verse after winnowing as barley grains are sifted by means of a winnowing basket, then men of equal knowledge understand the meaning (contained in the verse); in their verses (speech) blessed glory is enshrined."[80]

Sanskrit literature is characterized by the polysemous use of words as memory-aiding devices to maintain rhythm and tone. This led to the evolution of Sanskrit poetics which is rich in metaphorical descriptions (*alaṅkāra*). Sanskrit poetics does not demarcate poetry and prose–they both elucidate feelings or facts. This is in contrast to the popular notion that poetry is an art to provide aesthetic pleasure to the readers while prose is precise and more appropriate to the sciences.[81]

Poetic truth does not reveal itself to all: it unravels its mysteries and beauty only to connoisseurs. It requires the imagination of the readers; they need to recognize symbols presented through poetic expression and translate them into scientific facts. Writing science as poetry allowed gifted writers to express abstract scientific facts meaningfully and facilitate memorization of texts. Many ancient Hindu books were written in a language that had two or more meanings. One meaning was for the layperson, while the second was for scholars who were trained in the encryption system. To correctly identify the meaning of individual words, one must determine the reference type of a large segment of the text in which the particular word occurs.[82] This practice of writing in poetry allowed the preservation of the *Ṛgveda* for several thousand years in the oral tradition before the written records were prepared. Such intentional double meanings create serious hurdles for our comprehension of Hindu literature.

The most difficult aspect of Sanskrit poetics is the technical terminology which mostly uses the metaphorical meaning of common words. For example, *lal* is a term used for red color as well as for baby (mothers use them even for their adult children.). This means that the word could be used for appearance as well as for a person. Elizarenkova gives the example of the word *mánasa* which is defined as "inner emotion," "spirit," "mind," "rea-

[80] *Ṛgveda*, 10: 71: 2.
[81] Kane, 1961, p. 346.
[82] Elizarenkova, 1995, p. 285.

son," "thought," and "heart." Similarly, *hṛd-* or *hṛdaya* in the *Ṛgveda* meant not only heart (as part of body), but also the focal point of various emotions (as joy, fear, inspiration, etc.), and *hrd-* in the strictly anatomical sense often has the wider meaning of internal organs in general.[83] In some situations, these meanings are so divergent that multiple interpretations are possible. How do we know then what was the intended meaning of the author?

Synonymy has always been an outstanding feature of the Sanskrit vocabulary throughout its course of evolution, which is quite natural for a language with an abundant literary tradition. However, this metaphorical use of language, along with synonymy and polysemy, makes Hindu literature difficult for an uninitiated person.[84]

This has created a unique problem in the Sanskrit literature. Most Sanskrit texts are now translated into English and these translations serve as the only window to the richness of the literature to most scholars, especially for those who are not well versed in Sanskrit. Unfortunately, these translations of the same Sanskrit text vary considerably with each other, particularly between the native and foreign translators. It is common for some foreign scholars to bash the native scholars as biased and fundamentalists. These native scholars are accused of distorting or revising history. This has suppressed the translation efforts among the Hindus. For this reason, the popular secondary literature on Hinduism is produced mostly by non-practitioners of Hinduism. This is in contrast to Catholicism, Judaism, and Islam. This is certainly not a good situation.[85]

The hymns of sacred books of the Hindus have sentences with two meanings: one in the literal sense that makes it easy to memorize and, second, the encoded message. This poetic form of scientific writing, although time-consuming and cumbersome, was perhaps an excellent way to popularize tedious abstract scientific principles for the common people, especially in an oral culture. It was not due to the availability of free time that prompted

[83]Elizarenkova, 1995, p. 30.

[84]Elizarenkova, 1995, p. 74.

[85]Many of these native scholars do not have a global network and they at times cannot even present their counterarguments. In this book, being someone who has a marginal knowledge of Sanskrit, I struggled in choosing a particular translation. In most cases, I have consulted multiple translations to check the authenticity.

Hindu scholars to formulate their literature – even scientific – in poetry; it was done out of the need to preserve their knowledge for the following generations.[86]

Al-Bīrūnī, an Islamic scholar who lived in India for some thirteen years during the eleventh century, wrote of the importance of poetic literature in popularizing science: "By composing their books in metres [poetry] they [Hindus] intend to facilitate their being learned by heart, and to prevent people in all questions of science ever recurring to a written text, save in case of bare necessity. For they think that the mind of man sympathizes with everything in which there is symmetry and order, and has an aversion to everything in which there is no order. Therefore most Hindus are passionately fond of their verses, and always desirous of reciting them, even if they do not understand the meaning of words, and the audience will snap their fingers in token of joy and applause. They [Hindus] do not want prose compositions, although it is much easier to understand them."[87]

**Al-Bīrūnī**

Al-Bīrūnī (973 - 1051 A.D.) was born near Khwarizm near Persia, in modern Uzbekistan. His mother-tongue was Persian and his main language was Arabic, as is the case with most scholars in the region during the period. He also knew Sanskrit and translated several scientific texts from Sanskrit to Arabic. He had a good grasp of the Greek and Indian works. Al-Bīrūnī believed in Ptolemy's model of the geocentric earth and criticized Āryabhaṭa I for assigning motion to the Earth. (See Chapter 5)

He was a devout Muslim and wrote many books. Some of his popular books are: *Kitāb fi taḥqīq ma li'l Hind* (Book on India), *Nihāyāt al-amākin li-taṣḥiḥ masāfāt al-Masākin* (Determination of the Coordinates of Cities), and *Kitāb altafhīm li-awā'il sinā 'at al-tanjīm* (The Book of Instruction in the Art of Astrology). His book on India is valuable in

[86]Kane, 1961. Karpinski (1965, p. 42), perhaps in a frustrating moment, wrote that these Hindus "had a great deal of time to spend on thinking up the words." L. C. Karpinski (1878 - 1956) wrote a classic book on the history of arithmetic in which this comment was made. Yes, the ancient Hindus wrote hymns even in their mathematical texts.

[87]Sachau, 1964, vol. 1, p. 137.

knowing the stature of Hindu science during the eleventh century. This book also provides valuable cross-cultural perspectives on various sciences.

To understand the complexity, a list of words that can be substituted for numerals is compiled in Table 2.2.[88]

To understand the complexity of such a system for an English-speaking person, let me design a simplified analogy of such a system. The following numerical values are assigned to the consonants of the English language:

| Number | Letters |
|---|---|
| 1 | b, n, and z |
| 2 | c and p |
| 3 | d and q |
| 4 | f and r |
| 5 | g and s |
| 6 | h and t |
| 7 | j and v |
| 8 | k and w |
| 9 | l and x |
| 0 | m and y |

The vowels are not associated with any number, and are used as *fillers* in forming words with no magnitude assigned to them. For example, the number 2517 can be depicted by choosing various possible consonants for 2, 5, 1, 7, and inserting the vowels to complete the word. Thus, *Casanova* can represent number 2517 in this system. A large number of words can be chosen using this system, for a given set of numbers. The number 2516 can be represented as *pageant, paginate, cognate, or cogent.* Similarly, the number 28 can be expressed by the words *cow, paw, cake, or poke.*

[88]The list provided in Table 2.2 is not complete; a detailed list of such words can be found in Bag, 1979, p. 69; Datta and Singh, 1938, vol. I, p. 55; and Sachau, 1964, vol. I, p. 178, 179.

Table 2.2: Sanskrit Words for Various Numerals

| Number | Words |
|---|---|
| 0 | *śūnya* (nothing), *kha* (point), *gagana* (sky), *viyat* (heaven), *ākāśa* (sky), *ambra*, (heaven), *abhra* (heaven), *nabha* (sky), *bindu* (point), *viṣṇupada* (step of Lord Viṣṇu), *vyoma* (moving through air). |
| 1 | *candra* (Moon), *soma* (Moon), *sūrya* (Sun), *pṛthvī* (Earth), *indu* (Moon), *dharā* (Earth), *śaśi* (Moon), *bhūmi* (land and earth), *vasudhā* (Earth), *rūpa* (shape), *ek* or *ekam* (one), and Sītā (wife of Lord Rāma). |
| 2 | *Āśvin* (two celestial gods), *netra* (eyes), *nayana* (eye), *cakṣu* (eye), *bāhu* (arm), and *karṇa* (ear). |
| 3 | *Tri-kāla* (three times: past, present, and future), *tri-netra* (three eyes, it refers to the third eye of Lord Śiva), *loka* (worlds), Rāma (god), *agni* (fire), and *guṇa* (property). |
| 4 | *Veda* (four vedas), *śruti*, *varṇa* (caste), *yuga* (period), *diśā* (direction), *samundra* (ocean), *sāgar* (ocean), and *āśrama* (four stages of life). |
| 5 | *pañcabhūta* (five elements), *pāndava* (five brother kings), *indriya* (senses), *arth* (material or wealth), *pañca* (five) and *ratna* (jewel-stones). |
| 6 | *Aṅga* (body-part), *ṛitu* (season), *rāsa* (taste), *rāga* (melody), *ṣaṣtam* (six) and *śāstra* (holy book) |
| 7 | *ṛṣi* (sages), *parvata* (mountain), *nāga* (snake), *muni* (saint), *śaila*(stone or mountain), *svara* (voice), *saptam* (seven), and *dvīpa* (island). |
| 8 | *mūrti* (eight forms of Lord Śiva), *gaja* (elephant), *Maṅgala* (planet Mars),*vasu*, *ibha* (elephant), and *aṣṭa* (eight). |
| 9 | *aṅka* (numerals), *gṛha* (planets), *navam* (nine), *nanda*, *pavana*, *upendra*, and Durgā (goddess). |

With a little practice, a fairly large number of words can be chosen. For example, words such as *stared, stride, asteroid, stored, steroid, strode, shroud, shred, shard, and shared* can be chosen for the number 5643. Similarly, we can chose *aconite, cabotage, cantos, cants, cantus, counts, counties, cubits, iceboats, paints, panties, pants, peanuts, pintos, pints, points, and punts* for 2165. For 4264, we can choose *factor, raptor, rupture, rapture, richer, rector, erector,* and *archer.*

Let us try to create a sentence that can have literal as well as scientific meaning: "The magnitude of the acceleration due to gravity is equal to *Luke* over *bum.*" Here Luke represents 98 and bum represents 10. Thus, 98 over 10 = 9.8 which is the acceleration due to gravity in MKS system. Another sentence could be, "*Is* and *it* equals to *none.*" Here "is" equates to 5 while "it" signifies 6. The sum of the two is 11, signified by "none." As one can notice, both are imperfect examples as the literal meanings do not make much sense. To define someone's age who was born on February 1953, one can make the following poem:[89]

If you want to find the year of my birth,
find an *obelisk* and take way its *age.* [1958 - 5 = 1953, the year of my birth]
The days of my birth month are baked in a *cake,* [28 days = February]
And the month of my birth is mostly frozen in *ice,* [2, the second month]
This makes my age as *six* minus *one.* [59 - 1 = 58 in year 2011.]

It is difficult to write such poems where you have two meanings. However, this practice was helpful in preserving knowledge.

When the textual riches of Alexandria, China, Baghdad, and Rome were put to flame at different times in different contexts, knowledge of these centers and cultures disappeared like the smoke in the sky. In contrast, the ancient Hindus could develop a system that was fire-proof – a system that could survive for several thousand years.[90]

---

[89]The following poem is written by Ann Hughey, one of my students. I have made minor modifications.

[90]Montgomery and Kumar, 2000; Patel, 1996.

## 2.8 Discovery and Invention for *Mokṣa*

The connection between religion and science in Hindu literature is an interesting, and perhaps unexpected, variation from the Western model. The sciences of the ancient Hindus were an essential and integral part of their religion, as it was for the Mayans, Arabs, and Egyptians. The disciplines of astronomy, mathematics, chemistry, physics, yoga, and medicine were all practiced to meet the needs of religion, as well as to fulfill natural curiosity.

As mentioned earlier, Nārada considered astronomy (*Nakṣatra-vidyā*), mathematics (*rāsi-vidyā*), logic, history (*itihāsa-Purāṇa*), grammar (*vednāṁ vedaṁ*), and fine arts useful to know the ultimate truth.[91] Thus, astronomy, logic, mathematics, and history became tools to achieve *mokṣa* (salvation). Elsewhere in *Muṇḍaka-Upaniṣad*, when Śanunka went to Aṅgiras to find out "by knowing which a man comes to know the whole world," the reply included astronomy, along with several other disciplines.[92]

Like *Chāndogya-Upaniṣad* and *Muṇḍaka-Upaniṣad*, *Agni-Purāṇa* also treats various natural sciences as sacred, along with other domains of knowledge: "May the holy *Veda* such as *Ṛg*, etc., with their six branches of kindred sciences, the books of history, the *Purāṇa*, the medical sciences, the science of music and war, the science of proper pronunciation, ritual, grammar, lexicon, astronomy, and prosody, the six schools of philosophy . . . grant thee peace."[93] In such an intellectual environment in which religion suggested that people practice science, natural science was bound to prosper among the Hindus. It is important to mention here that this harmony between science and religion has always prevailed in history among the Hindus.

Progress in science has never been a hindrance to spiritual growth in the history of Hinduism. This is in contrast to periods in some religions where people had to make a choice between science and religion. The history of Europe provides examples of such confrontations between science and religion. For example, Bruno[94] was burned to death at the stake and Galileo[95]

---

[91] *Chāndogya-Upaniṣad*, 7: 1: 2, 4.
[92] *Muṇḍaka-Upaniṣad*, 1: 1: 5
[93] *Agni-Purāṇa*, 219: 58-62.
[94] Boulting, 1972; Yates, 1964.
[95] Finocchiaro, 1989.

was imprisoned when their scientific beliefs were in conflict with the Roman Catholic Church during the period of the Inquisition.

The natural philosophy of the Hindus fulfills the religious needs of people. For example, Āryabhaṭa I (born 476 A.D.) studied astronomy and other natural sciences in order to achieve *mokṣa*. "Whoever knows *Daśgītika Sūtra* [ten verses] which describes the movements of the Earth and the planets in the sphere of the asterisms passes through the paths of the planets and asterisms [stars] and goes to the higher Brahman [God]," the highest goal for any Hindu.[96] About a millennium before Āryabhaṭa I, Kaṇāda, proponent of the atomic theory of matter, suggested in his *Vaiśeṣika-Sūtra* that a study of physics leads to salvation.[97]

Astronomy was not the only science that was considered sacred by the Hindus; all the other sciences were also considered divine. Mastery of any of the sciences was considered to be a pathway to salvation – a goal desired by all Hindus. The *Agni-Purāṇa* suggests that knowledge of the human anatomy can also lead to salvation. "Said the God of Fire: Now I shall describe the system of veins and arteries [*Nāḍī-cakra*] that are to be found in the human body. A knowledge of these [arteries and veins] leads to a knowledge of the divine Hari [God]."[98]

Thus, as suggested in the sacred books of the Hindus, astronomy, physics, chemistry, medical science, and biology were sacred disciplines and any person well versed in these disciplines attained salvation.

Since science was a prescription to *mokṣa*, it became imperative for scientists to find true knowledge. Thus, science could grow independently and scientists could investigate whatever they deemed fit. This led al-Mas'ūdī (d. 957 A.D.), an Islamic historian during the tenth century, to write that science and technology were established without the aid of religious prophets in India. In his opinion, wise men could deduce the principles without the need of religion. It is not the prophets who dictated the domain of science, it was the logic, intuition, and experience of diligent observers who contributed

[96] *Āryabhaṭīya*, *Daśgītika*, 13.
[97] *Vaiśeṣika-Sūtra*, 1: 1 - 4
[98] *Agni-Purāṇa*, 214: 1 - 5.

to the domain of science. Al-Mas'udī considered India as the land of "virtue and wisdom."[99]

Knowledge is procured by logic (*tark*), intuition (*anubhūti*), and experience (*anubhava*), which are common to both religion and science. Science was a part of religion in the more comprehensive definition of religion. The religion of the Hindus was dynamic, in which contrary ideas could coexist with each other. As *mokṣa* was the prime goal for the ancient Hindus, natural philosophy, with its branches in physics, chemistry, mathematics, biology, and medicine, grew under the shelter of religion; these disciplines of science became the religion for the ancient Hindus. People could explore and practice these disciplines of science with logic, intuition, and experience. They could modify existing theories when confronted with new explanations. This allowed al-Mas'udī to claim that the Hindu science was not under the influence of religion.

As explained in this chapter, the ancient Hindus were obsessed with the virtues in life and tried their best to reach their fullest potential. This was crucial even for an illiterate person to reach her or his fullest potential as life was not a single step; it is a stage in a long series of reincarnations that eventually lead to *mokṣa*. The transmigration of soul theory encouraged people to excel. This excellence was in their conduct, physical health, and intellect. They tried hard to excel in science. Respect for the guru was an outcome of this necessity to excel. Science was bound to prosper among the ancient Hindus.

[99]Khalidi, 1975, p. 102 - 106.

# Chapter 3

# The Hindu Numerals

Human societies, even the most rudimentary ones, need a counting system. For primitive people, small numbers were sufficient to meet their daily needs. People did not conduct money transactions using currency, silver, or gold coins, as the coins came much later. Counting was mostly restricted to the use of fingers–to designate the size of the family, the number of cows or sheep a person inherited, etc.

People bartered different products of everyday necessities during the ancient world – the exchange of food for cotton or the exchange of wood for water, a visual estimate of volume or a feel for the weight of an object was good enough to finalize a transaction in the absence of proper weight standards. Eventually, currency transactions, systems of weight and measures, arithmetic, and algebra evolved to meet the needs of trade. In different parts of the world, attempts were made to invent a number system: the Greeks, Egyptians, Babylonians, Mayans, Chinese, and Hindus devised their own counting systems. The Hindu system of counting prevailed due to its ingenuity – the decimal (or base 10) system of counting with place-value notations. Historians of mathematics are in general agreement that the base 10 numeral system was invented in India and later transmitted to the rest of the world via Arabia. Thus, these numerals are popularly called Hindu-Arabic numerals.

The system of counting invented by the Hindus is so simple that it is difficult to realize its profundity and importance.[1] Young children in most countries are generally taught this counting system first and the alphabet of their native language later. Children are taught to write eleven as one and one (11) written side-by-side which they learn without much difficulty. "Our civilization uses it unthinkably, so to speak, and as a result we tend to be unaware of its merits. But no one who considers the history of numerical notations can fail to be struck by the ingenuity of our system, because its use of the zero concept and the place-value principle gives it an enormous advantage over most of the other systems that have been devised through the centuries."[2]

"It is India that gave us the ingenious method of expressing all numbers by means of ten symbols, each symbol receiving a value of position as well as an absolute value; a profound and important idea which appears so simple to us now that we ignore its true merit. But its very simplicity, the great ease which has lent to all computations, put our arithmetic in the front rank of useful inventions; and we shall appreciate the grandeur of this achievement more, when we remember that it escaped the genius of Archimedes and Apollonius, two of the greatest men produced by antiquity," wrote Laplace, a noted French mathematician and astronomer during the eighteenth century.[3]

## 3.1 The Decimal System

The decimal system represents numbers as linear combinations of powers of ten. Several civilizations such as the Egyptians, Chinese, and Romans used the additive decimal systems although their symbols were different. In their system, multiple repetition of symbols were used and added to increase magnitude. Thus, in the Roman system, X means 10 and XX becomes (10 + 10) = 20. Also, XVI amounts to 10 + 5 + 1 = 16. Romans also used subtractive symbolism where IX and IL represents 9 (as 10 - 1) and 49 (as

[1] Ifrah, 1985, p. 428. Ifrah's book is still one of the most engaging and scholarly book on numerals in various ethnic cultures. The other noted books are by Calinger, 1999; Cook, 1997; Katz, 1993; Smith, 1925; Smith and Karpinski, 1911; Suzuki, 2002.

[2] Ifrah, 1985, p. 428.

[3] Kadvany, 2007; Srinivasiengar, 1967, p. 5. This quotation originally appeared in Halstead's book (1912). Since then, it has been reproduced in several history texts. The original reference is unknown.

Table 3.1: Words and Their Numeric Equivalent in the *Vedas*

| Number | Word Numeral | Number | Word Numeral |
|---|---|---|---|
| 1 | *eka* | 20 | *viṁśati* |
| 2 | *dvi* | 30 | *triṁśat* |
| 3 | *tri* | 40 | *caturiṁśat* or *catvāriṁśat* |
| 4 | *catur* | 50 | *pañcāśat* |
| 5 | *pañca* | 60 | *ṣaṣṭi* or *ṣaṣṭyā* |
| 6 | *ṣaṣṭa* | 70 | *saptati* or *saptatyā* |
| 7 | *sapta* | 80 | *aṣīṭi* or *aṣīṭyā* |
| 8 | *aṣṭa* | 90 | *navatini* or *navatyā* |
| 9 | *nava* | 100 | *śata* |
| 10 | *daśa* | 1000 | *sahasra* |

50 - 1), respectively.

### 3.1.1 The Word-Numerals

In an oral tradition, it is reasonable to choose words for numbers since it is easier to use them in narrative. Many cultures of the ancient period used word numerals before they used symbols to represent numbers. In this system, as was common in several cultures, specific words were assigned to all numbers starting from one to ten, in the decimal systems (Egypt, Greece, China, and India) After ten, different words were assigned to 20, 30, 40, . . . 90, 100, 1000, 10000. No separate words were assigned to 11, 12, . . . 18, 19, 21, 22, . . . 98, 99, 101, 102, etc.; these numbers were built up from a combination of previously defined numbers. Table 3.1 provides these words that were used during the Vedic-period and their equivalent numeric values.

The word-numerals in the Sanskrit language are written similar to the way numbers are written in the German language. The English language uses the Roman system in word-numerals. For example, 34 is called four and thirty (*vier und dreizig*) in German while thirty and four in the English system. Similarly, fourteen is called *vier-zehn* (four and ten) and fifty-three is called *drei und fünfzig* (five and fifty). The *Ṛgveda* mentioned 12 as two and ten,[4] 34 as four-thirty,[5] 53 as three and fifty,[6] 77 as seven and seventy.[7] just like the German system.

In case of compound numerals, the number of higher order was placed as qualifier and the lower as qualified. For example, eleven is defined as ten qualified by the addition of one, thus giving *eka-dasa*. The number 720 is denoted as seven-hundred-twenty[8] and the number 60,099 is written as sixty-thousand-ninety-nine[9] in the *Ṛgveda*. Similarly, consistent with the practice in the *Ṛgveda*, the *Baudhāyana-śulbasūtras* expressed the number 225 as (200 and 5 and 20)[10] and the number 187 as (100 + 7 + 80).[11] Similarly, in the *Atharvaveda*, eleven is counted as one and ten.[12] The numbers 12, 13, . . ., are counted as two and ten, three and ten. . ., respectively.

The following hymns of the *Ṛgveda* provide a multiplication table for 2 up to 10 and a base-10 system of numeration: "Come, Indra, when invoked, with two horses, or with four or with six, or with eight, or with ten, to drink the Soma nectar: object of worship, the nectar is poured out; do not make a mistake. Come to our presence, Indra, having harnessed His car with twenty, thirty, or forty horses; or with fifty well trained steeds; or with sixty or with seventy, Indra, to drink the Soma nectar. Come to our presence, Indra, conveyed by eighty, ninety, or hundred horses: this Soma has been poured into the goblet, Indra, for his exhilaration."[13]

Near Bombay, in the city of Nāneghāt, inscriptions are written in the Brāhmī alphabet that are from the pre-Christian era (around 1st century B.C.). In word numerals, 12 is written as (10 + 2); 1101 as (1000 + 100 + 1); 189 as (100 + 80 + 9).[14]

---

[4] *Ṛgveda*, 1: 25: 8.
[5] *Ṛgveda*, 1: 162: 18; *Ṛgveda*, 10: 55: 3.
[6] *Ṛgveda*, 10: 34: 8.
[7] *Ṛgveda*, 10: 93: 15.
[8] *Ṛgveda*, 1: 164: 11.
[9] *Ṛgveda*, 1: 53: 9.
[10] *Baudhāyana-śulbasūtras*, 16: 8.
[11] *Baudhāyana-śulbasūtras*, 11: 2.
[12] *Atharvaveda*, 19: 23: 1 - 7.
[13] *Ṛgveda*, 2: 18: 4 - 6.
[14] A list of such numbers is provided by van Nooten, 1993.

For the numbers provided in the *Vedas* and *Śulbasūtras*, the following points are clearly established:

1. The numbers from one to ten have specific names.
2. After ten, specific words are designated to 20, 30, . . . 100. All other numbers in between are defined as a combination of these words.
3. After 100, new words are assigned to 1,000, 10,000, 100,000, etc.
4. In word numerals, numbers between 10 and 20 were defined as a number above ten and then 10. For example, 12 is defined as 2 and 10 (*dvi-daśa*). A similar practice was made for other numbers. For example, 99 is written as (9 + 90), 76 as (6 + 70), etc.

The above summary demonstrates that numbers were counted in tens during the *Ṛgvedic* period, implying a decimal base for the counting system.[15]

If the Egyptians, Greeks, and Romans also used the decimal systems, what was so unique about the Hindu system? Well, the Hindus used a decimal system with place-value notations. This system has a unique advantage in arithmetic operations, and is the basis of modern computational devices. The place-value notation were not used by the Egyptians, Greeks, and Romans, which made mathematical calculations quite cumbersome.

## 3.2 The Place-value Notations

The Greeks, Egyptians, and Romans did not use place-value notations in writing numbers while the Babylonians, Chinese, Mayans, and Hindu did use place-value notations. The current place-value system (also called position-value system) that is base-10 is Hindu in origin. This is certain. The uniqueness of the Hindu system lies in the fact that the position of a number qualifies its magnitude. Tens, hundreds, or thousands were not represented by different signs; they are represented by using digits in different positions. Notice that the one is in the second place in 10 (ten), in the third place in

[15] A decent summary of the terms is provided by Gupta, 1983b.

100 (hundred), and in the fourth place in 1,000 (thousand).

In a positional- or place-value system, a number, represented as $x_4x_3x_2x_1$ can be constructed as follows:

$$x_1 + (x_2 \times 10^1) + (x_3 \times 10^2) + (x_4 \times 10^3)$$

Where $x_1$, $x_2$, $x_3$, and $x_4$ are nonnegative integers that have magnitudes less than the chosen base (ten in our case). As you may have noticed, the magnitude of a number increases from right to left. For example, the number 1234 will be written as

$$4 + (3 \times 10^1) + (2 \times 10^2) + (1 \times 10^3)$$

A large number, one million two-hundred thirty-four thousand five-hundred sixty-seven is written as 1234567 in this system. It is equivalent to $[7 + (6 \times 10) + (5 \times 10^2) + (4 \times 10^3) + (3 \times 10^4) + (2 \times 10^5) + (1 \times 10^6)]$.

Similarly,

$$1.2345 = 1 + \frac{2}{10} + \frac{3}{10^2} + \frac{4}{10^3} + \frac{5}{10^4}$$

Such a positional number system does not have to be necessarily based on ten. This system can be designed for any base: Babylonians used a base-60 system with just two symbols while the Mayan used a base-20 system, also with just two symbols. In the Babylonian systems, number $x_4x_3x_2x_1$ is equivalent to:

$$x_1 + (x_2 \times 60^1) + (x_3 \times 60^2) + (x_4 \times 60^3)$$

Although the place-value system that was base-sixty was in use in Babylonia during the ancient period, the system was not popular. "It remained alien to Babylonian everyday life until it was replaced by the Hindu system."[16]

[16]Saidan, Ahmad S., Numeration and Arithmetic, in the book by Rashed, 1996, vol. II, p. 334.

In India, Āryabhaṭa I (born 476 A.D.) used a positional value system in describing numbers, did not even bother to explain much about it, and claimed it to be ancient knowledge. This indicates that the system was prevalent then.[17]

The earliest written record of the place-value notation comes from Vāsumitra, a leading figure of Kaniska's Great Council. According to Xuan Zang (also known as Hiuen Tsang, 602 - 664), Kusāna King Kaniska (144 - 178 A.D.) called a convocation of scholars to write a book, *Mahāvibhāsa*. Four main scholars under the chief monk, named Pārśva, wrote the book in 12 years. Vāsumitra was one of the four scholars. In this book, Vāsumitra tried to explain that matter is continually changing as it is defined by an instant (time), shape, mass, etc. As time is continually changing, therefore, matter is different in each situation although its appearance and mass do not change. He used an analogy of place-value notation to emphasize his point. Just as location of digit one (1) in the place of hundred is called hundred (100) and in place of thousand (1,000) is called thousand, similarly matter changes its state (*avasthā*) in different time designations.[18]

New explanations are generally given in terms of known and established facts. Thus, the very reason Vāsumitra used place-value notation as an example establishes that the place-value notation was considered as an established knowledge during the early Christian era.

In modern perspective, just imagine reading the values of various stocks in a newspaper. In a quick scan, you can recognize easily that 1089 is greater than 951. All you need to see is that the first number has four digits while the second number has only three. This is enough for a quick comparison. In contrast, in the Roman numerals, XC (90) is five times more in magnitude than XVIII (18). This is not easy to figure out in a quick glance. Also, mathematical operations of multiplication, division, addition and subtraction become much simpler in a place-value notation.

The shape of the Hindu numerals has gone through numerous changes over time. The most recent change came after the electronic counting de-

---

[17] *Āryabhaṭīya*, Gola, 49 - 50.

[18] Ruegg, 1993; Sinha, 1983, p. 130.

vices. Different fonts provide different shapes to the numerals, though quite similar to each other. A number written by hand is different from the numeral printed by a machine. The shape of numerals is not so important and one can choose any symbol for a numeral. Al-Uqlidīsī (920 - 980 A.D.), a scholar from the Middle East, in his book *Kitāb al-Fusūl*, noticed this fact: "The people of this craft have agreed on them. Thus the letters are conventional, and if they were to change form that will not affect the essence."[19] The shapes of the numerals were called *huruf al-Hind* in Arabic, meaning the Hindu letters.[20]

In India, varied symbols were used in different parts of the country, as noticed by al-Bīrūnī: "As in different parts of India the letters have different shapes, the numerical signs too, which are called *anka* [*aṅka*], differ. The numerical signs which we use [in Arabia] are derived from finest forms of Hindu signs."[21] This provides a clear view that, in al-Bīrūnī's mind, the so-called Arabic numerals were of Hindu origin.

The pronunciation used for numbers in various languages is similar to Sanskrit pronunciation, indicating a common origin. Table 3.2 compiles a list of these words in different languages. While the similarities of words for numerals in various cultures indicate a common origin, the antiquity of the Sanskrit language establishes the source. The words used for numerals in Sanskrit have their origin in the *Ṛgveda* which is about 3500 years old.

The ancient Greeks as well as the Arabs avoided the use of big numbers in their calculations. The Greeks used *myriad* ($10^4$) as the limit of counting while the Romans defined *mille* ($10^3$) as the largest number. The Hindus carried the decimal numeration, naming the successive powers of a number (usually 10), far beyond any other civilization of the past.

Al-Bīrūnī criticized the Hindus for their passion for large numbers: "I have studied the names of the orders of the numbers in various languages with all kinds of people with whom I have been in contact, and have found

[19] Saidan, 1978, p. 186.

[20] Saidan, Ahmad S., Numeration and Arithmetic, in the book by Rashed, 1996, vol. II, p. 335. Al-Uqlidīsī means "the Euclidean Scribe" as he must have been a Euclidean expert and earned his living by scribing Euclid's books.

[21] Sachau, vol. 1, p. 174.

Table 3.2: Numbers as Words in Different Languages

| Numeral | Sanskrit | Latin | Greek | German | Spanish | French | Russian |
|---|---|---|---|---|---|---|---|
| 1 | *eka* | *unus* | *hen* | *ein* | *uno* | *un* | *odin* |
| 2 | *dvi* | *duo* | *duo* | *zwei* | *dos* | *deux* | *dva* |
| 3 | *tri* | *tres* | *treis* | *drei* | *tres* | *trois* | *tri* |
| 4 | *caturtha* | *quattuor* | *tettares* | *vier* | *cuatro* | *quatre* | *tcheijre* |
| 5 | *pañca* | *quinque* | *pente* | *funf* | *cinco* | *cinq* | *pjat* |
| 6 | *ṣaṣṭha* | *sex* | *hex* | *sechs* | *seis* | *six* | *śest* |
| 7 | *saptam* | *septem* | *hepta* | *sieben* | *siete* | *sept* | *sem* |
| 8 | *aṣṭa* | *octo* | *okto* | *acht* | *ocho* | *huit* | *vosem* |
| 9 | *navam* | *novem* | *ennea* | *neun* | *nueve* | *neuf* | *deviat* |
| 10 | *deśam* | *decem* | *deka* | *zehn* | *diez* | *dix* | *desiat* |
| 100 | *satam* | *centum* | *hekaton* | *hundert* | *ciento* | *cent* | *sto* |

that no nation goes beyond thousand. The Arabs, too, stop with the thousand, which is certainly the most correct and the most natural thing to do. Those, however, who go beyond the thousand in their numeral system are the Hindus, at least in their arithmetical technical terms."[22] I am sure if al-Bīrūnī were to write his book today, he would not criticize the Hindus for their fondness of large numbers. Al-Bīrūnī mentioned $10^{19}$ as the largest number used by the Hindus. The Babylonians did use numbers beyond one thousand. However, for some reason, the Arabs later on restricted themselves to 1000 only.

During the first century B.C., in *Lalitavistara*, a book on the life of Lord Buddha, in a dialog between Lord Buddha and Arjuna, *tallaksana* ($10^{53}$) is defined.[23] As the story goes, when Gautama reached manhood, he courted Gopa, the daughter of King Daṇḍāpaṇī. In the tradition of *svayaṁbara* (a tradition where the young men demonstrate their abilities in the presence of the bride and her family. The bride, after consultation with the family,

[22]Sachau, 1964, vol. 1, p. 174.

[23]Bays, 1983, p. 224. A detailed list of large numbers used by the Hindus is compiled by Hayashi, 1995, p. 66 - 70.

identifies the man of her choice for marriage.), Gautama demonstrated a public proof of his scholarly abilities by debating (*śāstrārtha*) with Arjuna, the state mathematician of King Daṇḍāpaṇī. Arjuna asked Buddha to count numbers. Buddha defined *koti* ($10^7$) and in steps of 100 defined numbers all the way up to *tallaksana* ($10^{53}$).

The *Ṛgveda* mentions "three thousand and three hundred and thirty-nine (3339)"[24] as the count of people in a *yajna*, a holy gathering where worshipping is done around fire. *Atharvaveda* defines a hundred, thousand, myriad, hundred-million.[25] Table 3.3 provides a list of words for various numbers, as given in *Taittirīya-Saṁhitā*,[26] *Āryabhaṭīya*,[27] and *Vāyu-Purāṇa*.[28] In *Vālmīki-Rāmāyaṇa*, written around 200 B.C., Śukra, a Rāvaṇa's spy, defined large numbers in words, as given in Table 3.4.[29]

Some of these word numerals for various powers of ten were later used in China. In a fifth or sixth century document in the China, *koti*, *ayuta*, and *niyuta* in Sanskrit were called by a phonetically similar words *juzhi*, *ayuduo*, and *nayouta*, respectively.[30]

## 3.3 Transformation of Hindu Numerals to Arabic Numerals

As mentioned earlier, the Hindu numerals were generally called Arabic numerals from the sixteenth century to the beginning of the twentieth century. Afterward, these numerals were popularly called as Hindu-Arabic numerals. The reason lies in the fact that Hindu numerals were introduced to the western world by the Arabs when they controlled Spain in the beginning of the seventh-century. Through trade and invasions, the Hindu-Arabic numerals spread to the whole of Europe. "Arabic literary tradition, as generally interpreted, declares that the nine numerals with zero and place value were

[24] *Ṛgveda*, 10: 52: 6.
[25] *Atharvaveda*, 8: 8: 7.
[26] *Taittirīya-Saṁhitā*, 7: 2: 20: 1.
[27] *Āryabhaṭīya*, *Gaṇitpada*, 2.
[28] *Vāyu-Purāṇa*, 101: 93 - 99.
[29] *Vālmīki-Rāmāyaṇa* 6: 28
[30] Martzloff, 1997, p. 97

Table 3.3: Words for Numbers in the Power of Ten

| Number | *Taittirīya-Saṁhitā* | *Āryabhaṭīya* | *Vāyu-Purāṇa* |
|---|---|---|---|
| $10^0$ | *Eka* | *Eka* | *Eka* |
| $10^1$ | *Daśa* | *Daśa* | *Daśa* |
| $10^2$ | *Śata* | *Śata* | *Śata* |
| $10^3$ | *Sahasra* | *Sahasra* | *Sahasra* |
| $10^4$ | *Ayuta* | *Ayuta* | *Ayuta* |
| $10^5$ | *Niyuta* | *Niyuta* | *Niyuta* |
| $10^6$ | *Prayuta* | *Prayuta* | |
| $10^7$ | *Arbuda* | *Koti* | *Koti* |
| $10^8$ | *Nyarbuda* | *Arbuda* | *Arbuda* |
| $10^9$ | *Samudra* | *Vrnda* | *Abja* |
| $10^{10}$ | *Madhya* | | *Kharva* |
| $10^{11}$ | *Amta* | | *Nikharva* |
| $10^{12}$ | *Parārdha* | | *Śaṁkhu* |
| $10^{14}$ | | | *Samundra* |
| $10^{15}$ | | | *Anta* |
| $10^{17}$ | | | *Parārdha* |

Table 3.4: Powers of Ten in *Vālmīki-Rāmāyaṇa*

| | |
|---|---|
| Hundred-*Lākh* ($10^5$) = | 1 *Koti* ($10^7$) |
| Hundred-thousand *Koti* = | 1 *Śaṅkhu* ($10^{7+5} = 10^{12}$) |
| One *lākh-Śaṅkhu* = | 1 *Mahā-Śaṅkhu* ($10^{12+5} = 10^{17}$) |
| One *lākh-Vrnda* = | 1 *Mahā-Vrnda* ($10^{22+5}=10^{27}$) |
| One *lākh-mahā-Vrnda* = | 1 *Padma* ($10^{27+5} = 10^{32}$) |
| One *lākh-Padma* = | 1 *Mahā-Padma* ($10^{37}$) |
| One *lākh-mahā-Padma* = | 1 *Kharva* ($10^{42}$) |
| One *lākh-Kharva* = | 1 *Mahā-Kharva* ($10^{47}$) |
| One thousand-*mahā-Kharva* = | 1 *Samudra* ($10^{50}$) |
| One *lākh-Samudra* = | 1 *Oudh* ($10^{55}$) |
| One *lākh-Oudh* = 1 | *Mahā-Oudh* ($10^{60}$) |

invented by the Indians, and they were adopted by the Arabs during the last quarter of the eighth century A.D."[31] Owing their gratitude to the Hindus, numerals were always called *arqam hindiya* in Arabic, meaning the Hindu numerals.[32]

These numerals were known as Hindu numerals throughout the medieval period in the Arab world and are known so even today. Al-Jāḥiz, (ca. 776 - 868 A.D.), al-Khwārizmī (ca.800 - 847 A.D.), al-Uqlīdisī (ca. 920 - 980 A.D.) and Ibn Labbān (ca. 971 - 1029 A.D.), all from the Middle East or nearby regions, have testified to the Hindu origin of the so-called Arabic numerals.

Interestingly, in conformity with Arabic tradition, these numerals were called Hindu all through the medieval and early Renaissance periods in Europe by their top scholars. Adelard of Bath (1116 - 1142 A.D.) and Roger Bacon (1214 - 1292 A.D.) in England, Leonardo Fibonacci (1170 - 1250 A.D.) in Italy, Ṣā'id al-Andalusī (1029 - 1070 A.D.) and Ibn Ezra (11th century A.D.) in Spain, and Voltaire (1694 - 1778 A.D.) in France called them Hindu numerals. They were labeled Arabic only after the sixteenth century by some.[33] This was the beginning of the colonial period. Why the Hindu numerals got changed to Arabic numerals is a mystery in the history of mathematics. Did it happen due to a coincidence or was it a result of complex political ambitions of the West? This is a curious mystery that is not yet resolved. Today, most textbooks call these numerals the Hindu-Arabic numerals.

The ancient Hindus wrote numbers using a tablet (*takht*) with a layer of sawdust or sand. Such tablets were commonly used for computational works throughout in India and Arabia. The use of such tablets to perform computations was called *Hisāb al-Gubār* (dust-board Arithmetic) in the Arab-world.[34] Unfortunately, since such a practice does not leave a permanent record, we are left only with meager information about these tablets.

Al-Jāḥiz, (776 - 868 A.D.), a well-known Arab theologian and natural

[31]Clark, 1929, p. 217.
[32]Sarton, 1950.
[33]Clark, 1929, p. 217.
[34]Salem and Kumar, 1991, p. 14.

philosopher, attributed the origin of numerals to the Hindus.[35] Using Hindu numerals, Kūshyār Ibn Labbān (ca. 971 - 1029 A.D.) wrote a book on Hindu arithmetic, *Principles of Hindu Reckoning*,[36] which became quite popular during the eleventh-century Islamic world. Kūshyār ibn Labbān was primarily an astronomer from Jilan, a village in Persia on the south of Caspian Sea, and wrote his manuscripts in Arabic. He started his book with the following sentence: "In the name of Allāh, the Merciful and Compassionate. This book on the principle of Hindu arithmetic is an arrangement of Abu al-Hasan Kushyār bin [Ibn] Labbān al-Jīlī, may God have mercy on him."[37]

In the time of communal tensions around the world, Labbān's book sets an example for all–a Muslim who wrote a book on the Hindu arithmetic with the mercy of Allāh. It is completely in line with the doctrines of Prophet Mohammad who suggested that his followers travel around the world for knowledge: "You should insist on acquiring knowledge even if you have to travel to China."

Ibn Sīnā (980 - 1037 A.D.), a noted Persian scholar of the eleventh century mentioned that the Hindu method of calculation was common in Persia during his period. Even merchants with small businesses used the Hindu mathematical methods for calculations. Ibn Sīnā himself learned such methods from a vegetable merchant, Mahmūd al-Massāhī.[38] Ibn Sīnā was a prolific writer with books on medicine, metaphysics, natural philosophy, and mathematics. He explained the Greek works of Aristotle, Plato, and Plotinus within the framework of Islam.

## 3.4 From *Śūnyatā* and *Neti-Neti* to Zero

Zero is perhaps one of the grandest symbols ever invented in mathematics. It is the last numeral invented by the Hindus that prompted the natural philosophers to dump the abacus. It became much easier to perform calculations on a tablet or paper. "In the simple expedient of cipher [zero], which was permanently introduced by the Hindus, mathematics received one of the

[35] Pellat, 1969, p. 197; Levy and Petruck, 1965, p. 6.
[36] Levy and Petruck, 1965.
[37] Levy and Petruck, 1965.
[38] Gohlman, 1974, p. 21 and 121.

most powerful impulses."[39]

The presence of zero indicates a specific absence of the symbols of 1, 2, . . . , 9 at that location. Zero is thus a sign about no value or a missing value, a meta-symbol. Zero is a numeral and it has the same status as any other numerals, all in the absence of any magnitude. Zero defines the boundary between positive and negative numbers. It also defines the starting point in measurements, such as the coordinate axes, meter sticks, stop watches, and thermometers.

In digital electronics, the two different levels of voltage, high (1) and low (0), represent "on" and "off" states. In computers, these "on" and "off" states represent numbers 0 and 1. It is the location of these numbers that defines a large number.

Zero is a "place holder" in a string of digits. It moves other digits to its left by one place and increases their magnitude by a factor of ten. Zero get its meaning from digits that are on its left. Thus, zero plays the role of a number and at the same time signifies the metaphysical reality of absence of substance (emptiness). In Hindu philosophy, the terms *śunyata* and *neti-neti* define emptiness which evolved into a mathematical reality in the form of zero, as explained later.

The symbol zero was used by Pingala in *Candāḥ-sūtra*. In a short aphorism, to find the number of arrangements of long and short syllables in a meter containing $n$ syllables, "[Place] two when halved, when unity is subtracted then (place) zero . . . multiply by two when zero . . ."[40]

King Devendravarman of Kalinga, Orissa inscribed his deed on a copper plate in 681 A.D. This deed provides an archaeological evidence of place-value

[39] Cajori 1980, p. 147; Accounts on the invention of zero are provided by Bronkhorst, 1994; Datta, 1926; Gupta, 1995; Kak, 1989, 1990; Ruegg, 1978. Ruegg has provided perhaps the best review that includes philosophical insights and historical developments.

[40] Pingala's *Chandāḥ-sūtra*, 7: 28, 29, 30; Datta and Singh, 1938, vol. 1, p. 75. This quotation does not indicate the origin of number zero but provides a testimony that the zero was prevalent in India. Similar accounts of zero are also made elsewhere in the same manuscript (*Chandāḥ-sūtra*, 3: 2 and 17; 4: 8, 11, and 12; 18: 35, 44, 48 and 51).

notation. It lists twenty as two and zero (20) in a place-value notation.[41] The seventh-century inscriptions in Cambodia (Sambor) and Malaysia (Kota Kapur) list zero as a big dot or small circle using scripts that were derived from the Indian system. The *Bakhshālī manuscript* mentions *śūnya* for zero at several places in the text.[42]

By the eighth-century, the concept of zero was already in China. In Chapter 104 of the *Kaiyun Zhanjing* (*Astronomico-astrological Canon*), a book written during the reign of Kaiyun (713 - 741 A.D.), zero is mentioned and symbolized as a dot: "Each individual figure is written in one piece. After the nine, the ten is written in the next row. A *dot is always written in each empty row* and all places are occupied so that it is impossible to make a mistake and the calculations are simplified."[43] The dot is an obvious reference to zero. This book was written by Qutan Xida (Gautama Siddārtha) between 718 A.D. and 729 A.D..[44]

When the Arabs learned about zero, they literally translated the Sanskrit word *śūnya* (empty) into *sifr* (empty) in Arabic. "While the Arabs, as we have learned, did not invent the cipher [zero], they nevertheless introduced it with the Arabic numerals into Europe and taught Westerners the employment of this most convenient convention, thus facilitating the use of arithmetic in everyday life . . . al-Khwārizmī . . . was the first exponent of the use of numerals, including the zero, in preference to letters. These numerals he called Hindu, indicating the Indian origin."[45] Leonardo Fibonacci called zero as *zephir*, in Latin, in his book, *Liber Abaci*.[46] Adelard of Bath used the term *cifrae* in his translation of al-Khwārizmī's *Zīj al-Sindhind*.[47]

---

[41]Filliozat, 1993.

[42]Hayasi, 1995, p. 210, 213; also see Hoernle, 1888. The Bakhshālī manuscript consists of seventy fragmentary leaves of birch bark and is presently preserved in the Bodleian Library at Oxford University. The original size of a leaf is estimated to be about 17 cm wide and 13.5 cm high, containing mathematical writings. The manuscript was accidentally found in 1881 near a village, Bakhshālī, that is now near Peshawar in Pakistan. In his detailed analysis, Hayasi assigns the seventh century A.D. as the date when this manuscript was written.

[43]Martzloff, 1997, p. 207

[44]Yoke, 1985, p. 83

[45]Hitti 1963, p. 573.

[46]Horadam, 1975; Sigler, 2002, p. 17.

[47]Neugebauer, 1962, p. 18.

The word zero in the English language evolved from the terms in Latin and Italian.

The invention of zero was an important philosophical triumph in the progress of abstract mathematics; zero has played a crucial role in place-value notations, in analog computer circuits, and in other applications. The status of zero is different from that of the other numerals. Zero is the denial of number that gains its importance only in the place-value system.

### 3.4.1 Epistemology of Zero

Zero is used with 100% certitude in mathematics while, in science, this certitude is less than 100% and this metaphysical reality does not always exist in nature. For example, in nature, there is no zero temperature on the Kelvin temperature scale, there is no zero energy for a particle in a box in quantum mechanics and there is no zero pressure anywhere in the universe. Thus, zero is a rational number in mathematics but only an intuitive perception in science.

When we subtract any number by the same number, we get nothing (zero). The significance of this result is understood; however, the importance of the symbol and its epistemology is not so well known. To have a better understanding, let us perform a few elementary calculations:

$$5 \pm 0 = 5$$
$$5 \times 0 = 0$$
$$5 \pm 0 = 6 - 1$$
$$5 \times (6 - 6) = 0 \text{ since } 6 - 6 = 0$$

We can conclude:

1. Multiplication of any number by zero results in zero.

2. When zero is added or subtracted to any number, it does not change the number.

3. Zero is a number with no magnitude.

Using an analogy, zero is like a shadow. A shadow allows us to know about the object that is obstructing light. However, shadow is the absence of light which creates nothingness in a given space. Thus, the space with nothing, in turn, tells us a lot about the object that is creating emptiness in a given space. Similarly, zero defines nothing and it occupies a space and tells us about the magnitude of the number.

Zero, as a "void," is an integral part of the Hindu philosophy. The *nirguṇa-rūpa* (nonmanifestated-form, *amūrta-rūpa*) of God is worshipped by the Hindus. In this form, no attributes can be assigned to God. Thus, the Hindus can be said to worship emptiness. Nothingness (*śūnyatā*, emptiness or void) as such, in Hindu tradition, has a substance; it is not an absence of everything with 100% mathematical certitude; it is an absence of all manifestated forms that are within the realm of *māyā* (illusion of the manifestated world).

In the *nirguṇa* manifestation, God is beyond any attribute of nature yet the source of all. He is nowhere and yet everywhere. The mathematical symbol zero has similar qualities. Zero has no magnitude and, therefore, is present in every number as $a + 0 = a$. It shows its presence when associated with a number in the decimal system.

The concept of *neti-neti* (not this, not that; essential for *nirguṇa-rūpa*, as we cannot assign any particular attribute to God) dates back to the *Vedic* and *Upaniṣadic* periods. The *Bṛhadāraṇyaka-Upaniṣad* explains the Supreme God (ultimate reality) by defining God as the absence of all the attributes (*neti-neti*).[48]

Using an approach that is similar to *neti-neti*, J. Robert Oppenheimer, the father of the atom bomb, described the property of an electron: "If we ask, for instance, whether the position of the electron remains the same, we must say "no"; if we ask whether the electron's position changes with time, we must say 'no'; if we ask whether the electron is at rest, we must say 'no'; if we ask whether it is in motion, we must say 'no.'[49] Whatever is left is the reality for an electron. It is documented that Oppenheimer learned the style

---

[48] *Bṛhadāraṇyaka-Upaniṣad*, 3: 9: 26.

[49] Oppenheimer, 1954, p. 40

of expression from the studies of ancient Hindu scriptures, as acknowledged in his writings.[50]

Several thousand years before Oppenheimer, in the *Bṛhadāraṇyaka-Upaniṣad*, Yājñavalkya explains the nature of Brahmā to Gārgī: "It is neither course nor fine, neither short no long . . ." Nothingness is not the opposite of existence or being; it is perhaps the indetermination of attributes. In mathematics, it has the quality of "not being there" for any existence. Nothingness is not to be perceived as vacuum; rather, it is the source of all creation. After all, the universe was created from nothing, as indicated in the *Ṛgveda*. (See Chapter 5)

The absence of a number in place-value notation has a meaningful function. It does not have a similar usefulness in any other system that is not place-value. For example, the location of two digits 5 and 4 can be fifty four (54), five hundred four (504), five hundred forty (540), etc., depending on the location of the two digits. It is only natural to assign a symbol – a circle or a dot – for this absence of a number, for convenience. This is how zero became so central to mathematics.

Though the philosophical presence of zero is known to be present in ancient India during the *Upaniṣad*-period, its existence as a mathematical entity is not so certain. It would be inappropriate to conclude the presence of zero as a mathematical entity to the *Upaniṣad*-period. We do know that zero was used as mathematical entity in *Chandāḥ-sūtra* at the beginning of the Christian era.

In meditation, we concentrate on a single thought (for example, God). This is the first stage of meditation. If successful, we go beyond this single thought and reach a stage of thoughtlessness. It is this thoughtlessness which is aspired by all yogis, a state of bliss, and the dividing line between reincarnation and *mokṣa*. It is this thoughtlessness that is the key to salvation. It is this nothingness that has everything of value to a seeker.

The reality of an object is defined by the attributes it has. For example, a table can be defined by its shape, color, weight, and by the type of wood used.

[50]Oppenheimer, 1954, p. 66.

Therefore, we can have some idea about the table with just four attributes. Is it enough? Perhaps not. If we want to know more about the table, we might want to know its age, manufacturer, and current owner. This list will continue to grow as we become more and more inquisitive about table. Can we form a complete list of information? The attributes can never be listed in completeness. If such a small object is so difficult to explain, what about the discipline of science? What about defining the God? The attributes will be way beyond the grasp of any person.

The *mūrta-rūpa* (or *saguṇ-rūpa*; manifested form) of God has infinite manifestations that are impossible to count. The Vedic seers realized their inability to provide all attributes of God. That is the reason God is also named as *anant* (infinite).

"The importance of the creation of the zero mark can never be exaggerated. This giving to airy nothing, not merely a local habitation and a name, a picture, a symbol but helpful power, is the characteristic of the Hindu race whence it sprang."[51] Similarly, the metaphysical reality of zero and the basic interpretation of God in *neti-neti* are a curious connection. There is no historical document that defines this philosophy as the birth place of zero as the mathematical reality. However, there is no doubt that the ancient Hindus knew the concept of infinity and zero in the philosophical sense.

## 3.5 The Binary Number System

A system in which numbers are represented as linear combinations of powers of two (2) is called the binary number system. This is a positional-numeral system employing only two kinds of binary digits, namely 0 and 1. The importance of this system lies in the convenience of representing decimal numbers using a two-state system in computer technology. A simple "on-off" or "open-closed" system can effectively represent a number. The word, bit, used in information theory, is a short form of binary digits. A row of lights with a set of "on" and "off" can easily represent a number, using the system. Similarly, a set of condensers "charged" or "not-charged" can represent a number, or a set of two different voltages on a device can effectively represent a number. For this reason, the binary number system is popular

[51]Halstead, 1912, p. 20.

Table 3.5: Decimal Numbers and Their Binary Equivalent

| Decimal Numbers | Binary Equivalent |
|---|---|
| $1 = (0 \times 2^1) + 1$ | 01 |
| $2 = (1 \times 2^1) + 0$ | 10 |
| $3 = (1 \times 2^1) + 1$ | 11 |
| $4 = (1 \times 2^2) + (0 \times 2^1) + 0$ | 100 |
| $5 = (1 \times 2^2) + (0 \times 2^1) + 1$ | 101 |
| $6 = (1 \times 2^2) + (1 \times 2^1) + 0$ | 110 |
| $7 = (1 \times 2^2) + (1 \times 2^1) + 1$ | 111 |
| $8 = (1 \times 2^3) + (0 \times 2^2) + (0 \times 2^1) + 0$ | 1000 |
| $9 = (1 \times 2^3) + (0 \times 2^2) + (0 \times 2^1) + 1$ | 1001 |
| $10 = (1 \times 2^3) + (0 \times 2^2) + (1 \times 2^1) + 0$ | 1010 |
| $11 = (1 \times 2^3) + (0 \times 2^2) + (1 \times 2^1) + 1$ | 1011 |
| $15 = (1 \times 2^3) + (0 \times 2^2) + (1 \times 2^1) + 1$ | 1111 |
| $17 = (1 \times 2^4) + (0 \times 2^3) + (0 \times 2^2) + (0 \times 2^1) + 1$ | 10001 |

in electronic circuitry.

The presence of a binary number system is an example of the place-value notational system. In this system, the number 3 can be written as $(1 \times 2^1 + 1)$. In the binary system, we write 3 as 11. In this way, every integer is the sum of distinctive powers of 2 in a combination. To understand the system, a few examples are provided in Table 3.6:

Using this system, a somewhat larger number 444 is represented as 110111100 $[(1 \times 2^8) + (1 \times 2^7) + (0 \times 2^6) + (1 \times 2^5) + (1 \times 2^4) + (1 \times 2^3) + (1 \times 2^2) + (0 \times 2^1) + 0]$.

To convert a number into a binary number, divide the number by two. If there is one as remainder, then write one (1). If it is divisible, then write zero. For example, to write 45 in the binary system, let us divide by two:

$45 \div 2$. The resultant is 22 and the remainder is 1. Let us write 1.
$22 \div 2$. The resultant is 11 and the remainder is 0. Let us write 0.
$11 \div 2$. The resultant is 5 and the remainder is 1. Let us write 1.

Table 3.6: Decimal Numbers and Their 7-based and 12-based Equivalent

| Decimal Number | Seven-based System | Twelve-Based System |
|---|---|---|
| 1 | 1 | 1 |
| 2 | 2 | 2 |
| 3 | 3 | 3 |
| 4 | 4 | 4 |
| 5 | 5 | 5 |
| 6 | 6 | 6 |
| 7 | 10 | 7 |
| 8 | 11 | 8 |
| 9 | 12 | 9 |
| 10 | 13 | $ |
| 11 | 14 | # |
| 12 | 15 | 10 |
| 13 | 16 | 11 |
| 14 | 20 | 12 |
| 15 | 21 | 13 |
| 16 | 22 | 14 |
| 17 | 23 | 15 |
| 18 | 24 | 16 |
| 19 | 25 | 17 |
| 20 | 26 | 18 |

5÷ 2. The resultant is 2 and the remainder is 1. Let us write 1.
2÷ 2. The resultant is 1 and the remainder is 0. Let us write 0.
2÷ 1. The resultant is 0 and the remainder is 1. Let us write 1.

By taking the first quotient 1 and the remainders in the reverse order, we can write 45 in binary number: 101101. This number implies $(1 \times 2^0) + (0 \times 2^1) + (1 \times 2^2) + (1 \times 2^3) + (0 \times 2^4) + (1 \times 2^5)$. This converts to $(1 + 4 + 8 + 32)$ and equals to 45. Interestingly, this rule is provided in *Pingala-Sūtra* (8: 24 – 25) in a cryptic language which is nicely explained by Barend A. van Nooten using history documents that provided commentary to Pingala's verse.[52]

To understand the binary system better, let us define the seven-based system and twelve-based system. As we need twelve symbols for the twelve-based system, we have chosen $ and # symbols for ten and eleven, respectively. These two systems are not far away from the realities as the Mayans and the Babylonians use base-20 and base-60 systems, respectively.

The ancient Hindus carefully defined the methodology to write hymns which involved the study of language and prosody. This allowed a meter to have specific rhythm in chanting. The first comprehensive treatise that is known to us was written by Pingala, called *Chandāḥ-sūtra* (or Candāh) or *Chandāḥ-śāstra.*[53]

The birth, death, and place of Pingala and *Chandāḥ-sūtra* are not known. We do not even know if the manuscript written by him is his original contribution or that of his school. Fortunately, Pingala's treatise is mentioned by Śabra in his manuscript *Mīmāṁsā-sūtra* (1: 1: 5). We do know that the work of Śabra was written around the 4th century A.D.. Varāhamihira, a famous astronomer of the early 6th century A.D. mentioned Śabra's work.[54] It is also suggested that Pingala was the younger brother of Pānini[55] or his

---

[52]van Nootan, 1993. Barend A. van Nooten was a professor of Sanskrit at the University of California at Berkeley. He is known for his co-authored books, *Rigveda: A Metrically Restored Text* and *The Sanskrit Epics.*

[53]Weber, 1863; taken from van Nooten, 1993, also reproduced in Rao and Kak, 1998.

[54]van Nooten, 1993.

[55]Agarwala, 1969, p. 16; taken from Singh, 1985.

maternal uncle.[56] Based on the existing literature, Agarwala concluded[57] that Panini lived between 480 B.C. and 410 B.C. Datta and Singh assigns 200 B.C. to the period of Pingala.[58]

In Sanskrit, the science of versification (prosody) consists of verse feet, *padas*, that are composed of syllables: light (*laghu*) and heavy (*guru*). The basic unit in Sanskrit meters consist of a letter having a single *mātrā* (syllabic instant) called *laghu* (short or light). A light syllable consists of a short vowel followed by one consonant. All other syllables are *guru* (heavy). Each verse usually consists of a set of four quarter verses, called *padas*. All verses contain the same number of syllables. For example, most verses in the *Vedas* are either 8-, 11-, or 12-syllabic. Using a complex procedure described by van Nooten[59] and Dutta and Singh,[60] which requires a sound knowledge of Sanskrit, a matrix, called *prastāra*, was defined where *laghus* and *gurus* are listed horizontally. John Kadvany studied the positional value and mathematics in Sanskrit language. In studying Sanskrit grammar, he concludes that "word compounding is the structural feature facilitating the formation of positional number words and is the heuristic center of the Sanskrit language."

Pingala provided mathematical formula that defined the position of verse meter within the *prastāra* and reconstructed a metrical pattern. This pattern allowed the metrists to write Vedic hymns. As van Nooten has described, "[T]he metrist's task is to discover the parameters within which that variation takes place, to classify the meters and to organize them into larger categories." This process is similar to that of finding the binary representation of a decimal number.

Pingala gave the following rule: "(Place) two when halved" "when unity is subtracted then (place) zero"; "multiply by two when zero"; "square when halved."[61]. It was van Nootan's expertise in Sanskrit metrics that allowed him to discover this unique binary system. In this system, each syllable is

---

[56]Vinayasagar, 1965; taken from Singh, 1985.

[57]Agarwala, 1969, p. 463 - 476.

[58]Datta and Singh, 1962, p. 75.

[59]van Nooten 1993; van Nooten in the book by T. R. N. Rao and Subash Kak, 1998.

[60]Datta and Singh, 1962.

[61]Datta and Singh, 1962, vol. 1, p. 76; van Nooten has provided a slightly different translation. However, both systems provide similar results.

Table 3.7: Pingala's Rule with Six Syllables

| | Number | Result |
|---|---|---|
| Number of syllables | 6 | |
| Halve the number | 3 | separately place 2 |
| Since 3 cannot be halved, subtract 1 | 3 - 1 = 2 | separately place 0 |
| Halve the number | $2 \div 2 = 1$ | separately place 2 |
| Since one cannot be halved, subtract 1 | 1 - 1 = 0 | separately place 0 |

assigned a numerical value, based on its position in the meter.

As *Gāyatrī mantra* contains six syllables, the above rule gives us results that are defined in Table 3.7. In the last column, we have written "2" for each operation of halving and "0" for each operation of subtraction. This gives us a set of (2, 0, 2, 0). Let us start from the last number 0 and apply Pingala's rule by placing two there. The next is "2" and we need to square the previous number. This gives us $2^2$ (= 4). The next number is "0", meaning that we need to multiply by 2. This gives us $2^3$ (= 8). The next number is "2". This means we need to square the number. This gives us $2^6$ (= 64). This is the total number of ways two items can be arranged in six places.

Mathematically, we can write it as:

| | | | |
|---|---|---|---|
| 2 | 0 | 2 | 0 |
| $2^6$ | $2^3$ | $2^2$ | 2 |

This is the total number of places one can arrange for two objects in 6 places (64).

To understand the above operation better, let us take another example with 7 syllables, as provided in Table 3.8.

The process ends and we get a set of number:

Table 3.8: Pingala's Rule with Seven Syllables

| | Number | Result |
|---|---|---|
| Number of syllables | 7 | |
| Since 7 cannot be halved, subtract 1 | 7 - 1 = 6 | separately place 0 |
| Halve the number | $6 \div 2 = 3$ | separately place 2 |
| Since 3 cannot be halved, subtract 1 | 3 - 1 = 2 | separately place 0 |
| Halve the number | $2 \div 2 = 1$ | separately place 2 |
| Since one cannot be halved, subtract 1 | 1 - 1 = 0 | separately place 0 |

| | | | | |
|---|---|---|---|---|
| 0 | 2 | 0 | 2 | 0 |
| $2^7$ | $2^6$ | $2^3$ | $2^2$ | 2 |

The number two raised to power seven ($2^7$) gives us 128. This is the number of arrangements one can have with two objects for seven different places. Similarly, with 9 syllables, it gives us $2^9$ (= 512) possible combinations. The work of Pingala in *Candāḥ-sūtra* definitely shows that he knew the place-value system of numeric notations and used a binary numerical base, and not base-10.[62].

The discovery of binary numbers is generally attributed to Gottfried Leibniz (1646 - 1716 A.D.) at the end of 17th century. Leibniz is said to have come up with the idea when he interpreted Chinese hexagram depictions of Fu Hsi in *I-Ching* (*The Book of Changes*) in terms of a binary code.[63] Pingala, in van Nooten's view, did not provide the further applications of the discovery. However, this knowledge was available to Sanskrit scholars of meterics.[64] "Unlike the case of the great linguistic discoveries of the Indians which directly influenced and inspired Western linguistics, this discovery of the theory of binary numbers has so far gone unrecorded in the annals of the

---

[62] van Nooten, 1993

[63] Loosen and Vonessen, 1968, p. 126-131; reference taken from van Nooten, 1993.

[64] van Nooten, 1993.

West," remarks van Nooten.

## 3.6 Diffusion of Hindu Numerals to Other Cultures

As indicated in Section 3.3, the Hindu numerals became known to the Arabs during the earliest period of the Abbasid dynasty (750 - 1258 A.D.). As the kingdoms in the Middle East later made contact with their European counterparts, these numerals also became known to Europeans. This section deals with the diffusion of the Hindu numerals, which started from India and later engulfed the entire world.

The Muslims arrived in the Southern parts of India as merchants and missionaries during the late sixth century A.D.. In 711 A.D., the Umayyad-dynasty controlled the Sind region and channeled the Hindu sciences to Baghdad.[65] Arab merchants learned the Hindu system of numeration, used by Hindu merchants in their business dealings, and introduced them to Arabia. Many Hindu books on wars, weapons, and medicines were translated from Sanskrit to Persian or Arabic.[66] "India acted as an early source of inspiration [for Arabia], especially in wisdom, literature and mathematics," noted by P. K. Hitti, a noted scholar on the Middle-East.[67]

Since the Greek numerals were already in use in Arabia when the Hindu numerals were introduced, both numeral systems competed with each other. Lobbies were formed where the superiority of one system over the other was debated/discussed. Severus Sebokht (died, 662 A.D.), a Syrian natural philosopher mentioned of such a rivalry between the Greek and Hindu numerals. [68]

---

[65]al-Hasan and Hill, 1986, p. 31

[66]al-Hasan and Hill, 1986, p. 2.

[67]Hitti, 1963, p. 307. Philip Khuri Hitti (1876 - 1978) was a Syrian who mostly lived in America. He received his Ph.D. from Columbia University and taught at the American University of Beirut and later at Princeton University and Harvard University. He was a scholar of Islam and the Middle East and his book is quite popular and sets a milestone in our understanding of the history of the Arabs.

[68]Wright, 1966, p. 137. Sebokht was born in Nisibis in Persia and taught Greek philos-

"I will omit all discussion of the science of the Hindus, a people not the same as Syrians, their subtle discoveries in the science of astronomy, discoveries which are more ingenious than those of the Greeks and the Babylonians; their valuable method of calculation; their computing that surpasses description. I wish only to say that this computation is done by means of nine signs. If those who believe, because they speak Greek, that they have reached the limits of science, should know these things, they would be convinced that there are also others [Hindu] who know something."[69] This quotation is a proof that the Hindu numerals were in practice in Arabia by the seventh-century. Also, by the seventh century, people started to discard the Greek numerals in favor of the Hindu numerals in Arabia.

Al-Uqlidīsī (920 - 980 A.D., also written as Uklidisi, meaning 'the Euclid-man' for his role as copyist of Euclid's work) in his book, *Kitāb al-Fusūl fī al-hisāb al-Hindī* defined the Hindu numeral system and explained Hindu arithmetic: "I have looked into the works of the past arithmeticians versed in the arithmetic of the Indians [Hindus, in the original] . . . We can thus dispense with other works of arithmetic. The reader who has read other works will realize this fact and thus adhere to this work and prefer it to others, new or old, for it contains more than any other books of this kind."[70] The title of the above book indicates that the book contained the mathematics of the Hindus. According to al-Uqlidīsī, "even the blind and the weak sighted will find in our explanations and summaries something to benefit from without toil or cost, by the will and help of God."[71] He mentioned the use of the dust-board abacus for the Hindu arithmetic [*hisāb al-Ghubār*, *hisāb al-takht* (board), *al-Hisāb al-Hindī* or *Hisāb al-turāb* (dust)].[72] This kind of board became so popular for mathematical calculations, arithmetic was called *Hisāb*

ophy in Syria. Later, he became the bishop of the convent of Ken-neshre (Qenneshre), a center of Greek learning in Upper-Euphrates in West-Syria. He had a good grasp of various sciences and wrote on astronomy, geography, and mathematics. His books in manuscript form are preserved in museums in London and Paris. Sebokht believed in the universal nature of science and was of the opinion that many cultures, not just the dominant Greeks, contributed to the Arab scientific pool.

[69]Datta and Singh, pt. 1, p. 96; Clark, 1929, p. 220.

[70]Saidan, 1978, p. 35.

[71]Saidan, 1978, p. 189. This book was composed in Damascus in 952 - 953 A.D. and a 1186 A.D. manuscript has survived. Read Burnett, 2006.

[72]Saidan, 1978, p. 12.

*al-Ghubār*, meaning mathematics of the dust-board. This term was common in the medieval Spain.[73] On the practicality of a dust board in mathematical calculations, al-Uqlidīsī wrote, "If some persons dislike it because it needs the *takht* (board), we say that this is a science and technique that needs a tool."[74]

In the Middle East, al-Khwārizmī wrote a book in the Arabic language on the use of Hindu numerals, *Kitāb al-hisāb al-Hindī*. Although the original book in the Arabic language has been lost, its Latin translation by Robert of Chester has survived.[75] A thirteenth century manuscript from Spain has been found recently which is more complete than the first Latin translation. This manuscript along with its German translation was published by Menso Folkerts in 1997.[76] Al-Khwārizmī's book is a compilation of methods used by the Hindus. He has acknowledged that his book will "explain the Indian [Hindu] method of calculation (*numerus*, Arabic *ḥisāb*) with nine digits (*littere* with which any number can be easily represented. With these digits arithmetic i.e. multiplication, division, addition and subtraction, may be simplified."[77]

Monk Vigilia of the monastery of Albelda in the Rioja, Asturias (an autonomous community in Spain), in 976 A.D., made a copy of the *Etymologies* written by Isidore of Seville. In this copy, the monk provided information about the Hindu numerals with the following remark: "We must know that the Indians have a most subtle talent and all other races yield to them in arithmetic and geometry and the other liberal arts. And this is clear in the nine figures with which they are able to designate each and every degree of each order (of numbers)."[78]

---

[73]Read, Salem and Kumar, 1991.

[74]Saidan, 1978, p. 35; Saidan has also compiled a list of books on Hindu-Arabic arithmetic that were in the Arab world. See, Saidan, 1965.

[75]Hughes, 1989.

[76]Folkerts, Menso, *Die älteste Lateinische Schrift über das Indische Rechnen nach al-Ḫwārizmī*, Übersetzung und Kommentar von M. Fokerts under Mitarbeit von Paul Kunitzsch, Bayerische Akademie der Wissenschaften, Philosophisch-historische Klasse, Abhandlungen, Neue Folge, Heft 113, 1997. Taken from Folkerts, 2001.

[77]Folkerts, 2001. Al-Khwārizmī was perhaps the most influential Islamic philosopher in Europe before the European Renaissance. The mathematical terms sine, algorithm, and zero are attributed to him.

[78]Burnett, 2006.

Yoshio Mikami, in his book, *The Development of Mathematics in China and Japan*, wrote of the Indian influence on Chinese mathematics: "Things Indian exercised supremacy in art and literature, in philosophy, in the mode of life and the thoughts of the inhabitants, in everything. It is even said, astronomy and calendrical arts had also felt their influence. How then could arithmetic remain unaffected? No doubt the Chinese studied the arithmetical works of the Hindoos [Hindus]."[79]

On the popularity of Indian literature in China, Mikami wrote: "we read in history, the Indian works were read in translation ten times more than the native [Chinese] classics, a fact that vividly tells how the Indian influence had swept over the country [China]."[80]

The biggest export of India to China is definitely the religion founded by Lord Buddha, Buddhism. When Buddhism was introduced to China, this country was already an old civilization with a powerful tradition and history. Philosophers such as Confucius guided Chinese society with their philosophies. Confucianism mostly deals with ethical rules that are applicable to this world; it does not deal with the spiritual life. Buddhism filled that niche to allow people to think about the "ultimate questions" of salvation, heaven, etc. There was a contrast in the intentions of these visitors who came from China in comparison to the Europeans. Most Chinese travelers visited India for their spiritual pursuits while Marco Polo and Vasco de Gamma visited to collect wealth. For example, Xuan Zang (b. 603 A.D., also known as Hiuen Tsang, Huan Chwang, Yuan Chwang, Hiouen Thsang, and Hsuan Tsang)[81] studied in Nālandā and carried books on twenty-two horses on his way back to China. He build a pagoda in Xian, China to house the books. Kumārjīva, an Indian scholar, went to Chang-an in 401 A.D. and served as "Grand Preceptor" of an enormous translation project involving thousands of monks and scholars that advanced the philosophy of the great Indian philosopher Nāgārjuna.[82] He was from the Andra Pradesh region and was the founder of the *Madhyamaka* (Middle Path) school of Buddhism. This school became popular in China under the name *Sanlun*. Nāgārjuna was based in Nālanda University.

---

[79]Mikami, 1913, p. 57.
[80]Mikami, 1913, p. 57.
[81]Bernstein, 2001, p. 22.
[82]Bernstein, 2001, p. 95

Ch'u-t'an Hsi-ta from Tang's Court translated a Sanskrit text into Chinese and introduced decimal notation and the arithmetic rules during the early eighth century. His work was continued further by Yijing (or I-tsing, 643 - 713 A.D.), under the order of the emperor.[83] (See Chapter 2)

The Muslims invaded Spain and later on captured the region. The Islamic religious requirement of Hajj, the pilgrimage to Mecca, prompted people to visit Mecca from Spain. These travels allowed the exchange of information and books. Europe learned not only about the Middle East but also about India and China.

Ṣā'id al-Andalusī (1029 - 1070 A.D.), a Spanish scholar of the eleventh century, wrote about the Hindu numerals and arithmetic: "That which has reached us from their [Hindus'] work on numbers is *Hisāb al-Ghubār* (dust board arithmetic) which was simplified by Abū Jā'far Muhmmad ibn Mūsā al-Khwārizmī. This method of calculating is the simplest, fastest, and easiest method to understand and use and has a remarkable structure. It is a testimony to the intelligence of the Indians [Hindus, in the original], the clarity of their creativity, and the power of their inventiveness."[84]

Ṣā'id al-Andalusī was not the only one in Spain to write about the dominance of India in the sciences. Rabbi Abrahm Ibn Ezra (1096 - 1167 A.D.), a poet, scholar, and author of numerous books on grammar, philosophy, medicine, astronomy, astrology and mathematics, wrote on the ways the Hindu numerals were introduced in Arabia. Ibn Ezra mainly lived in Spain and translated al-Bīrūnī's and al-Khwārizmī's work into Hebrew. Ibn Ezra mentioned the role of the Jewish scholars in bringing a Hindu scholar, named Kanaka, who taught place-value notations to the Arabs.[85]

Leonardo Fibonacci (1170 - 1250 A.D.) is well known for his contributions in arithmetic and geometry. He is also known for the Fibonacci Sequence. He learned the Hindu mathematics from the Arabs and popularized Hindu numerals in the western world. "Of all the methods of calculation,

---

[83]Sarton, 1927, vol. 1, p. 513.
[84]Salem and Kumar, 1991, p. 14.
[85]Goldstein, 1996.

he [Fibonacci] found the Hindu [method] to be unquestionably the best,"[86] concludes Cajori in his studies. Fibonacci attributed the numerals correctly to the Hindus.

[87] The book was based on Hindu numerals in which Fibonacci shared the knowledge he gained during his interactions with the Islamic scholars. Fibonacci wrote about the purpose of this book in the following words: "I pursued my study in depth and learned the give-and-take of disputation. But all this even, and the algorism, as well as the art of Pythagoras I considered as almost a mistake in respect to the method of the Hindus. Therefore, embracing more stringently the method of the Hindus, and taking stricter pains in its study, while adding certain things from my own understanding and inserting also certain things from the niceties of Euclid's geometric art, I have striven to compose this book in its entirety as understandably as I could, dividing it into fifteen chapters."[88]

Fibonacci wrote a statement in this book which erases any confusion about his knowledge of the Hindu numerals: "The nine Indian figures are: 9 8 7 6 5 4 3 2 1. With these nine figures, and with the sign 0 . . . any number may be written, as is demonstrated below."[89] He wrote: "The first place in the writing of the numbers begins at the right. The second truly follows the first to the left. The third follows the second. . . . the nine figures that will be in the second place will represent as many tens . . . the figure that is in the third place denotes the number of hundreds. . . if the figure four is in the first place, and the unit in the second, thus 14, undoubtedly xiiii will be denoted . . if the figure of the unit is in the first place, and the figure four in the second, thus 41, xli will be denoted."[90]

Fibonacci was fascinated with Hindu mathematics. ". . . the art of the nine Indian figures, the introduction and knowledge of the art pleased me so much above all else, and I learnt from them, whoever was learned in it, from nearby Egypt, Syria, Greece, Sicily and Provence, and their various methods, to which locations of business I traveled considerably afterward for

[86]Cajori, 1980, p. 121.
[87]Sarton, 1931, vol. 2, p. 7.
[88]Grimm, 1973. Taken from Horadam, 1975.
[89]Boncompagni, 1857 - 1862; Horadam, 1975.
[90]Sigler, 2002, p. 17 - 18.

much study, and I learned from the assembled disputations."[91] Thus, if *Liber Abaci* is the birthdate of European mathematics, as indicated by George Sarton, Hindu mathematics has played an important role in its inception.

Italian merchants carried Hindu mathematics to the neighboring country of Germany. The Italian word *cosa* (meaning *thing*) was used for the unknowns in algebraic equations. This led to the German word *cossists* for people known for their mathematical expertise. Leonardo used the mathematical knowledge of the Hindus to convert money from one currency to another, investment of money, calculation of simple and compound interest, and in defining the rules of barter. Fibonacci found the prevalent Roman numerals inferior to the Hindu numerals. The purpose of writing his book *Liber Abaci* was to introduce the Romans to the very best available tools of mathematical calculations and their applications.

Fibonacci started the book with the following sentence: "You, my Master Michael Scott, most great philosopher, wrote to my Lord about the book on numbers which some time ago I composed and transcribed to you; whence complying with your criticism, your more subtle examining circumspection, to the honor of you and many others I with advantage corrected this work. In this rectification I added certain necessities, and I deleted certain superfluities. In it I present a full instruction on numbers close to the method of the Indians whose outstanding method I chose for this science."[92]

Fibonacci's sources of information were largely Islamic. He copied several problems verbatim from the books of al-Khwārizmī, Abū Kamil, and al-Karajī.[93] Fibonacci called zero as *zephir* which was derived from of the Arabic word *sifr*, indicating nothing.[94] Zero was used as a place-holder to facilitate the reading of numbers.

**Leonardo Pisano Fibonacci**

Leonardo Pisano Fibonacci (1170 - 1250 A.D.) was born in Italy and

[91]Sigler, 2002, p. 15 - 16.
[92]Sigler, 2002. Michael Scot was Leonardo's mentor.
[93]Katz, 1993, p. 282.
[94]Sigler, 2002, p. 17.

educated in Algeria where his father held a diplomatic assignment as a secretary of the Republic of Pisa. The struggle between the Papacy and the Roman Empire by the end of twelfth century had left many Italian cities as independent republics, including the Republic of Pisa. The town of Pisa is known for its famous leaning tower.

Fibonacci learned the Hindu numerals in Algeria and went on business trips to Egypt, Syria, Sicily and Greece. He learned various techniques of mathematics by learning Arabic and studying under several Islamic teachers. He published several well-known books: *Liber Abaci* in 1202, *Practica Geometriae* in 1220, and *Liber Quadratorum* in 1225. *Liber Abaci*, which he revised in 1228, deals with the Hindu numerals and methods of reckoning. Fibonacci is also known to have popularized the Hindu numerals in Europe.

He posed practical problems in his books and solved them using Hindu-Arabic numerals and the Hindu arithmetic. These problems popularized the book in Europe as the readers found a quick way to calculate profits and currency conversions. Eventually, bankers, traders, scientists, and merchants replaced the Roman numerals with the Hindu numerals. Fibonacci sets an example of multiculturalism for today's culture-centric intellectuals. He kept his scholarly integrity and gave due recognitions to Islamic and Hindu cultures.

Fibonacci had such a large impact on mathematics that a journal, *The Fibonacci Quarterly*, is dedicated to his work. The Fibonacci Association, an international organization, is the meeting ground of all scholars who are interested in the mathematics of Fibonacci. This demonstrates the significant impact of this thirteenth century man. It is interesting that, despite such popularity, his most important work, *Liber Abaci*, the book where Fibonacci numbers were introduced, was not translated into English for about thousand years. Therefore, the mathematicians and scientists that were not conversant with Latin could not read the book for so long. It was translated for the first time into English in 2002, thanks to the work of Sigler.

Adelard of Bath (1116 - 1142 A.D.), an English translator, traveled extensively in Asia Minor, France, Sicily, Egypt, Syria, and perhaps in Spain. He learned Arabic in Sicily and translated Arabic manuscripts, including Abū Masher's book on astronomy. Adelard wrote a book in Latin, *Liber Ysagogarum Alchorismi in artem astronomicam a magistro A. Compositus*, that details the Hindu mathematics and numerals. He wrote, ". . . since no knowledge goes forth if the doctrine of all the numbers is neglected, our tract begins with them, following the reasoning of the Indians."[95] Only the first three sections of the book are published in German.[96]

Roger Bacon (1214 - 1294 A.D.), a noted Franciscan natural philosopher from England, studied at Oxford University and the University of Paris. At the University of Paris several manuscripts about the Hindu mathematics were available in French that were read by Bacon.[97] Bacon knew the works of ibn al-Haytham (Alhazen), al-Battānī (Albategnius), Ibn Sīnā, and al-Khwārizmī who wrote about the Hindu numerals. Bacon emphasized the role of mathematics for Europe in his *Opus Majus* and deemed it "necessary to the culture of the Latins"[98] Fibonacci's *Liber Abaci* was published thirteen years before the birth of Bacon, and was known to him. Bacon suggested people should study mathematics from "all the sages of antiquity" and chastised them for being "ignorant" of its "usefulness".[99] Bacon felt that mathematics is "the gate and key" to learn other sciences. In his view, a study of mathematics was essential for a study of philosophy that, in turn, was essential for the study of theology. Therefore, mathematics "has always been used by all the saints and sages more than all other sciences,"[100] and is essential for theology and to know about the creatures and other creations.[101] Thus, mathematics, in the mind of Roger Bacon, was a tool to know more about theology or to know about God's creations. This is similar to how Nārada viewed mathematics to achieve salvation (Chapter 2).

Roger Bacon contacted Pope Clement IV and asked for his support to

[95] *Dictionary of Scientific Biographies*, vol. 1, p. 62.

[96] Curtze, 1898.

[97] Smith, a chapter, The Place of Roger Bacon in the History of Mathematics, in the book by Little, 1914, p. 156.

[98] Bacon, 1928, vol. 1, p. 27.

[99] Bacon, 1928, vol. 1, p. 27.

[100] Bacon, 1928, vol. 1, p. 116.

[101] Bacon, 1928, vol. 1, p. 195.

compile the works of the ancients. The Pope gave his approval to the project and, thus, Bacon wrote *Opus Majus*. He calculated the radius of earth to be 3,246 miles and discussed the possibility of a voyage from Spain to India in his book. It was this information that led Christopher Columbus to write a letter to Ferdinand and Isabella[102] to fund his expedition. The voyage was funded and Columbus landed in America, assuming it to be India.[103]

The origin of the actual source for the numerals was known in intellectual circles in Europe even in the eighteenth century. Voltaire (1694 - 1778 A.D.), a French philosopher and historian who traveled extensively in Europe, Arabia, Persia, China, and India wrote: "The arithmetical figures we use, which the Arabians imported into Europe about the time of Charlemagne, are originally derived from India."[104] Charlemagne (Charles the Great, 742 - 814 A.D.) was a king of France who later became an emperor of the Roman empire. Voltaire assertion that the number system came to Europe in the eighth century is in tune with other documents.

The triumph of the Hindu numeral system over the Roman numerals precipitated neither rapidly nor without conflict. It was not easy to discard the Roman numerals as they were deeply rooted in the system. The new Hindu system was mixed with the Roman by some. For example, 1482 was written as M.CCCC.8II, whereas 1089 was written as I.0.VIII.IX.[105] In the twelfth and early thirteenth century, it was common to write numbers below 360 in Arabic, Greek or Latin symbols and the numbers above 360 in Hindu numerals.[106] In 1299 A.D., the City Council of Florence forbade the use of Hindu numerals in official accounting books. Similarly, as late as the fifteenth century, the mayor of Frankfurt issued an order to the master cal-

[102]Smith, a chapter in the book, The Place of Roger Bacon in the History of Mathematics, by Little, 1914, p. 180. As one can notice, a mistake in radius by Roger Bacon led to a shortage of more than four thousand miles on the surface of the Earth for Columbus.

[103]He considered the native people of America as Indian. Later, they were called Red Indian. Today, this term is considered as a slur in popular conversation.

[104]Voltaire, 1901, vol. 24, p. 40.

[105]Gupta, 1983a; Menninger, 1970, p. 287. In the number 1482, an amalgamation of the Roman and Hindu system is practiced. The number 1 represents 1000 and M is the corresponding symbol. Similarly, C represents 100. The second number 1089 is written in place-value notations with Roman symbols for different numbers. As mentioned earlier, symbols can easily be changed without compromising quality.

[106]Burnett, 2006.

culators to abstain from the use of the Hindu numerals.[107] Some even tried to create a parallel system with the first nine Roman numerals. For example, H. Ocreatus, a student of Adelard of Bath, tried to create this system which could have been appropriate scientifically. However, it did not catch on among the scholars.[108] As late as the eighteenth century, the National Audit Office (*Cour des Comptes*) in France was still using Roman numerals in their accounting.[109]

### 3.6.1 Support of Pope Sylvester II

Science is international in nature; it fosters international cooperation that is conducive to its intellectual growth. Scientists travel from place to place, help their fellow scientists across the globe, and seek for truth. The issue of race, gender, and culture do not come in the picture. Prophet Mohammad told his disciples to even visit China, considered as a far land that was culturally different. This is not true only of Islam. All religions have suggested similar measures to seek truth. It is this international character of science that allowed Pope Sylvester II to become instrumental in popularizing the Hindu numerals, in his life as Gerbert d'Aurillac (945 - 1003 A. D.), his name before he became a pontiff.

Gerbert d'Aurillac was born as an obscure peasant from a poor family in France and was first educated at Aurillac in France, later moving to Catalonian Spain. He visited the cities of Barcelona, Seville, and Cordova in Spain which were leading centers of learning during the period. Gerbert became the headmaster of the cathedral school at Reims, later became the archbishop of Reims and then of Ravenna in Italy. With the patronage of Otto III of Saxony, in 999 A.D., he was elected as pontiff and chose Pope Sylvester II as his new name. It was an important period for Christianity as it was about to complete the year one thousand. Pope Sylvester II is known for his efforts to find and translate Greek and Arabic texts on natural philosophy into Latin. He raised the profile of natural philosophy and reinforced intellectual aspects of theology in the mainstream activities on the Church.

---

[107]Gazalé, 2000, p. 48.
[108]Burnett, 2006
[109]Sarton, 1950.

Pope Gerbert had a unique training: monastic life from Christian teachers, pagan life from Latin classics, and academic learnings (astronomy and mathematics) from the Muslim teachers of tenth century Spain. His interests included literature, music, philosophy, theology, mathematics and the natural sciences. He became familiar with the Hindu numerals during his stay in Spain. Later, he wrote and taught in his own school in Rheims about the Hindu numerals with its positional notations and rules related to the arithmetical operations.[110] He used the abacus for mathematical computations.

Gerbert used the nine signs of the Hindus to construct his abacus and "gave the multiplication or the division of each number, dividing and multiplying their infinite numbers with such quickness that, as for their multiplication, one could get the answer quicker than he could express in words.".[111] Later, after several years of teaching in Rheims, he was designated as abbot of St. Columban of Bobbio with the help of Otto II. With the premature death of Otto II, the empire was now controlled by Otto III, a three-year-old child. This child needed the help of his loyal people. Gerbert provided this loyalty and played an important role in keeping the empire safe for Otto III, particularly from the hands of Henry of Bavaria. In one letter, he wished Otto III to have a long life in terms of numbers counted in an abacus: "Let the highest of the figures of the abacus prescribe the limit of your life."[112] Otto III used his influence and helped Gerbert to become pope on April 9, 999. With his education in France and Spain in Latin grammar, science and mathematics, his untiring efforts, and his political astuteness, he eventually became the hundred forty-sixth pope.

Gerbert wrote five books dealing with science and mathematics that are now lost. What we know about Gerbert is through the writings of his students, particularly Richer, the son of a French nobleman[113] and through his letters.[114] Richer was a disciple of Gerbert. He became a monk of St. Rémy of Rheims and wrote a popular book on the history of France, *Historia Francorum.*[115]. We also know about Gerbert due to the extensive scholarship of

[110]Lattin, 1961.
[111]Darlington, 1947.
[112]Darlington, 1936, taken from Darlington, 1947.
[113]Darlington, 1947.
[114]Lattin, 1961.
[115]Richer, 1964 - 67.

Russian scholar Nikolai Bubnov who throughout his life studied the works of Gerbert and published a book about his contributions to mathematics, in Latin.[116] This book is yet to be translated in English. Another book on the life of Gerbert was published by Pierre Riché in French.[117]

In one letter that was written to Constantine, monk of Fleury in 980 A.D., Gerbert explained the Hindu numerals and mathematics in a letter to his monk friend using the abacus that was popular in Spain. Following are the excerpts of the letter: "Do not let any half-educated philosopher think they [Hindu numerals and mathematics] are contrary to any of the arts or to philosophy. . . For, who can say which are digits, which are articles, which the lesser numbers of divisors, if he disdains sitting at the feet of the ancient?"[118]

Richer described Gerbert's abacus in the following words: "in teaching geometry, indeed, he expended no less labor [than in teaching astronomy]. As an introduction to it he caused an abacus, that is, a board of suitable dimensions, to be made by a shield maker. Its length was divided into 27 parts [columns] on which he arranged the symbols, nine in number, signifying all numbers. Likewise, he had made out of horn a thousand characters, which indicated the multiplication or division of each number when shifted about in the 27 parts of the abacus. [He manipulated the characters] with such speed in dividing and multiplying large numbers that, in view of their very great size, they could be shown [seen] rather than be grasped mentally by words. If anyone desires to know this more thoroughly let him read his book which he wrote to Constantine, the grammaticus; for there he will find these matters treated completely enough."[119]

Another disciple of Gerbert, Bernelinus, has described his abacus as consisting of a smooth board with a uniform layer of blue sand. For arithmetical purposes, the board was divided into 30 columns, of which 3 were reserved for fractions. These 27 columns were grouped with three columns in each group. These were marked as C (*cenlum*), D (*decem*), and S (*singularis*) and M (*monas*). Bernelinus gave the nine numerals and suggested that the

[116]Bubnov, 1899.
[117]Riché, 1987.
[118]Lattin, 1961, p. 45.
[119]Lattin, 1961, p. 46.

Greek letters could also be used instead.[120]

Pope Sylvester lived in a period that is before the times of Fibonacci of Italy and Roger Bacon of England, both credited for the introduction of the Hindu numerals in Europe. Unfortunately, Pope Sylvester II died in 1003, less than four years after he was crowned as Pope. He was a champion of scientific scholarships along with his usual duties as a religious leader.

[120]Cajori, 1980, p. 116.

# Chapter 4

# Mathematics

"Of the development of Hindu mathematics we know but little. A few manuscripts bear testimony that the Indians had climbed to a lofty height, but their path of ascent is no longer traceable," wrote Cajori in his book, *A History of Mathematics*.[1] This was the status of scholarship about Hindu mathematics in 1893 when Cajori's book was first published. The book is still a landmark in the history of mathematics. However, the scholarship of the last hundred years has allowed us to know a lot more about Hindu mathematics.

The history of mathematics is a field that has not evolved for a variety of reasons.[2] In the words of George Sarton, "if the history of science is a secret history, then the history of mathematics is a doubly secret, as secret within a secret."[3] This is also, incidentally, a discipline where the non-Western cultures have made considerable contributions. The disciplines of mathematics owe a lot to their Hindu and Middle Eastern roots.[4]

Roger Bacon (c. 1214 - 1292 A.D.), a noted Franciscan natural philosopher from England, considered mathematics essential for the study of natural

[1]Cajori, 1980, p. 83.

[2]Dauben, Joseph W., "Mathematics: a Historian's Perspective," a chapter in the book by Chikara, Mitsuo and Dauben, 1994.

[3]Taken from Dauben, Joseph W., Mathematics: a Historian's Perspective, a chapter in the book by Chikara, Mitsuo and Dauben, 1994, p. 1.

[4]Bag, 1979; Colebrooke, 1817; Datta and Singh, 1938; Ifrah, 1985; Joseph, 1991; Rashed, 1996; Rouse Ball, 1908; Smith, 1925; Srinivasiengar, 1967. Ifrah's and Joseph's books are written in a quite engaging style.

philosophy. He also considered it "absolutely necessary" in the study of "sacred science".[5] This idea was shared in his book that was written for Pope Clement IV to propagate Christianity.

Prior to Roger Bacon, al-Khwārizmī (ca. 800 - 847 A.D.), from the Middle East, considered mathematics to be useful in the dealing with inheritance, legacies, partition, lawsuits, trade, land measurements, digging canals, and geometrical computations.[6] Elsewhere, the Egyptians considered mathematics as the "divine source", as Plato had stated this in *Phaedrus* (274 c - d). As has been mentioned earlier (Section 2.6), mathematics was considered helpful in achieving salvation for the Hindus, as defined in *Chāndogya-Upaniṣad*.

In the Sanskrit language, mathematics is called *gaṇita*–meaning the science of calculation. *Aṅka-gaṇita* or *pāṭī-gaṇita* (mathematics by means of algorithms) is the Sanskrit word for arithmetic, the science of numbers. *Bīja-gaṇita* is the word used in Sanskrit for algebra, which literally translates into "mathematics (*gaṇita*) of the seeds (*Bīja*)." This name has evolved from an analogy of algebraic equations (*samīkaraṇa*) as multiple branches of a tree that lead to seeds, which represented the mathematical solutions. Since algebra deals with unknown quantities in mathematical equations, Hindus also used the expression *avyakta-gaṇita* for algebra, which means the mathematics of the unknowns. Geometry is called *Rajju* (rope) or *rekhā-gaṇita* (line-mathematics) in the Sanskrit language. Like the Egyptians, the Hindus used ropes or strings in various geometrical constructions. This prompted them to call geometry by the name *Rajju*, meaning rope.

## 4.1 Arithmetic (*Aṅka-gaṇita*)

Arithmetic deals with the arts of mensuration, numerical computation, and the elementary aspects of the theory of numerals. In comparison to other cultures "in all fundamental aspects the Hindu arithmetic corresponds to the modern subject much more closely than the same subject as developed by any other people before the year 800 A.D.,"[7] concludes Karpinski in his analysis of the ancient contributions of various civilizations.

[5]Bacon, 1928, p. 195.

[6]Rosen, 1986, p. 3 - 4.

[7]Karpinski, 1965, p. 46. L. C. Karpinski was a known historian of mathematics.

Al-Bīrūnī (973 - 1050 A.D.), a Muslim natural philosopher from the Middle East, compared the knowledge of arithmetic in India with that of Arabia, and accepted the authority of the Hindus. "The Hindus use the numeral signs in arithmetic in the same way as we do. I have composed a treatise showing how far, possibly, the Hindus are ahead of us in the subject [of arithmetic].[8]

al-Khwārizmī wrote two texts on arithmetic: one on Hindu arithmetic that has come to us in a Latin translation and the other work is mentioned in Arabic bibliographies, *al-Jamʿ wa al-Tafriq* (*Augmentation and diminution*).[9] Though, al-Khwārizmī indicated that his book is the compilation of Hindu mathematics, we cannot assume that the Latin version[10] of his book covers all the Hindu arithmetic that went to Islam. We can also not assume that everything in this book is from India, although al-Khwārizmī indicated that the book has Hindu mathematics at the outset in his book. "However, in fairness to all, we can tell for sure that the decimal scale with the operations of addition, subtraction, multiplication, division and square root extraction, of whole numbers and fractions, as well as the same operations in the scale of sixty, were taken over from India,"[11] concludes Saidan in his analysis of the history of mathematics.

The *Yajurveda* provides the successive addition of 2 to 1 (1, 3, 5, 7, . . .) and a multiplication table of four up to forty-eight.[12] The *Ṛgveda* provides the multiplication table of number two up to 10 and the multiplication table of ten up to hundred.[13] The *Atharvaveda* gives the multiplication table of

[8]Sachau, 1964, vol. I, p. 177.

[9]Saidan, Ahmad S., Numeration and Arithmetic, in the book by Rashed, 1996, vol. II, p. 331.

[10]Vogel 1963.

[11]Saidan, Ahmad S., Numeration and Arithmetic, in the book by Rashed, 1996, vol. II, p. 336-7; also see Saidan, 1965.

[12]"May my One [*Eka*] and my Three, and my Three and my Five, and my Five and my Seven, and my seven and my nine . . . [the addition continues to thirty-three] prosper by sacrifice." (*Yajurveda* (white), 18: 24)

"May my Four and my Eight; my eight and my Twelve; my twelve and my sixteen; my sixteen and my twenty, . . . my twenty-four. . . my twenty-eight. . . my thirty-two. . . my thirty-six my forty and my forty-four. . . my forty-eight prosper by sacrifice." *Yajurveda* (White), 18: 25.

[13]"Come, Indra, when invoked, with two horses, or with four or with six, or with eight,

number eleven up to ninety-nine.[14] The *Ṛgveda* mentions one-fourth and three-fourth parts which can only be done using division.[15]

The *Śatapatha-Brāhmaṇa* provides the division of 1,000 with three, and mentions one as a remainder.[16] The *Śatapatha-Brāhmaṇa* gives the division of seven hundred twenty by two, three, four, five, six, seven, eight, nine, ten, twelve, fifteen, sixteen, eighteen, twenty, and twenty four.[17] *Āpastambā-Śulbasūtra* defines a large number of fractions[18] and one is provided as an example: "*anuka* is one-fourth (of a *purusa*), aratni one-fifth (of a *purusa*), and so is urvasthi (one-sixth of a *purusa*.).[19] The *Kātyāyana-Śulbasūtra* provides the fraction $14\frac{3}{7}$) = 14 + (3 divided by 7).[20]

When the rules of Hindu arithmetic reached Europe, these methods were called *algorism*.[21] This term was perhaps named after al-Khwārizmī who wrote about the Hindu arithmetic and was known as Algorismus in the Latin world. Interestingly, the word could also be a metamorphosed form of its Arabic term for arithmetic that was equally common in Europe, *Ḥisāb al-ghubār* or dust-board-arithmetic.[22] *Ḥisāb* means mathematics and *al-ghubār* means dust-board.

---

or with ten, to drink the Soma juice: object of worship . . . Come to our presence, Indra, having harnessed thy car with twenty, thirty, or forty horses. Come thou with fifty well trained coursers, Indra, sixty or seventy, to drink the Soma. Come to us hitherward, O Indra, carried by eighty, ninety, or hundred horses." *Ṛgveda*, 2: 18: 4 - 6.

[14]Thy ninety-nine examiners, O night, who look upon mankind, eighty and eight in numbers, or seventy-seven are they. Sixty-six, O opulent, fifty-five and, O happy one, forty-four and thirty-three are they, O thou enriched with spoil. Twenty-two hast thou, O night, eleven, yea, and fewer still." *Atharvaveda*, 19: 47: 3 - 5.

[15]"So mighty is his greatness; yea, greater than this is *Puruṣa*. All creatures are one-fourth of him, three-fourth eternal life in heaven. With three-fourths *Puruṣa* went up: one-fourth of him again was there." *Ṛgveda*, 10: 90: 3, 4.

[16]"For Indra and Viṣṇu divided a thousand cows into three parts, there was one left . . . If anyone were to divide a thousand by three, one would remain over." *Śatapatha-Brāhmaṇa*, 3: 3: 1: 13.

[17]*Śatapatha-Brāhmaṇa*, 10: 4: 2: 4 - 17.

[18]*Āpastambā-Śulbasūtra* 11: 3; 15: 8; 18: 3; 19: 1 - 8.

[19]*Āpastambā-Śulbasūtra*, 11: 3.

[20]*Kātyāyana-Śulbasūtra* , 6: 2.

[21]Høyrup, 1994, p. 128.

[22]Salem and Kumar, 1991, p. 14

### 4.1.1 The Fibonacci Sequence

Leonardo Fibonacci's book *Liber Abaci* played an important role in the growth of mathematics in Europe. Fibonacci came in contact with Hindu mathematics during his stay in Bugia, located on the Barbary Coast of Africa. In Fibonacci's own account, his father was a public official there to help the visiting Pisan merchants there. His father wanted Fibonacci to learn mathematics for "a useful and comfortable future." He arranged some lessons in mathematics for the young Fibonacci from well known scholars. This is how Leonardo learned about Hindu numerals. Leonardo recognized their superiority over the Roman numeral system which was used in his native land, now known as Italy, and decided to write a book on the Hindu numerals to explain to the Italians their use and applications.

The Fibonacci sequence is connected with cumulative growth, and plays a role in various number games and natural phenomena, including a botanical phenomenon called phyllotaxis where the arrangement of leaves on a system is studied.[23] The seeds distributed in a sunflower have the Fibonacci sequence.

The following is the problem that Fibonacci posed in *Liber Abaci* that is well known today as the Fibonacci sequence:

"A certain man had one pair of rabbits together in a certain enclosed place, and one wishes to know how many are created from the pair in one year when it is the nature of them in a single month to bear another pair, and in the second month those born to bear also. Because the above written pair in the first month bore, you will double it; there will be two pairs in one month. One of these, namely the first, bears in the second month, and thus there are in the second month 3 pairs; of these in one month 2 are pregnant, and in the third month 2 pairs of rabbits are born, and thus there are five pairs in the month; in this month 3 pairs are pregnant, and in the fourth month there are 8 pairs, of which 5 pairs bear another 5 pairs; these are added to the 8 pairs making 13 pairs in the fifth month; these 5 pairs that are born in this month do not mate in this month, but another 8 pairs are pregnant, and thus there are in the sixth month 21 pairs; to these are added the 13 pairs that are born in the seventh month; there will be 34 pairs in

[23] For more information, see Hoggatt, 1969.

this month; to this are added the 21 pairs that are born in the eighth month; there will be 55 pairs in this month; to these are added the 34 pairs that are born in the ninth month; there will be 89 pairs in this month; to these are added again the 55 pairs that are born in the tenth month; there will be 144 pairs in this month; to these are added again the 89 pairs that are born in the eleventh month; there will be 233 pairs in this month. To these are still added the 144 pairs that are born in the last month; there will be 377 pairs, and this many pairs are produced from the above written pair in the mentioned place at the end of the one year."[24]

Based on the assumptions, at the beginning of the second month, there will be two pairs. After the second month, there will be three pairs. In the third month there will be 5 pairs. In the consecutive month 8, 13, 21, 34, 55, 89, 144, 233, and 377 pairs will be there. Thus, 1, 2, 3, 5, 8, 13, 21, 34, 55, 89, 144, 233, and 377 pairs of rabbits will be available. As one can notice, in this number sequence,

$$x_n = x_{n-1} + x_{n-2}$$

Also,

$$x_{n+1} \times x_{n-1} = x_n^2 + (-1)^n$$

It was the French mathematician Édouard Lucas who gave the name "the Fibonacci sequence" to this mathematical series in the nineteenth century. Later, an Oxford botanist, A. H. Church, recognized that the number of seeds in the spiral patter of sunflower head match with the Fibonacci sequence numbers. In 1963, an International Fibonacci Society was formed to study related topics.[25]

The work of Leonardo Fibonacci and his personal ethics sets an example of multiculturalism. He was from Pisa, a strong hold of Christianity. His interactions were with several noted Muslim scholars who taught him Hindu mathematics. In all these intellectual endeavors, he kept his scholarly integrity and gave due recognitions to Islamic as well as Hindu cultures.

---

[24] Sigler, 2002, p. 404 - 405.

[25] Gies and Gies, 1969, p. 81 - 83.

Leonardo was quite clear in his book that the contents assembled were based on the mathematics of the Hindus and Muslims. However, some information in the book could be his own contribution. On the Fibbonacci sequence, Leonardo has not provided any clear indication of its origin.

In India, this sequence appeared in the science of hymn-composing, or metrics, just like the binary numbers that were discussed in the previous chapter. In one particular category, the number of variations of meters having 1, 2, 3, . . . *morae* (syllabic instant) are 1, 2, 3, 5, 8, 13, . . ., respectively, the so-called Fibonacci numbers. Ācārya Virāhanka (lived sometime between 600 - 800 A.D.), Gopāla (before 1135 A.D.) and Hemachandra (ca. 1150 A.D.) had provided this sequence before it was suggested by Leonardo Fibonacci. To understand this metrical science, a knowledge of Sanskrit is required. Readers can get more information on this from the work of Permanand Singh that is published in a prestigious journal *Historia Mathematica.* He concludes that "the concept of the sequence of these numbers in India is at least as old as the origin of the metrical sciences of Sanskrit and Prakrit poetry.[26]

### 4.1.2 The Square-root Operation

The approximate value of the square root of the number 2 can be calculated using Pythagorean theorem of the ancient Hindus in arithmetic form, as explained later in Section 4.4. If we apply the expression to a square of side 1 in any unit system, we can find the value of $\sqrt{2}$. According to this theorem, the diagonal of a right-angle triangle with two equal sides ($a$) is

$$L = a + \frac{a}{3} + \frac{a}{3 \times 4} - \frac{a}{3 \times 4 \times 34}$$

For $a = 1$, this gives us the approximate value of $\sqrt{2}$.

$$L = \sqrt{2} = 1 + \frac{1}{3} + \frac{1}{3 \times 4} - \frac{1}{3 \times 4 \times 34} = \frac{577}{408}$$

$$\sqrt{2} = 1.41$$

[26]Singh, 1985.

This is the accepted value. If one tries to get higher accuracy in the results, the value is correct to the fifth decimal place, 1.41421. After the fifth decimal place, the value given in this formula is slightly higher (1.4142156) than the actual value (1.4142135). John F. Price of the University of New South Wales provided the rationale of this formula.[27] We know the value of the $\sqrt{2}$ is between 1 and 2. If we equate $\sqrt{2}$ to 1 or 2 and square both sides we get 1 or 4 on one side while 2 on the other side in both cases. Using similar considerations, we know that $\sqrt{2}$ will be less than $(1 + \frac{1}{2})$ and more than $(1 + \frac{1}{3})$, Therefore, our first initial approximation is

$$\sqrt{2} \approx 1 + \frac{1}{3}$$

If we improve this value by adding a small term ($x$) to our value of $\sqrt{2}$, square both sides, assume $x^2$ to be quite small and neglect it, we go through the following sequence:

$$\sqrt{2} = 1 + \frac{1}{3} + x$$

$$2 = \left(\frac{4}{3}\right)^2 + \frac{8}{3}x + x^2$$

$$x = \frac{1}{3 \times 4}$$

In our next approximation,

$$\sqrt{2} \approx 1 + \frac{1}{3} + \frac{1}{3 \times 4} + y$$

This reduces to

$$\sqrt{2} = \frac{4}{3} + \frac{1}{3 \times 4} + y$$

Again, square and ignore $y^2$ term. The next step yields

---

[27]Price, in Gorini, 2000, p. 46 - 55.

$$2 = \frac{16}{9} + \frac{1}{9 \times 16} + \frac{2 \times 4}{9 \times 4} + y\left(\frac{2 \times 4}{3} + \frac{2}{3 \times 4}\right)$$

If we simplify this equation and calculate $y$, we get

$$y = -\frac{1}{3 \times 4 \times 34}$$

If we continue further, the next approximation will be

$$L = 1 + \frac{1}{3} + \frac{1}{3 \times 4} - \frac{1}{3 \times 4 \times 34} - \frac{1}{3 \times 4 \times 34 \times 2 \times 577}$$

This will give us $\sqrt{2} = 1.414213562374$, a value accurate up to the thirteenth place, as was done by David W. Henderson of Cornell University[28] and Price[29]. It is interesting that the *Śulbasūtra* does mention that the value is only approximate and a little bit higher than the actual value (*saviśeṣa*).[30] This process is called the method of successive approximations in most mathematics books, a modern technique.

This process continues with more terms added. It will always provide only an approximate value. However, it loses it practical relevance at the stage that was defined in *Śulbasūtra*. The fifth place of the decimal was good enough for most measurements for the ancient Hindus. Henderson as well as Price also tried to compare the popular "divide-and-average" method, also called the Newton's method, with the Baudhayāna's method. The Newton's method starts with an initial approximation $a_0$ for the square-root of the number $N$. The next approximation is the average of $a_0$ and $\frac{N}{a_0}$. In general, if $a_n$ is the $n^{\text{th}}$ approximation, then

---

[28]Read, Square Roots in the Śulba Sūtras by Henderson, in Gorini, 2000, p. 39 - 45.

[29]Price, John F., Applied Geometry of the *Śulba Sūtras*, in the book by Gorini, 2000, p. 46 - 55.

[30]Read, Square Roots in the Śulba Sūtras by Henderson, in Gorini, 2000, p. 39 - 45.

$$a_{n+1} = \frac{a_n + \frac{N}{a_n}}{2}$$

Henderson concluded that Baudhayāna's method "uses significantly fewer computations" than the Newton's method to reach the same order of accuracy.

Āryabhaṭa I gave rules for the square-root[31] and the cube-root.[32] For simplicity, let us explain the square-root operation: "(Having subtracted the largest possible square from the last odd place and then having written down the square-root of the number subtracted in the line of the square-root) always divide the even place (standing on the right) by twice the square-root. then, having subtracted the square (of the quotient) from the odd place (standing on the right) set down the quotient at the next place (*i.e.*, on the right of the number already written in the line of the square-root). This is the square-root." (Repeat the process if there are still digits on the right.)

The above statement reduces to the following steps:

1. Define odd and even digits, starting from the right to the left.
2. Subtract the largest possible square from the number that is at the last odd place (count from right to left). If the last place is not odd, consider the number with even and odd places.
3. The remainder from the subtraction and the next even digit will form the next number. Divide this number by twice of the number that was squared in the preceding step.
4. The remainder and the next odd digit will form the next number. Subtract the square of the quotient in the preceding step.
5. Continue this process as long as there are digits left.

Let us understand this process by taking a few examples:

[31] *Āryabhaṭīya, Gaṇitpada*, 4. Also, Shukla and Sarma, 1976, p. 36.
[32] *Āryabhaṭīya, Gaṇitpada*, 5.

Example 1: Find the square-root of 1369.

We will write the number as

*eoeo*
1369

Remember, we are looking for the last odd place and not the number. In our case, this is the third place where we have number 3. Therefore, we need to find the largest possible square for 13. We know that $3^2 < 13 < 4^2$. Therefore, subtract $3^2$ from 13. This leaves us the remainder 4.

Let us borrow the next digit (6). This makes the next number 46. Divide this number by six ($2 \times 3 = 6$, six is twice the number that was squared.)

This gives us the quotient 7 and the remainder is 4 ($6 \times 7 = 42$, $46 - 42 = 4$). Borrow the next digit 9. This makes the number 49. Subtract the square of the quotient 7. We end up with zero. Therefore, the square root is 37. If the square-root is not exact, we will be left with remainder.

Example 2: Find the square-root of 66564.

*oeoeo*
66564

The digit at the last odd place is 6, at the fifth place. Therefore, we will square 2 and subtract from 6. This gives us the remainder 2. This makes our next number 26. Divide 26 by 4 (as twice of 2). The quotient is 5 and the remainder is 6. Borrow 5 which makes the next number as 65. Subtract $5^2$. This gives us 40. Borrow 6. This makes the next digit 406. Divide it by 50 ($25 \times 2$) (First number and the quotient multiplied by two). We end up with the quotient 8. The remainder is 6. The next number is 64. Subtract the square of 8. This gives us zero. Therefore, the square-root is 258.

As you may have noticed. When we divided 26 by 4, we could have taken 6 as the quotient. However, that would leave us 2 as the remainder. This leaves us zero in the next step. We must always avoid such a situation where we end up with zero or negative numbers.

Example 3: Find the square-root of 41692849.

*eoeoeoeo*
41692849

The first number is 41. We will take the square of 6 (= 36) and subtract. The next number is 56. We will divide is by twice of 6, meaning 12. We get the quotient 4. The next number is 89. We will subtract it by the square of 4 (= 16). The next number is 732. We will divide it by twice of 64 (= 128). We get the quotient 5. The next number is 928. We will subtract $5^2$ (= 25). The next number is 9034. We will divide this number by twice of 645 (= 1290). The quotient is 7. The next number is 49. Subtract the square of 7. We get zero. Therefore, the square-root is 6457.

## 4.2 Algebra, Algorithm, and al-Khwārizmī

The Arabs were certainly well versed in algebraic techniques and may have invented some techniques or improved upon the techniques of algebraic calculations from India; but they certainly did not invent algebra. Algebraic techniques were in a more advanced state among the Hindus much before they flourished in Arabia and Europe. The Hindu influence of al-Khwārizmī's work was unmistakable.[33] He used *jidhr* as the term to denote the first power of the unknown. This is a Hindu term that was used for some time in India.[34]

al-Khwārizmī, Muhammad ibn Mūsā is one of the earliest known astronomers and mathematicians from the glorious period of Baghdad or Islam. He was native to the Khwarizm Region in Persia, as the name suggests. He later moved to Baghdad and served in the court of al-Mamūn (813 - 833 A.D.). His best known works are: *Kitāb al-jabr wa'l Muqābala* (*The Book of Manipulation and Restoration*), *Kitāb al Hisāb al-Hindī* (*Book of Indian Mathematics*), and *Zīj al-Sindhind* (*Astronomy Table from India*). The last two books are obviously the works of the Hindus as the titles suggest. Even the third book was based on the work of the Hindus. The impact of al-Khwārizmī was so great in Europe that the title of his one book became synonymous with the theory of equations. The word "algebra" stemmed from the word *al-jabr* which appeared in the title of al-Khwārizmī's book.

[33]Rosen, 1986; Hughes, 1989.
[34]Høyrup, 1994, p. 95.

*Al-jabr* literally means bone-setting, indication manipulation of equations. Al-Khwārizmī played a crucial role as a disseminator of science. Latinization of his name took many forms: from *algorizmus* to *algoritmus* to *algorithmus*. The mathematical term, algorithm, has possibly stemmed from his Latinized name.[35] It may also be a metamorphosed form of *Hisāb al-Ghubār* (dust board arithmetic) in which *al-Ghubār* became algorithm. Today, algorithm is used to denote any systematic computation, or to define a system of step-by-step instructions for solving a problem.

By at least the fifth century A.D., the following are some of the rules that were known to the ancient Hindus. All these rules are taken from *Āryabhaṭīya*. Āryabhaṭa I did not provide the derivation of the rules in most cases and called the knowledge ancient.

### 4.2.1 Unknown in an Algebraic Equation

"The difference of the known *amounts* relating to the two persons should be divided by the difference of the coefficients of the unknown. The quotient will be the value of the unknown, if their possessions are equal."[36]

This statement can be simplified in the following language: Two persons, who are equally rich, possess respectively, $a$ and $b$ times a certain unknown amount of money ($x$) plus $c$ and $d$ amounts of money in cash, respectively. Mathematically, it can be written as:

$$ax + c = bx + d$$

Where $x$ is the unknown amount. Then,

$$x = \frac{d - c}{a - b}$$

This result is correct.

[35] For more information, read Crossley and Henry, 1990; Hughes, 1989; King, 1983; Sesiano, in the book by Selin, 1997. As previously mentioned, this speculation is not based on historical documents. The term may have evolved from the term *al-ghubar*, dust-board mathematics that was used for these algorithms.

[36] *Āryabhaṭīya*, *Gaṇitpada*, 30.

### 4.2.2 Product of Two Numbers Using Squares

"One should subtract the sum of the squares of two factors from the square of their sum. Half the result is the product of two factors."[37] Mathematically, the meter can be written for two number $a$ and $b$ as:

$$ab = \frac{(a + b)^2 - (a^2 + b^2)}{2}$$

This is a correct formula.

### 4.2.3 Loan with Complex Interest Scenario

Āryabhaṭa I also gave the solution of a complex money-transaction problem: "Multiply the sum of the interest on the principal and the interest on this interest by the time and by the principal. Add to this result the square of the half of the principal. Take the square-root of this. Subtract half the principal and divide the remainder by the time. The result will be the interest on the principal."[38]

The problem can be written as follows: "A certain amount of money ($P$ for principal money) was given on loan for one month with unknown interest $x$. The unknown interest $x$ that was accrued in one month was again loaned for time $T$ months. On the completion of this period, the original interest $x$ and the interest on this interest, all together, became $I$. Find out the rate of interest $x$ on the amount $P$. The answer provided by Āryabhaṭa I is as follows:

$$x = \frac{\sqrt{IPT + (P/2)^2} - P/2}{T}$$

The solution provided is correct.

[37] *Āryabhaṭīya, Gaṇitpada*, 23.
[38] *Āryabhaṭīya, Gaṇitpada*, 25.

### 4.2.4 Sum of a Series

"The desired number of terms minus one, halved, plus the number of terms which precedes, multiplied by the common difference between the terms, plus the first term, is the middle term. This multiplied by the number of terms desired is the sum of the desired number of terms. Or the sum of the first and last terms is multiplied by half the number of terms."[39]

This rule can be mathematically written for a series with initial term $a$ and common difference between terms as $d$. mathematically, it can be written as

$$a + (a + d) + ...[a + (n - 1)d]$$

The sum of the series for $n$ terms is:

$$\begin{aligned} S_n &= n\left[\left(\frac{n-1}{2}\right)d + a\right] \\ &= \frac{n}{2}[a + (a + (n-1)d)] \end{aligned}$$

This is a correct result.

Similarly, in another verse, Āryabhaṭa I provides the solution to the following problem: "The number of terms (is obtained as follows): Multiply (the sum of the series) by eight and by the common difference, increase that by the square of the difference between twice the first term and the common difference, and then take the square root; then subtract twice the first term, then divide by the common difference, then add one (to the quotient), and then divide by two."[40]

To understand the above statement, let us assume $S$ be the sum of the series

$$a + (a + d) + (a + 2d) + (a + 3d) + ...\text{to } n \text{ terms}$$

[39] *Āryabhaṭīya, Gaṇitpada*, 19.
[40] *Āryabhaṭīya, Gaṇitpada*, 20.

In this case,

$$n = \frac{1}{2}\left[\frac{\sqrt{8ds + (2a-d)^2} - 2a}{d} + 1\right]$$

Again, this is a correct result.

### 4.2.5 Sum of a Series with $\Sigma n^2$ and $\Sigma n^3$

"The continued product of the three quantities viz, the number of terms, number of terms plus one, and twice the number of terms increased by one when divided by 6 gives the sum of the series of squares of natural numbers. The square of the sum of the series of natural numbers gives the sum of the series of cubes of natural numbers."[41]

Mathematically, the above verse translates as follows:

For the series $1^2 + 2^2 + 3^2 + ... + n^2$, the sum equals to $\frac{n(n+1)(2n+1)}{6}$. This provides the correct value for all values of $n$.

Similarly, for the series $1^3 + 2^3 + 3^3 + ... + n^3$, according to Āryabhaṭa I, the sum equals to $(1 + 2 + 3 + ... + n)^2$. This reduces to $\left[\frac{n(n+1)}{2}\right]^2$ which provides correct result.

### 4.2.6 Simultaneous Quadratic Equations

Āryabhaṭa I provided the solution of simultaneous quadratic equations:

$$x - y = d$$

$$xy = b$$

"Multiply the product [of two factors] by the square of 2 [meaning 4], add the square of the difference between the two factors, take the square

[41] *Āryabhaṭīya, Gaṇitpada*, 22.

root, add and subtract the difference between the two factors, and divide the result by two."[42]

This rule provides the value of $x$ and $y$ as:

$$x = \frac{\sqrt{(d^2 + 4b)} + d}{2}$$

$$y = \frac{\sqrt{(d^2 + 4b)} - d}{2}$$

### 4.2.7 System of Linear Equations

Āryabhaṭa I provided the solution of $n$ linear equation of the following kinds:[43]

$$(x_1 + x_2 + x_3 + ... + x_n) - x_1 = a_1$$

$$(x_1 + x_2 + x_3 + ... + x_n) - x_2 = a_2$$

$$(x_1 + x_2 + x_3 + ... + x_n) - x_3 = a_3$$

$$.$$

$$.$$

$$(x_1 + x_2 + x_3 + ... + x_n) - x_n = a_n$$

where $x_1, x_2, x_3, ...x_n$ are the unknown quantities and $a_1, a_2, a_3, ...a_n$ are the known values. Then,

$$(x_1 + x_2 + x_3 + ... + x_n) = \frac{a_1 + a_2 + a_3 + ... + a_n}{n - 1}$$

$$x_1 = \frac{a_1 + a_2 + a_3 + ... + a_n}{n - 1} - a_1$$

[42] *Āryabhaṭīya, Gaṇitpada*, 24.
[43] *Āryabhaṭīya, Gaṇitpada*, 29.

$$x_2 = \frac{a_1 + a_2 + a_3 + ... + a_n}{n-1} - a_2$$

.

.

$$x_n = \frac{a_1 + a_2 + a_3 + ... + a_n}{n-1} - a_n$$

The solution of linear equations, quadratic equations, and indeterminate equations of the first order, etc. were also known to the ancient Hindus.[44]

## 4.3 Geometry

Geometry is the study of shapes and forms. These studies of geometry are applicable to a variety of situations – from architecture to engineering, from politics to military warfare, and from religion to science.[45] The ancient Hindus used an elaborate knowledge of geometry for the construction of altars for religious purposes and for arranging various battalions of soldiers in wars.[46] The cities of the Indus-Sarasvatī Civilization demonstrate careful planning and a knowledge of geometry and mensuration in their city-planning.[47]

Tribal warfares are like fully frontal assaults where the larger armies with young people prevail. However, in scientific warfare, a smaller army can prevail with proper geometrical constructions in achieving their tactical goals. For example, Pāṇḍava in *Mahābhārata* used their much smaller army and

---

[44]A detailed analysis of such rules is provided by Datta and Singh, 1938 and Srinivasiengar, 1967.

[45]Gorini, 2000.

[46]Datta, 1932; Kulkarni, 1983; Sarasvati Amma, 1979; Sen and Bag, 1983; Staal, 1999; Thibaut, 1875.

[47]Kulkarni, 1978b, 1983.

these tactical geometrical constructions against a much larger army of Kaurava. *Cakravyūha* was an important geometrical construction of the placements of warriors to trap enemy warriors during the period of *Mahābhārata*. Abhimanyu, son of Arjuna, as mentioned in *Mahābhārata*, knew how to break and enter into the geometrical fortification of *cakravyūha*. However, he did not know the ways to come out of it. He was trapped and lost his life. In *Mahābhārata* war, multiple geometrical constructions, defined by their similarity to mostly animal or object shapes, were used in organizing soldiers to fight.

The word *śulba* means a chord, a rope, or a string. *Śulbasūtra* signifies geometry using strings. The Egyptians also used ropes in geometry, especially in the building of pyramids, as many scholars believe. The *Śulbasūtra* of Baudhayāna, Āpastambā and Kātyayāna provide descriptions and discussion on the use of geometrical knowledge in the construction of altars. The *Śulbasūtra* are not the books of geometry or mathematics; these books deal mostly with rituals. The knowledge of geometry and mathematics is used to perform the rituals. One can consider them as the "first applied geometry text in the world" to "combine geometry and numerical techniques."[48]

Elaborately designed altars were used extensively for longevity of life, good luck, and victory in wars. *Baudhayāna-Śulbasūtra* suggests, "He who desires heaven may construct the falcon shape altar."[49] Or, "He who wishes to conquer the world of Brahman is to construct the altar for *yajna* in the shape of tortoise, such is the tradition."[50] Varieties of fire-altars were constructed in different shapes to achieve desired specific goals: falcon, heron, Alaja bird, isosceles triangle, square, rhombus, chariot wheel, trough, circle, pyre, and tortoise. These shapes were constructed not by simply artistic representations; they were constructed with different parts in exact proportion using geometry and other mathematical tools.

The ancient Hindus knew many basic rules of geometry in which they were ahead of other civilizations. A glimpse of their knowledge is provided here.

[48]Henderson, in Gorini, 2000

[49]*Baudhayāna-Śulbasūtra*, 3: 1

[50]*Baudhayāna-Śulbasūtra*, 3: 270

### 4.3.1 Area of a Triangle

"The product of the perpendicular (dropped from the vertex on the base) and half the base gives the measure of the area of a triangle," suggests *Āryabhaṭīya*.[51]

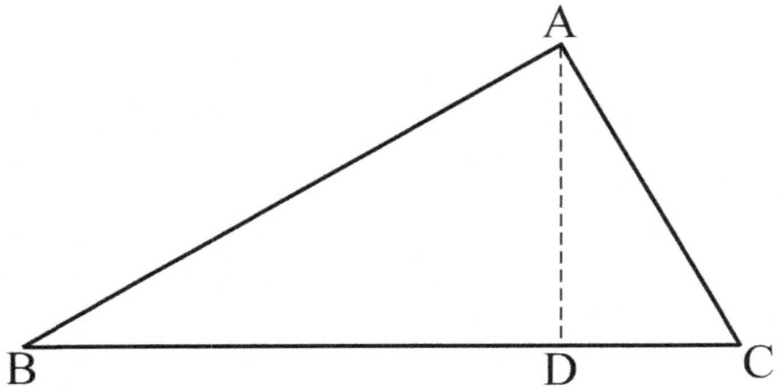


Figure 4.1: Area of a Triangle (Designed with the help of Morgan Gelfand and David Valentino).

Thus, Area of the triangle $= \frac{1}{2}$(length of BC $\times$ length of AD). This gives the correct value of area. (Figure 4.1)

### 4.3.2 Transforming a Square into a Circle

"If it is desired to transform a square into a circle, (a cord of length) half the diagonal (of the square) is stretched from the center to the east (a part of it lying outside the eastern side of the square); with one-third (of the part lying outside) added to the remainder (of the half diagonal), the (required) circle is drawn."[52]

Let PQRS be the given square and O is the center of the square (see Figure 4.2). The half diagonal OP is drawn and an arc PAQ is formed, implying

[51] *Āryabhaṭīya, Gaṇitpada*, 6.

[52] *Baudhayāna-Śulbasūtra*, 2: 9.

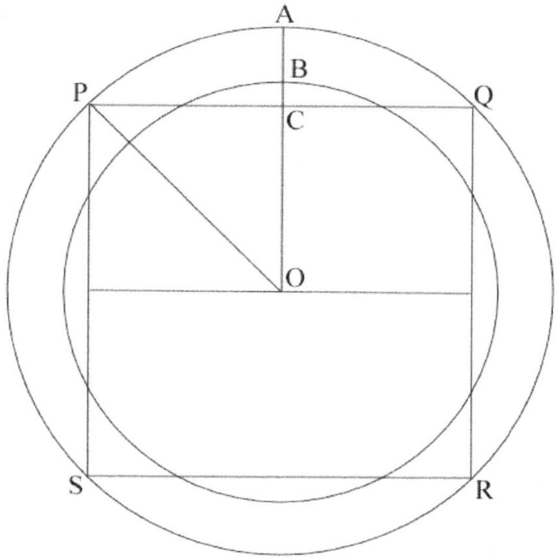


Figure 4.2: Transforming square into a circle of the same area (Designed with the help of Morgan Gelfand and David Valentino).

OP = OA. A circle of radius OB ($r$) is drawn where OB = OC + CB and CB = $\frac{1}{3}$CA.

This implies that OB = OC + $\frac{1}{3}$CA.

This equals OB = OC + $\frac{1}{3}$ (OA - OC).

Let $2a$ be the side of square PQRS,

$$r = a + \frac{1}{3}(\sqrt{2}a - a)$$

$$r = a[1 + \frac{1}{3}(\sqrt{2} - 1)]$$

$$r = \frac{a}{3}(2 + \sqrt{2})$$

The area of the circle is $\pi r^2 = 3.14[\frac{a}{3}(2 + \sqrt{2})]^2 = 4.1a^2$. This is in rough agreement with the area of the given square ($4a^2$).

There is a distinct difference between Greek and Hindu geometry. While the Greeks were rigorously logical and systematic in their approach, the Hindus were abstract, empirical and symbolic. Hindus' mathematical books did not provide axioms and postulates, as the Greek's books do; the mathematical rules are based on experimental verification. Seidenberg tried to assign a possible period for the general use of geometry among the Hindus but realized, "no matter how far we go back in 'history', we find geometrical rituals."[53]

### 4.3.3 Height of a Tall Object

"(When two gnomons of equal height are in the same direction from the lamp-post), multiply the distance between the tips of the shadows (of the two gnomons)[CD] by the (larger [GD] or shorter [EC]) shadow and divide by the larger shadow diminished by the shorter one [GD - EC]: the result is the upright (*i.e.*, the distance of the tip of the larger [BD] or shorter shadow [BC] from the foot of the lamp-post). The upright multiplied by the height of the gnomon and divided by the (larger or shorter) shadow gives the base (*i.e.* height of the lamp-post)."[54]

Mathematically,

$$\text{BD} = \frac{\text{CD} \times \text{GD}}{\text{GD - EC}}$$

and

$$\text{AB} = \frac{h \times \text{BC}}{\text{EC}} = \frac{h \times \text{CD}}{\text{GD} - \text{EC}}$$

The proof of this theorem is as follows (see Figure 4.3):[55] Let AB be the unknown height that we want to find out. We place a gnomon of height $h$ at point E. This gives us the shadow EC. Now we place the same gnomon or another one of the same size at point G. This gives us shadow GD.

Using similar triangles,

[53]Seidenberg, 1962.

[54]*Āryabhaṭīya*, *Gaṇitapada*,16. For more information, read Mishra and Singh, 1996; and Shukla and Sarma, 1976, p. 58.

[55]Professor George Baloglou, formerly at SUNY Oswego, helped me in deriving this equation.

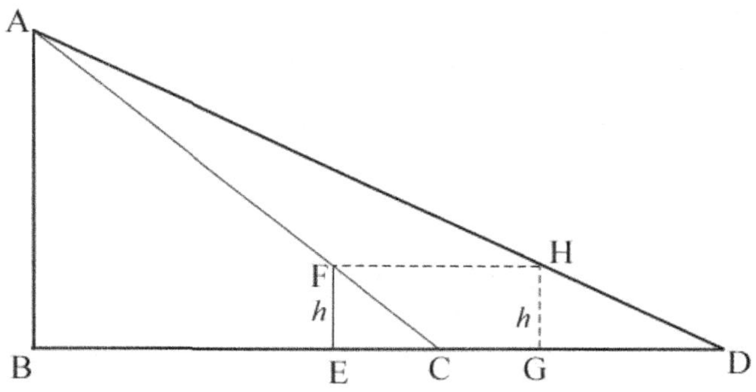


Figure 4.3: Determining height of a tall object using a small stick (Designed with the help of Morgan Gelfand and David Valentino).

$$\frac{\text{AB}}{h} = \frac{\text{BC}}{\text{EC}} = \frac{\text{BD}}{\text{GD}}$$

$$\text{AB} = \frac{\text{BC} \times h}{\text{EC}}$$

As given before,

$$\frac{\text{BC}}{\text{EC}} = \frac{\text{BD}}{\text{GD}}$$

Since, BD = BC + CD,

$$\frac{\text{BC} + \text{CD}}{\text{GD}} = \frac{\text{BC}}{\text{EC}}$$

$$\frac{\text{BC} + \text{CD}}{\text{BC}} = \frac{\text{GD}}{\text{EC}}$$

$$1 + \frac{\text{CD}}{\text{BC}} = \frac{\text{GD}}{\text{EC}}$$

$$\frac{\text{CD}}{\text{BC}} = \frac{\text{GD}}{\text{EC}} - 1$$

$$\frac{\text{CD}}{\text{BC}} = \frac{\text{GD} - \text{EC}}{\text{EC}}$$

$$\text{BC} = \frac{\text{CD} \times \text{EC}}{\text{GD} - \text{EC}}$$

Therefore,

$$\text{AB} = \frac{h \times \text{BC}}{\text{EC}} = \frac{h \times \text{CD}}{\text{GD} - \text{EC}}$$

Thus, by measuring the lengths of shadows at two different locations for a stick, one could measure the height of the unknown object. Or, by aligning the stick in such a way that one could see the top of the object in line with the top of stick while looking from the ground level from two different locations, one could measure the height of the unknown object.

### 4.3.4 The Value of $\pi$

The Greek letter $\pi$ (pronounced as *pi*) indicates the ratio of the circumference of a circle to its diameter that is a constant for any circle. Hindu books generally provided two values of $\pi$: one for rough calculations and second for precise measurements. The knowledge of $\pi$ was useful in the construction of altars, wheels of a cart, the metallic rims of a wheel, and in geometry.

The *Baudhayāna-Śulbasūtras* provided an approximate value of $\pi$ to be three; "The pits for the Pupas have a diameter of one *pada*. The periphery of the base of the Yupas is three *pada*."[56]

---

[56] *Baudhayāna-Śulbasūtra*, 4: 15.

The *Mānava-Śulbasūtra* provided the value of $\pi$ to be 3.2: "The fifth part of the diameter added to the three times the diameter gives the circumference (of a circle). Not a hair of length is left over."[57] This provides the value of circumference, $C$, from diameter, $D$, as,

$$C = \frac{D}{5} + 3D = \frac{16}{5}D = 3.2D$$

This gives a value of $\pi$ which is close to the actual value of 3.14. The readers must remember that the purpose of these books was to prepare altars and these calculations were good enough for that purpose because they worked for simple designs.

Āryabhaṭa I gave the value of $\pi$ that is correct to the fourth decimal place: "Add four to hundred, multiply by eight, and add sixty two thousand. The result is approximately the circumference of a circle [$C$] of which the diameter [$D$] is twenty thousand."[58]

Mathematically, we know that $C = \pi D$. Therefore, the value of $\pi$, based on Āryabhaṭa I's method, is equal to:

$$\pi = \frac{[(4 + 100) \times 8] + 62,000}{20,000}$$

$$= \frac{62,832}{20,000} = 3.1416$$

This is equal to the presently accepted value of 3.1416, for up to 4 decimal places. The readers must notice that Āryabhaṭa I suggested this value to be approximate. This makes sense since the value of $\pi$ can only be determined approximately since the ratio of circumference to diameter is not evenly divisive; it can have an innumerable number of significant figures. It is an endeavor for many mathematicians to calculate a more precise value of *pi*. The value of $\pi$ to a large number of significant figures is commonly used to check the speed, efficiency, and the accuracy of computers.

---

[57] *Mānava-Śulbasūtra*, 11: 13.

[58] *Āryabhaṭīya*, *Gaṇitapada*, 10. Also, Read Hayashi *et al*, 1989 and Kulkarni, 1978a. The work of Hayasi *et al* has a good review of the later developments on the issue in India. For example, Madhav (14th century) calculated the value of $\pi$ that was correct to eleven places.

## 4.4 The Pythagorean Theorem

The so-called Pythagorean theorem connects the three sides of a right-angle-triangle with the relation,

$$a^2 + b^2 = c^2$$

where $a$, $b$, and $c$ are the base, perpendicular, and hypotenuse, respectively of a right-angle triangle (see Figure 4.4). Pythagoras, a Greek philosopher is said to be the originator of the theorem sometime during sixth century B.C.

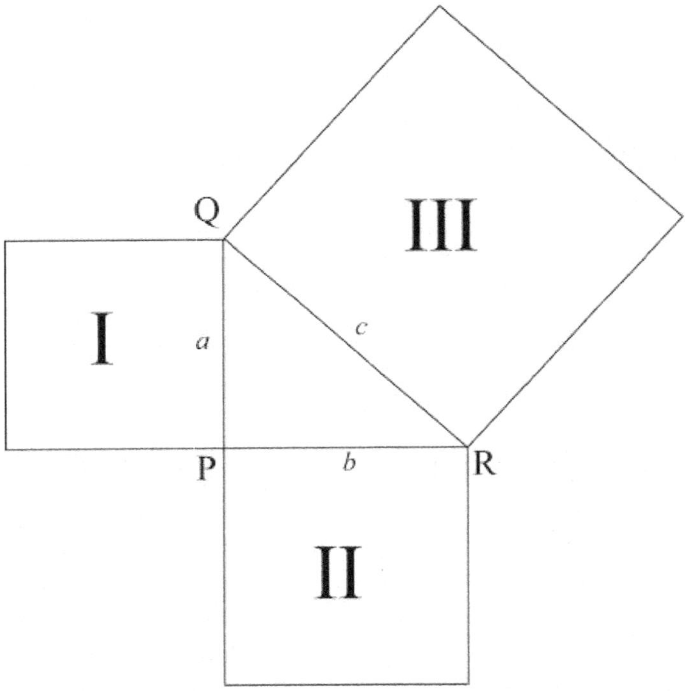


Figure 4.4: A right-angled triangle, PQR (Designed with the help of Morgan Gelfand and David Valentino).

*Śulbasūtras* provides the following specific mathematical relationship between base, perpendicular, and hypotenuse for right-angled triangles:

122 and 92 = 152[59]
32 and 42 = 52[60]
362 and 152 = 392[61]

This means that a right-angle triangle with two sides of 122 and 92 in any arbitrary unit will have hypotenuse of 152 in that unit. These values of the hypotenuse for a given base and perpendicular are approximately correct.[62]

Let us provide a few more statements from the *Śulbasūtra* that indicate the understanding of the Pythagorean theorem among the Hindus:

"The diagonal of a square produces double the area (of the square). It is $\sqrt{2}$ (*dvikaraṇī*) of the side of the square (of which it is the diagonal)."[63] The word *dvikaraṇī* literally means "that which produces 2," implying the twice area of the square constructed using *dvikaraṇī* as one side in comparison to the original square.

"The diagonal (of a right angle triangle) of which the breadth is *pada* and the length 3 *padas* is $\sqrt{10}$ *padas*."[64]

"The diagonal (of a right triangle) of which the breadth is 2 *pada* and the length is 6 *pada* is $\sqrt{40}$ *pada*."[65]

The *Baudhayāna-Śulbasūtra* provides another rule for a right-angle triangle: "The areas (of the squares) produced separately by the length and the breadth of a rectangle together equal the area (of the square) produced by the diagonal."[66]

Similar rules are given elsewhere in the Hindu literature: "The areas (of

[59] *Kātyāyana-Śulbasūtra*, 2: 5.

[60] *Āpastambā-Śulbasūtra*, 1: 13; 5: 3.

[61] *Baudhayāna-Śulbasūtra*, 1: 13 and *Āpastambā-Śulba-Sūtra*, 5: 4.

[62] A large number of similar mathematical calculations can be seen from Sen and Bag, 1983, p. 9 - 10.

[63] *Āpastambā-Śulbasūtra*, 1: 5.

[64] *Kātyāyana-Śulbasūtra*, 2: 4.

[65] *Kātyāyana-Śulbasūtra*, 2: 5.

[66] *Baudhayāna-Śulbasūtra*, 1: 48.

the squares) produced separately by the length and the breadth of a rectangle together equal the area (of the square) produced by the diagonal. By the understanding of these (methods) the construction of the figures as stated (is to be accomplished)."[67]

"The diagonal of a square produces a square twice as large (that is, the diagonal equals $\sqrt{2}$ times the side of the square).[68]

This indicates that the square of the hypotenuse (diagonal) is equal to the sum of the square of the two sides. Baudhayāna wrote that he had tried various possible combinations of the sides and found this rule to be valid.[69]

The above rules from the *Śulba-sūtra* provide ample examples of the so-called Pythagorean theorem in a variety of examples using multiple expressions. It is obvious that the ancient Hindus understood this theorem well. It is interesting to point out that Pythagoras, as the story goes, did not define his theorem in the form $a^2 + b^2 = c^2$; he defined this theorem in the form of the area of a square drawn on each side, as we know from the following Greek writers. There is a curious similarity between Baudhayāna and Pythagoras on the so-called Pythagorean theorem. It is important to understand that Baudhayāna's work was compiled at least around 1700 B.C., while the Pythagorean theorem was compiled over a millennia later, i.e. around sixth-century B.C.[70] There are indications that Pythagoras perhaps came in contact with India or Indian wisdom. A detailed discussion on the issue of his possible contact with India is provided in Appendix A-1.

The Pythagorean theorem is mostly given in terms of geometry in most cultures. However, among the Hindus, they used a mathematical expression for the Pythagorean theorem which is unique to them. No other culture used a similar mathematical expression for this theorem. Several *Śulbasūtras* provided a arithmetical method to find the diagonal of a right-angled triangle: "The measure is to be increased by its third and this (third) again by its own fourth less the thirty-fourth part (of the fourth); this is (the value of)

[67] *Āpastambā-Śulbasūtra*, 1: 4.

[68] *Kātyāyana-Śulbasūtra*, 2: 8.

[69] *Baudhayāna-Śulbasūtra*, 1: 49.

[70] Seidenberg, 1962. It is likely that the date assigned by Seidenberg will be revised with new scholarship.

the diagonal of a square (whose side is the measure).[71]

To understand, let us assume a square of 4 m length. The length of the diagonal ($L$) is,

$$L = 4 + \frac{4}{3} + \frac{4}{3 \times 4} - \frac{4}{3 \times 4 \times 34}$$

This is equal to (4 + 1.33 + 0.33 - 0.01) = 5.65 m. When we apply the Pythagorean theorem, we get the same value for the diagonal.

Let us take another example with a larger number, thirty-five. The diagonal will be:

$$L = 35 + \frac{35}{3} + \frac{35}{3 \times 4} - \frac{35}{3 \times 4 \times 34}$$

This gives us the diagonal, $L$ = 49.50, which is an approximate value accurate to two decimal places. This is the same value we get from the Pythagorean theorem. This rule is applicable to any right-angled triangle with equal sides. As mentioned before, this kind of mathematical formula is unique to the ancient Hindus only.

## 4.5 Trigonometry: From *Jyā* to Sine

Trigonometry is a branch of science which deals with specific functions of angles and their application to calculations in geometry. The sine function, as defined in trigonometry, is essential to the study of geometry. This function also plays an important role in the physical sciences – including mechanics, the study of electromagnetism, optics, acoustics, and astronomy. It allows people to compute the height of a distant mountain and the distance between two lakes, and to solve a multitude of similar problems. Astronomers use trigonometry to locate heavenly objects in the sky, travelers use it in navigation, engineers use it in construction, and scientists use it in studying periodic phenomena. Architects, surveyors, navigators also use trigonometry

[71] *Baudhayāna-Śulbasūtra*, 2: 12; *Āpastambā-Śulbasūtra*, 1: 6; and *Kātyāyana-Śulbasūtra*, 2: 9.

in their work.

Although known to some readers, let us provide a brief description of this trigonometric function. For a right-angle-triangle, if $\theta$ is the acute angle of a right triangle, the *sine* of $\theta$ is the ratio of the side opposite ($b$) and the hypotenuse ($c$). Mathematically, $\sin\theta = \frac{b}{c}$.

Why do we call this function *sine*? Who chose this word for the scientific community? What is the meaning of this word? These are simple questions that intrigue curious minds when they first learn about this trigonometric function.

"The most significant contribution of India to medieval mathematics is in trigonometry. For a circle of unit radius the length of an arc is a measure of the angle it subtends at the centre [center] of the circle. The Greeks, to facilitate calculations in geometry, tabulated values of the chord of arcs. This method was replaced by Hindu mathematicians with half chord of an arc, known as sine of the angle . . . No influence on the West was exerted by the development in India . . . Thus methods had been known in India were not rediscovered until 1624 by the French mathematician Claude-Gaspar Bachet, sieur de Méziriac,"[72] suggests the *Encyclopaedia Britannica.*

Trigonometry was introduced to the Arab world by India.[73] al-Battānī, Abu ʻAbd Allāh Muhammad ibn Jābir ibn Sinān al-Raqqī al-Harrānī al-Sābiʻ (858 - 929 A.D.) used the half-cord (that leads to *sine* function) instead of the chord in his book *Kitāb al-Zīj* (Book of Tables) following the examples of his predecessors who used the Hindu method rather than the Greek method.[74] al-Battānī's *Zīj* was translated into Latin by Robert of Ketton and Plato of Tivoli during the twelfth century. This table was also translated into Spanish under the patronage of Alfonso X.[75] Adelard of Bath, an English philosopher, mathematician, and scientist, translated al-Khwārizmī's astronomical tables, that had been previously revised by al-Majrīṭī (d. 1007 A.D.), from Arabic to Latin. These tables included tables of *sines*. This way the *sine* function

[72] *Encyclopaedia Britannica*, vol. 23, p. 605, 1989.
[73] Needham and Ling, 1959, vol. III, p. 108.
[74] *Dictionary of Scientific Biographies*, vol. I, p. 508
[75] Julio Samsó in Selin, 1997

was introduced to the Latin world.[76]

The trigonometric function *sine* is ubiquitous in science and mathematics. The function has a rich history that transcends the boundaries of one country, one culture, or even one period. Its history provides an excellent example of international cooperation, multiculturalism in science, and science/mathematics as a dynamic discipline. This fascinating history is not discussed much in the introductory textbooks in science or mathematics.

In the Western tradition, the trigonometric function is attributed to Hipparchus (fl. in the second century B.C.). There is not a single book of Hipparchus available today that used this trigonometric function. What we know about Hipparchus' contribution to trigonometry is from a comment by Theon (about 365 A.D.) who, like Hipparchus, lived in Alexandria. Theon indicated that Hipparchus wrote a treatise in 12 books on the calculation of chords in a circle.[77] Neither the contemporary natural philosophers nor later authors have provided any details/commentaries of Hipparchus' methods. As popularly believed, Hipparchus used the length of an arc as a measure of the angle it subtends at the center of the circle, as shown in Figure 4.5a. He took a circle and tabulated the chords for angles with increments of 7.5°

Āryabhaṭa I used the half chord on an arc, defined the *sine* function and gave a table of *sines* of the angles (Fig. 4.5b). In his book, the circumference of a circle was divided into $360 \times 60 = 21,600$ equal segments. He also divided the quadrant of a circle into 24 equal parts. The smallest is thus $225'$ or $3°45'$. Āryabhaṭa I defined *jyā* (or $\sin\theta$) for a circle of radius, $R$, of 3438 units. We know that the half-chord equals $R\sin\theta$ in the modern system. Thus, Āryabhaṭa I's values were 3438 times the modern values for a given angle.[78] The radius 3438 may have come from the division of $\frac{180°}{\pi}$. This gives us $3438'$. In Table 4.1, a comparison of Āryabhaṭa I's values with the modern values is provided. The trigonometric function *sine* is called *jyā* in the Sanskrit language.

In the tenth and eleventh chapter of *Almagest*, Ptolemy related the chord

---

[76] Sarton, 1931, vol. 2, p. 167.

[77] Bond, 1921.

[78] For more details, read Achar, 2002; Clark, 1930; Hayasi, 1997; and Shukla and Sarma, 1976.

Table 4.1: Āryabhaṭa I's and Modern *sine* values

| Angle | Āryabhaṭa I's Value | Modern *sine* value × 3438 |
|---|---|---|
| 3°45′ | 225 | 225 |
| 7°20′ | 449 | 449 |
| 11°15′ | 671 | 671 |
| 15°0′ | 890 | 890 |
| 18°45′ | 1105 | 1105 |
| 22°30′ | 1315 | 1315 |
| 26°15′ | 1520 | 1521 |
| 30°0′ | 1719 | 1719 |
| 33°45′ | 1910 | 1910 |
| 37°30′ | 2093 | 2092 |
| 41°15′ | 2267 | 2266 |
| 45°0′ | 2431 | 2431 |
| 48°45′ | 2585 | 2584 |
| 52°30′ | 2728 | 2728 |
| 56°15′ | 2859 | 2859 |
| 60°0′ | 2978 | 2977 |
| 63°45′ | 3084 | 3083 |
| 67°30′ | 3177 | 3176 |
| 71°15′ | 3256 | 3256 |
| 75°0′ | 3321 | 3321 |
| 78°45′ | 3372 | 3372 |
| 82°30′ | 3409 | 3409 |
| 86°15′ | 3431 | 3431 |
| 90°0′ | 3438 | 3438 |

to an arc. He defined the length of the chord corresponding to an arc of angle $\alpha$ in a circle of radius 60 units up to $180°$ in steps of $\frac{1}{2}°$. In comparing Ptolemy and Āryabhaṭa I, a main difference is that Ptolemy used a full chord while Āryabhaṭa I used a half-chord. Āryabhaṭa I's procedure could be considered to lead to modern trigonometry.

### 4.5.1 Spread of the Sine Function to Other Cultures

The trigonometric term *jyā*, as defined by Āryabhaṭa I, was called *ming* in China.[79] In the section, *Tui yueh Chien Liang Ming* (On the prediction of the moon's positions) of *Chiu-Chih-li*, an astronomical book that is based on Sanskrit works, *chia* is the term used in this book to define the trigonometric function sine which is an obviously a minor modification of the term *jyā*.[80] The term *ming* may have been adopted from the title of this book later as the book became popular. In 718 A.D, a new calender, called *Jiuzhi li* (Nine Upholder Calendar), was compiled in China.[81] This calendar is based on Varāhamihira's *Pañca Siddhānta* and contains a table of sines and the Hindu numeral zero in the form of a dot (*bindu*).[82]

In Arabia, al-Khwārizmī introduced this function in his book that is now extant, like the work of Hipparchus. However, the book was revised by al-Majrīṭī of Cordoba, Spain, during the eleventh century. This revised Arabic translation of al-Majrīṭī was translated into Latin by Adelard of Bath (England) during the early 12th century. Further, this Latin translation was translated into English by O. Neugebauer.[83] It is this English translation that is our major source of al-Khwārizmī's knowledge in astronomy and trigonometry.

Adelard used the term *elgeib* for the trigonometric function *sine*, following the word used by al-Khwārizmī who use *geib* or *jaib* for this term.[84] Rabbi Ben Ezra (born c. 1093), of Toledo, Spain, also used a similar term, *al-geib*,

---

[79] Martzloff, 1997, p. 90 and 100.
[80] Sen, 1970
[81] Yoke, 1985, p. 162
[82] Yoke, 1985, p. 162.
[83] Neugebauer, 1962.
[84] Neugebauer, 1962, p. 44, 45.

for this trigonometric function.[85]

The Arabic term *geib* or *jaib* is a metamorphosed form of the term *jyā* used by the Hindus. The Arabic word *jaib* has multiple meanings: pocket, fold, or bosom. Due to this, Gherardo of Cremona (ca. 1114 - 1187 A.D., sometimes also spelled as Gerard or Gerhard, born in Carmona, Spain and not Cremona, Italy) literally translated the Arabic term into Latin and used the term *sinus* to define the operation.[86] The term *sinus* means bosom, fold, or pocket in Latin. Pocket or bosom has nothing to do with the trigonometric function. However, this term has been in use for about a millennium.

Using Hindu trigonometric method of half-chord, Abu-Wefa (ca. 940-998 A.D., lived in Baghdad) improved Ptolemaic tables.[87] Two popular astronomical tables, *the Toledan Tables* (11th century A.D.) and *the Alfosine Tables* (13th century A.D.), were also compiled in Europe where this Hindu trigonometric operation was used. Both tables were quite popular in the medieval Europe in astronomy.

Robert of Chester (12th century, England) and Leonardo Fibonacci (13th century, Pisa, Italy) also used the term sinus to define the operation which later on evolved into *sine*.[88] These multiple examples over a long period all over the world prompted Boyer and Merzbach to write that, "there is no doubt that it [*sine*] was through the Hindus, and not the Greeks, that our use of half chord has been derived; and our word "*sine*," through misadventure in translation."[89]

In 1631, Richard Norwood (1590 - 1675), a British mathematician and surveyor, published a book on trigonometry where the word *sine* was used for this trigonometric function and the symbol *s* was used to depict the function in his mathematical equations.[90] In 1634, the French mathematician Pierre Hérigone (1580 - 1643) became the first person to use the symbol sin for

[85] Goldstein, 1996.
[86] Smith, 1925, vol. II, p. 616.
[87] Bond, 1921.
[88] Bond, 1921 and Boyer and Merzbach, 1991,p. 252.
[89] Boyer and Merzbach, 1991., p. 209.
[90] Smith, 1925, vol. II, p. 618.

*sine* and the practice has continued since then.[91] Hérigone's book, *Cursus mathematicus, nova, brevi, et clara methodo demonstratus, per notas reales et universales, citra usum cujuscunque idiomatis intellectu faciles,* was published in six volumes from 1634 to 1637. This work used a large number of symbols for various mathematical operations and was written in French and Latin. Due to the large number of symbols, his work was criticized by some. However, the second edition was printed in 1644 which is rather unique for books published during this period. David Eugene Smith has provided a review of sine function and symbol.[92]

This chapter provides a brief sketch of the contributions of the Hindus which certainly, by any standard, is not an in-depth study. It reminds us of the mathematical tools that were perfected by the Hindus more than a thousand years ago and which have become a part of mainstream mathematics. "Think of our notation of numbers, brought to perfection by the Hindus, think of the Indian arithmetical operations nearly perfect as our own, think of their elegant algebraic methods, and then judge whether the Brahmins on the banks of the Ganges are not entitled to some credit," questions Cajori.[93] The answer is definitely in the affirmative.

[91]Smith, 1925, vol. II, p. 618.
[92]Smith, 1925, vol. II, p. 614 - 619.
[93]Cajori, 1980, p. 97.

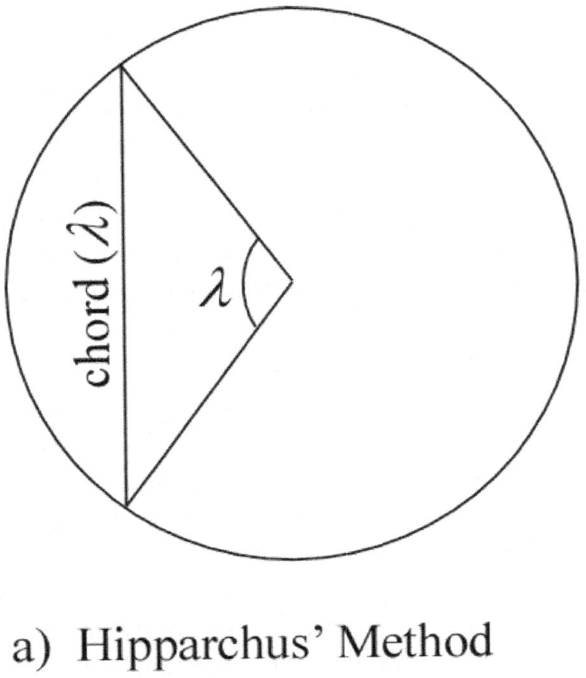


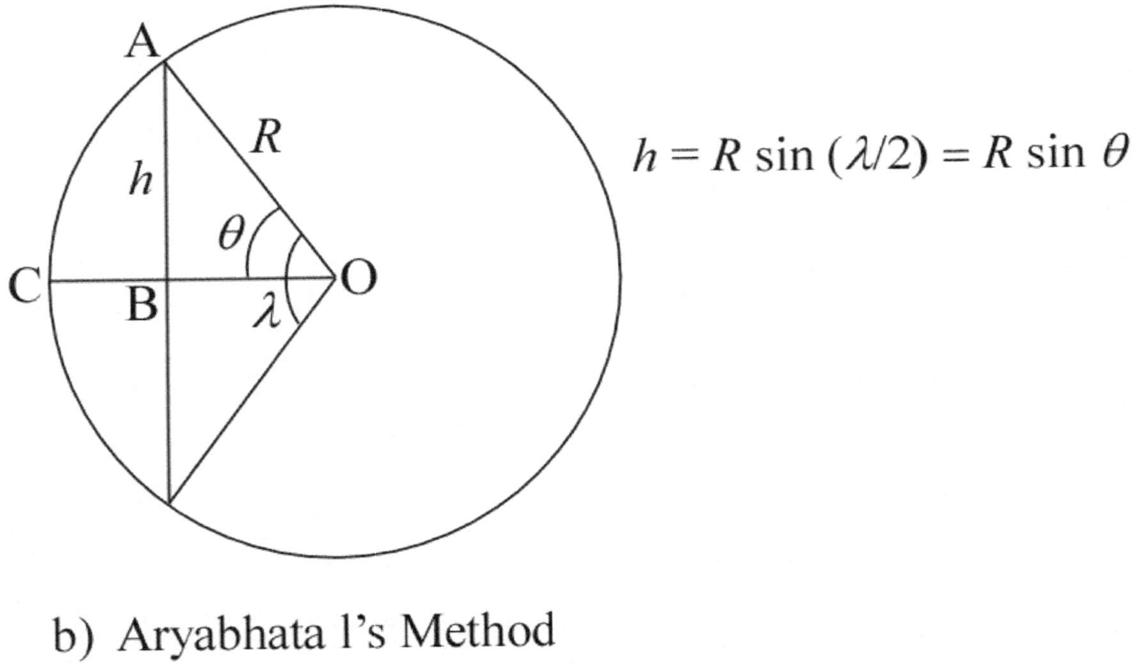


Figure 4.5: Hipparchus' method and Āryabhaṭa I's method to define an arc (Designed with the help of Morgan Gelfand and David Valentino).

# Chapter 5

# Astronomy

The sun gives us light that allows us to see the beauty of flowers, allows us to sense the magnitude of a mountain range and ocean, and, most important of all, gives us warmth. In contrast, dark nights bring us cold temperatures, insecurities, and helplessness. It was only natural for the ancients to know more about the sun, its periodicity, its radiation, and its relationships to the other luminaries in the sky. The continually changing phases of the moon on different nights raises curiosities to inquisitive minds. The relationship of the northward or southward motions of the sun with changing seasons, and the occurrences of eclipses of the sun and the moon were some of the natural phenomena that were observed and explained by the ancients.

Astronomers in India, China, Egypt, Greece, Mesoamerica, and Arabia made careful observations of the night sky. They observed and named clusters of stars, constellations, and planets, and fabricated mythological stories involving these luminaries. Astronomical observations allowed government officials and religious priests to choose and define the day and time for a religious or official activity. Astronomy served many practical needs, as Roger Bacon (1214 - 1292 A.D.) noted in the thirteenth century A.D.: ". . we learn the suitable time for sailing, the time for ploughing, the dog star of summer, and the mistrusted storms of autumn."[1]

The ancient Hindus systematically observed the night sky and noticed changes in the positions of some luminaries (planets, meteors, comets) against

[1]Bacon, 1928, vol. 1, p. 199.

the fixed background of other luminaries (stars). They tried to know the shape of the Earth, looked for an explanation for the various phases of the moon, the changing seasons, and designed their own calendar which was unique and distinctive from those of other civilizations. Once scientific observations were made and conclusions were drawn, the ancient Hindus documented the facts into stories or verses. For example, a story was written in which Rohiṇī (Aldebaran; a star in constellation Taurus), the celestial female deer, was pursued by the stars (male) of Orion labeled as the celestial stag. Sirius in his role as a hunter pins Orion with his arrow to protect Aldebaran. The line that connects Sirius with Aldebaran go through the Belt of Orion, indicating the arrow's flight. This story is provided in the *Ṛgveda*.[2]

The discipline of astronomy is described by two Sanskrit words: *jyotiṣa-śāstra* or *khagola-śāstra*. As mentioned previously, Hindu astronomy partly flourished out of the need to perform religious rituals on proper days at particular times that were governed by the positions of the sun or the moon with respect to various constellations. The ancient Hindus performed rites at sunrise (*brahm-muhūrta*) and sunset (*sandhyā*), at the rising and setting of the moon, and at other well-defined entrances of the moon or the sun into particular constellations. Such rites were performed in public as well as within individual families. These needs required keen observations of the sky and a map of the heavens. A special class of priests (*khagola-śāstri*, scholars of the sky) made observations of astronomical events, including the motions of the planets, and documented them in their hymns.[3] The *Yajurveda* mentions *Nakṣatra-daraśa*[4] as the term for astronomer. This term is made up of two words: *nakṣatra* means constellation or a prominent star and *daraśa* means seer or observer, signifying a person who studies astronomy. The *Chāndogya-Upaniṣad* mentions *nakṣatra-vidyā* (knowledge of constellations or astronomy) as a discipline.

---

[2]Krupp, October 1996. Krupp is an internationally known astronomer and works at the Griffith Observatory in Los Angeles. Krupp failed to provide a proper reference in his journal article. I could not find it in the *Ṛgveda* since he used the scientific names in the article. However, knowing the reputation of Krupp, I have decided to share it with the reader.

[3]see Brennard, 1988; Burgess, 1893; Lishk, 1987; Paramhans, 1991; Roy, 1976; Saha and Lahri, 1992; Shukla and Sarma, 1976; Somayaji, 1971; Subbarayappa and Sarma, 1985.

[4]*Yajurveda*, 30: 10.

The ancient Hindus were keen observers of the sky and designed instruments to facilitate their observations. Yukio Ōhashi, a Japanese scholar, researched the instruments and came up with the conclusion that the Indians had the following instruments for astronomical observation during the period of Āryabhaṭa I:[5]

1. Chāyā-yantra (Shadow-Instrument)
2. Dhanur-yantra (semi-circle instrument)
3. Yaṣṭi-yantra (staff)
4. Cakra-yantra (circular instrument)
5. Chatra-yantra (umbrella instrument)
6. Toya-yantrāṇi (water instrument)
7. Ghaṭikā-yantra (clepsydra)
8. Kapāla-yantra(clepsydra)
9. Śaṅku-yantra (gnomon)

This list of the instruments were first compiled by a noted scholar K. S. Shukla from Lucknow who found the fragments of Āryabhaṭa I's work in some books written later.

The *Taittirīya-Brāhmaṇa* suggested the study of stars before sunrise to find the exact time of the rituals. "The position of an auspicious star [relative to the sun] has to be determined at sunrise. But when the sun rises, that star would not be visible. So, before the sun rises, watch for the adjacent star. By performing the rite with due time adjustment, one would have performed the rite at the correct time."[6]

Most Hindu festivals are governed by astronomical events. For example, *Saṁkrānti*, an important festival in India, is celebrated on the day when

[5]Ōhashi, 1994.
[6]*Taittirīya-Brāhmaṇa*, 1: 5: 2: 1.

the sun leaves (apparent motion) one *rāśi* (zodiac) and enters a new zodiac. There are 12 zodiac signs and, therefore, twelve *Saṁkrānti* festivals every year. The most popular of these festivals is *makar-Saṁkrānti.* This is the day when the sun moves from the *dhanu-* (Sagittarius) to *makar* (Capricorn) zodiac. On this day, the Hindus visit temples, fast for the day, donate money to needy people, feed hungry people, and some bathe in the holy rivers. This religious need required regular observations of the constellations and the apparent motion of the sun.

Because of the observation of stars, distant objects, and the curvature of the Earth, the ancient Hindus knew the shape of the Earth as spherical from the earliest periods. The *Śatapatha-Brāhmaṇa*, an ancient book of the Hindus, mentioned the spherical shape of the universe: ". . . womb is spherical and moreover this terrestrial world doubtless is spherical in shape."[7] In his book *Geography*, Strabo (ca. 63 B.C. - 25 A.D.), a Greek traveler and historian, mentioned that the Indians, like the Greeks, believed in the spherical shape of the Earth.[8]

Al-Bīrūnī (973 - 1050 A.D.) also affirmed this view: "According to the religious traditions of Hindus, the Earth on which we live is round . ."[9] The key word in this quotation is "traditions." There is about a millennium of time between Strabo and al-Bīrūnī. Throughout from the Vedic period, the Earth was considered to be spherical. Al-Bīrūnī quoted the Hindu astronomers to indicate that the size of the Earth was very small in comparison to the visible part of the universe:[10] "These are the words of Hindu astronomers regarding the globular shape of heaven and earth, and what is between them, and regarding the fact that the earth situated in the center of the globe, is only a small size in comparison with the visible part of heaven."

Āryabhaṭa I used an analogy of a kadamba flower to demonstrate the distribution of various life forms on the Earth: "Half of the sphere of the Earth, the planets, and the asterisms is darkened by their shadows, and half, being turned toward the sun, is lighted according to their size. The sphere of the earth, being quite round, situated in the center of space, in the middle

[7] *Śatapatha-Brāhmaṇa*, 7: 1: 37.
[8] Strabo, 15: 1: 59.
[9] Sachau, 1964, vol. 1, p. 233.
[10] Sachau, 1964, vol. 1, p. 269.

Figure 5.1: Kadamba flower

of the circle of asterisms [constellations or stars], surrounded by the orbits of the planets, consists of water, Earth, fire, and air. Just as the ball formed by a kadamba flower is surrounded on all sides by blossoms just so the Earth is surrounded on all sides by all creatures terrestrial and aquatic."[11] (Figure 5.1)

There are numerous accounts in the Hindu literature indicating that the moon gets its luminosity from the sun. The *Ṛgveda* tells us that "he [the moon] assumed the brilliancy of the sun,"[12] "He [the moon] is adorned with sun's arrowy beam . . .,"[13] or, "the Sun-God Savitar bestows his sunlight to his Lord, the moon."[14] The *Yajurveda* (white) also tells us that "the moon whose rays are the sun's ray ."[15]

Al-Bīrūnī (973 - 1050 A.D.), in stating the Hindu view that planets were illuminated by the sun, wrote: "the sun aloft is of fiery essence, self-shining,

[11] *Āryabhaṭīya*, *Gola*, 5 - 7.
[12] *Ṛgveda*, 9: 71; 9.
[13] *Ṛgveda*, 9: 76: 4.
[14] *Ṛgveda*, 10: 85: 9; *Atharvaveda*, 14: 1: 9.
[15] *Yajurveda* (white), 18: 40.

and 'per accidens' illuminates other stars when they stand opposite to him."[16] For the moon, al-Bīrūnī wrote that "the moon is watery in her essence, therefore the rays which fall on her are reflected, as they reflect from the water and the mirror towards the wall."[17]

Varāhamihira stated that, in the words of al-Bīrūnī: "The moon is always below the sun, who throws his rays upon her, and lit up the one half of her body, whilst the other half remains dark and shadowy like a pot which you place in the sunshine. The one half which faces the sun is lit up, whilst the other half which does not face it remains dark."[18] "If the moon is in conjunction with the sun [new moon], the white part of her turns towards the sun, the black part towards us. Then the white part sinks downward towards us slowly, as the sun marches away from the moon."[19]

The above statements from al-Bīrūnī demonstrate that the ancient Hindus knew that the moonlight in the night sky is actually the reflected sunlight. The night sky does not have the sun. If so, how could it illuminate the moon? Based on the statements provided, the ancient Hindus had some knowledge of the changing geometry of the Earth, the moon, and the sun. Information to support this argument is provided later, especially in relation to Rāhu and Ketu. The ancient Hindus knew that the sun does not dissolve after the sunset. The sun does not set or rise. (Section 5.1) Al-Bīrūnī states that "every educated man among Hindu theologians, and much more so among their astronomers, believes indeed that the moon is below the sun, and even below all the planets."[20]

The ancient Hindus used parallax techniques to find distances of the planets from the Earth. Al-Bīrūnī provides its description: "It is well known among all astronomers that there is no possibility of distinguishing between the higher and the lower one of two planets except by means of the occultation or the increase of the parallax. However, the occultation occurs only very seldom, and only the parallax of a single planet, viz. the moon, can be observed. Now the Hindus believe that the motions are equal, but the distances

[16]Sachau, 1964, vol. 2, p. 64.
[17]Sachau, 1964, vol. 2, p. 66.
[18]Sachau, 1964, vol.2, p. 66.
[19]Sachau, 1964, vol. 2, p. 67.
[20]Sachau, 1964, 2, p. 67.

are different. The reason why the higher planet moves more slowly than the lower is the greater extension of its sphere [orbit]; and the reason why the lower planet moves more rapidly is that its sphere or orbit is less extended."[21]

The above statement clearly defines different distances for different planets – some closer to the Earth while the others are farther away. The differences in the observed motions of these planets are due to their distances from the Earth.

Āryabhaṭa I explained the geometry of the solar system and the universe: "Half of the globe of the Earth, the planets, and the stars are dark due to their own shadows; the other halves facing the sun are bright in proportion to their sizes."[22] "Beneath the stars are Saturn, Jupiter, Mars, the sun, Venus, Mercury, and the moon, and beneath these is the Earth situated in the center of space like a hitching-post."[23] This observation of the heavenly objects, as given by Āryabhaṭa I, is correct for an observer on the Earth. Āryabhaṭa I knew correctly that the five planets, the sun, and the moon are below the stars.

The apparent path of the sun as viewed from the Earth against the background of stars is called the ecliptic. Actually this path results from the Earth's motion around the sun against the backdrop of the celestial sphere. The moon with its changing phases moves around the Earth in a plane that is close to the ecliptic plane. The plane of the moon lies 5° north or south of the ecliptic plane. The two points at which the moon's path crosses the elliptic plane are called nodes – the descending and ascending nodes. The nodal points are not fixed and move along the ecliptic plane of the Earth's motion. The ancient Hindus called them Rāhu and Ketu, respectively. The rotational periods of Rāhu and Ketu as defined in Hindu astrological charts, the horoscopes, are the same as that of these nodal points. These two nodes are diametrically opposite to each other and so are the Rāhu and the Ketu in the horoscope.

These nodal points describe the relation of the moon and the sun to the

[21]Sachau, 1964, vol. 2, p. 69.
[22]*Āryabhaṭīya*, *Gola*, 5.
[23]*Āryabhaṭīya*, *Kālakriyā*, 15.

Earth. Along with the individual effects of the sun and moon, there is also a collective effect exhibited by the sun and the moon, as noticed in lunar and solar eclipses. The solar eclipse and the lunar eclipse can be explained with the help of these nodal points. The ancient Hindus used this fact and tried to explain eclipses by the motion of Rāhu and Ketu, the two nodal points.

The *Ṛgveda* mentions solar eclipse[24] which is a common observation in most cultures. An explanation for the cause was warranted. In Hindu mythology, as given in the *Ādi-parva* of *Mahābhārata*, eclipses are described as the swallowing of the sun or the moon by two demons, Rāhu and Ketu. As the story goes, once the gods decided to churn the ocean to get gifts from it. During the churning, the elixir or divine nectar (*amṛta*) also came as a gift. All the gods wanted to drink elixir to become immortal. A demon, named Rāhu, took the guise of a god and tried to drink the elixir. When he was about to drink, the sun and the moon recognized him. They both warned Lord Viṣṇu about Rāhu's transgression. Lord Viṣṇu acted promptly and slit Rāhu's throat. By then, however, the elixir had already reached Rāhu's throat and, therefore, he could not die. His head, devoid of his rest of the body, went to the sky. Ever since there has been a long lasting feud between Rāhu's head and the sun and the moon. Rāhu chases them across the sky and tries to swallow them. Whenever Rāhu succeeds, an eclipse occurs. An eclipse is therefore symbolizes a momentary victory of Rāhu over the sun or the moon. In essence, it is a temporary victory of evil over good. Therefore, eclipse is considered as an inauspicious occasion in Hindu mythology. Hindus, therefore, pray during an eclipse and bathe in their holy rivers to purify themselves afterward. Astrologers predict possible misfortunes since eclipses are indicative of them. As an example, after Yudhiṣṭhira lost the game of dice to Duryodhana, Pāṇḍava had to live in the forest for 12 years. Draupadī, the wife of Pāṇḍavas, was also embarrassed in the process. During that period, "horrible meteors fell from the sky, and Rāhu swallowed the sun." Several wise people thus predicted the demise of Kaurava and Duryodhana, as explained in *Mahābhārata*.

The *Chāndogya-Upaniṣad* provided an account of the lunar eclipse: "From the dark, I go to the varicolored. From the varicolored, I go to the dark. Shaking off evil, as a horse his hairs; shaking off the body, as the moon

[24] *Ṛgveda*, 5: 40: 5 - 9.

releases itself from the mouth of Rāhu."[25] Here "varicolored" is indicative of the corona rings that appear during an eclipse. The mention of Rāhu is indicative of the role of the nodal points.

The *Atharvaveda* provides the cause of a solar eclipse: "Peace to us during lunar eclipse, peace to us during the period when Rāhu swallow the sun."[26] Āryabhaṭa I used scientific terms and explained the cause of solar and lunar eclipses as the moon blocks the sun or the Earth comes in between the sun and the moon. He also provided the method to calculate the area of the moon or the sun that would be affected during an eclipse: "The moon obscures the sun and the great shadow of the Earth obscures the moon."[27] This is a clear indication that the geometry of eclipse was known to Āryabhaṭa I.

"When at the end of the true lunar month the moon, being near the node, enters the sun, or when at the end of the half-month the moon enters the shadow of the Earth that is the middle of the eclipse which occurs sometimes before and sometimes after the exact end of the lunar month or half-month."[28]

"Multiply the distance of the sun by the diameter of the Earth and divide (the product) by the difference between the diameters of the sun and the Earth: the result is the length of the shadow of the Earth (i.e. the distance of the vertex of the Earth's shadow) from the diameter of the Earth (i.e. from the center of the Earth).[29]

Thus,

$$\text{Length of the Earth's Shadow} = \frac{\text{sun's distance} \times \text{Earth's diameter}}{\text{sun's diameter} - \text{Earth's diameter}}$$

"The difference between the length of the Earth's shadow and the distance of the moon from the Earth multiplied by the diameter of the Earth and divided by the length of the Earth's shadow is the diameter of the Earth's shadow (in the orbit of the moon)."[30]

[25] *Chāndogya-Upaniṣad*, 8: 13: 1.
[26] *Atharvaveda*, 19: 9: 10.
[27] *Āryabhaṭīya*, *Gola*, 37.
[28] *Āryabhaṭīya*, *Gola*, 38.
[29] *Āryabhaṭīya*, *Gola*, 39, taken from Shukla and Sarma, 1976, p. 152.
[30] *Āryabhaṭīya*, *Gola*, 40.

"Subtract the given time from half of the duration of the obscuration. Add this to the square of the celestial latitude. Take the square root. Subtract this from half the sum of the diameters. The remainder will be the obscuration at the given time."[31]

All these statements provide enough information to demonstrate that the ancient Hindus used a systematic approach to explain astronomical events that were based on astronomical observations.[32] Al-Bīrūnī confirmed that the cause of solar and lunar eclipse was known to the Hindus. "It is perfectly known to the Hindu astronomers that the moon is eclipsed by the shadow of the Earth, and the sun is eclipsed by the moon."[33]

## 5.1 Heliocentric Solar System

To understand the dimensions of the solar system, if we reduce the size of our solar system by a factor of one billion (1,000,000,000 or $10^9$), the sun would be a sphere of 1.4 m in diameter. Other planets and their distance from the sun are defined in Table 5.1.

By the fifth century, at the time of Āryabhaṭa I, the ancient Hindus knew the existence of only five planets: Mercury, Venus, Mars, Jupiter, and Saturn. Table 5.2 provides the names of these planets that are mentioned in the Hindu literature:[34]

The pre-Copernican astronomers had an internally consistent geocentric theory that described the universe that they observed: The earth was stationary. The stars appeared to move in circles. Do you see earth moving when you watch outside the window? Do you feel the earth moving like you feel on a merry-go-round? Do you observe a difference in range if you shot a cannon ball toward the east in comparison to toward the west? Thus, the geocentric theory was a thing of beauty, however, it later proved to be wrong.

---

[31] *Āryabhaṭīya*, *Gola*, 44.

[32] For more information, read Shukla and Sarma, 1976, p. 152 - 161.

[33] Sachau, 1964, vol. 2, p. 107.

[34] Kak, Subhash, Birth and Early Development of Indian Astronomy, p. 303 - 340 in Selin, 2000.

Table 5.1: Solar System, dimensions reduced by a factor of $10^9$

| Object | Diameter | Distance from the sun |
|---|---|---|
| Sun | 1.4 m | 0 |
| Mercury | 5 mm | 58 m |
| Venus | 12 mm | 108 m |
| Earth | 12 mm | 150 m |
| Mars | 7 mm | 228 m |
| Jupiter | 14.2 cm | 778 m |
| Saturn | 12 cm | 1427 m |
| Uranus | 5 cm | 2870 m |
| Neptune | 5 cm | 4497 m |
| Pluto | 3 mm | 5913 m |

Table 5.2: Five Planets and Their Sanskrit Names

| Planet | Sanskrit Names |
|---|---|
| Mercury | Budha, Saumya, Rauhiṇeya, Tuṅga (yellow) |
| Venus | Vena, Uśanas, Śukra, Kavi, Bhṛgu (white) |
| Mars | Aṅgāraka, Bhūmija, Lohitāṅga, Bhauma, Maṅgala, Kumāra, Skanda (red) |
| Jupiter | Bṛhaspati, Guru, Āṅgiras (yellow) |
| Saturn | Śani, Sauri, Manda, Paṅgu, Pātangi (black) |

Most cultures explained planetary motion in a geocentric system where the Earth remained stationary, like the Ptolemy's model. In this model, epicyclic motion of the planets was necessary to explain the planetary motions. However, Āryabhaṭa I came up with an alternative and innovative explanation of the solar system in which the Earth was provided with an axial motion. With the known spherical shape of the Earth, Āryabhaṭa I correctly theorized the diurnal motion of the Earth, and was able to explain the occurrence of day and night. He suggested that at the time of sunrise in Sri Lanka, it should be midnight in Rome. Obviously, Āryabhaṭa I knew the longitudes of Sri-Lanka and Rome: "Sunrise at Lanka (Sri-Lanka) is sunset at Siddhapura, mid-day at Yavakoti (or Yamakoti, Indonesia), and mid-night at Romaka (Rome)."[35]

The above statement is remarkable since it involves information that deals with simultaneous observation. Needless to say, Āryabhaṭa I had no way to make a phone call since the phones were not available. Similarly, all other means to contact with people in Rome was not possible for a person sitting in India. The only possibility is to predict it using geometry and astronomy in which the relative motion of the sun and the Earth, including its shape, is know to the person. This is similar to what Eratosthenes did when he made simultaneous measurements at Syene and Alexandria, both in Egypt, to measure the size of the Earth.[36] Roger Billard applied mathematical statistics and demonstrated that the ancient astronomical works in India were based on astronomical observations.[37]

### 5.1.1 Ujjain, Greenwich of the Ancient World

Sri Lanka was used as a reference point because the prime meridian of Ujjain, a city in India, (Longitude 75°43′E, Latitude 23°13′N) intersects the equator in Sri Lanka. The location of Ujjain played an important role in astronomy around the world. It was the Greenwich of the ancient world and most ancient astronomers in India and Arabia used the city as a reference point for their astronomical observations. Āryabhaṭa I defined the distance between Ujjain and Sri Lanka as one-sixteenth of the Earth's circumference in the

[35] *Āryabhaṭīya*, *Gola*, 13. The locations of Siddhapura and Yavakoti are not known as the names of these locations must have changed.

[36] Brown and Kumar, 2011.

[37] Billard, 1977.

north direction.[38] This gives the latitude of Ujjain as $\frac{360}{16} = 22°30'$ which is quite close to the actual number. A difference of 1° in latitude creates an error of 69 miles in distance. Therefore, the value given by Āryabhaṭa I for the latitude differs with the actual value by 43′ which is equivalent to less than 50 miles. People in most metropolitan cities would agree that this is a small distance. For example, suburban sites in New York or Los Angeles can be 50 miles away from the city center and still would be considered as a part of these cities. Since the settlements around Ujjain have evolved over the last thousand years, it could easily be the distance of modern city center and the ancient location of the observatory.

The *Viṣṇu-Purāṇa* suggested that sunrise and sunset are merely terms to explain our observations. Otherwise, the sun is always there: "Actually the sun neither rises or sets. . . When the sun becomes visible to people, to them He [the sun] is said to rise; when He [the sun] disappears from their view, that is called his [the sun] setting. There is in truth neither rising or setting of sun, for he is always there; and these terms merely imply his presence and his disappearance."[39] The *Viṣṇu-Purāṇa* is one of the oldest *Purāṇa*, written around the beginning of the Christian era. This knowledge became known to some in the Western world when, about 1500 years later, Copernicus introduced it.

There are several books of the Hindus indicating that the sun constantly illuminated the Earth. Sunrise or sunset at a place on the Earth represents the side of the globe illuminated by the sun at that particular instant. The following are representative statements from the ancient literature of the Hindus: "He [the Sun] never sets or rises. When [we] think that he is setting, he is only turning round, after reaching the end of the day, and makes night here and day below. Then, when [we] think he is rising in the morning, he is only turning round after reaching the end of the night, and makes day here and night below. Thus, he (the sun) never sets at all," suggests the *Āitareya-Brāhmaṇa*.[40]

"Then rising from these upward, he (the Sun) will neither rise nor set. He

[38] *Āryabhaṭīya*, *Gola*, 14; Shukla and Sarma, 1976, p. 123 - 126.
[39] *Viṣṇu-Purāṇa*, 2: 8.
[40] *Āitareya-Brāhmaṇa*, 14: 6; taken from Subbarayappa and Sarma, 1985, p. 28.

will remain alone in the middle. Never does this happen here. Never did the sun set there nor did it rise. O Gods! By this my assertion of the truth, may I not fall from Brahman. Verily, for him, the sun neither rises nor sets. He who thus knows this secret of the *Vedas*, for him, there is perpetual day,"[41] suggests the *Chāndogya-Upaniṣad*.

The above quotations clearly suggest that the sun is always there, illuminating one side of the Earth, creating day on that side and night on the opposite side.

### 5.1.2 Diurnal Motion of the Earth

Āryabhaṭa I assigned diurnal motion to the Earth and kept the sun stationary in his astronomical scheme. According to Āryabhaṭa I the motion of the stars that we observe in the sky is an illusion. To explain the apparent motion of the sun, Āryabhaṭa I used an analogy of a boat in a river. "As a man in a boat going forward sees a stationary object moving backward just so in Sri-Lanka a man sees the stationary asterisms (stars) moving backward exactly toward the West."[42]

**Āryabhaṭa I**

Āryabhaṭa I was the Head at the famous Nālandā University near modern Patna. He composed a book, *Āryabhaṭīya*, dealing with mathematics (*gaṇita-pada*), spherical astronomy (*gola-pada*), and time-reckoning (*kāla-kriyā-pada*). Āryabhaṭa I was only 23 years old when he composed the verses of his book. This means that the book was composed in 499 A.D. His work made a major impact in India for several centuries as the following writers like Brahmgupta (born around 598 A.D.) and Varāhamihira (505 - 587 A.D.) wrote extensive commentaries on his work. Al-Bīrūnī, an Islamic philosopher-scientist, commented on the

[41] *Chāndogya-Upaniṣad*, 3: 11: 1- 3; taken from Subbarayappa and Sarma, 1985, p. 28. Other translators have translated the text differently. In most translations, "sun never sets" is stated.

[42] *Āryabhaṭīya*, *Gola*, 9; to read more about Āryabhaṭa I, read Hooda and Kapur, 1997; and Shukla and Sarma, 1976.

work of Āryabhaṭa I and criticized him for assigning motion to the Earth in the eleventh century. Obviously, Al-Bīrūnī considered the Earth to be stationary.

*Āryabhaṭīya* is an invaluable document for the historians of science; it provides an account of the ancient sciences of the Hindus. The content of this book was not new knowledge created by Āryabhaṭa I. He labeled the content of this book as "old knowledge."

*Āryabhaṭīya* has 118 metrical verses subdivided into four chapters. The first chapter, *Daśa-gītikā* has ten verses and provided astronomical constants. The second chapter is on mathematics. The third chapter is on time-reckoning, and the fourth chapter concerns spherical astronomy. In this book, Āryabhaṭa I provided the value of $\pi$ as *approximately equal to* 3.1416, the solution of indeterminate equations and quadratic equations, theory of planetary motions, and calculations of the latitudes of planets. And most important of all, a millennium before Copernicus, he assigned axial motion to the Earth in his astronomical model and kept the stars stationary.

In Āryabhaṭa I's honor, India named its first artificial satellite after him that was launched in 1975. The International Astronomical Union named a lunar crater after Āryabhaṭa I in 1979.

The interpretation is that a person standing on the equator of the Earth, that rotates from the west to the east, would see the asterisms moving in the westward motion. The clear grasp of Āryabhaṭa I about Earth's rotational motion is splendidly explained in the analogy of a boat man given above.

Interestingly, about one millennium after Ārybahaṭa I, Copernicus used a similar argument to assign motion to the Earth. "For when a ship is floating calmly along, the sailors see its motion mirrored in everything outside, while on the other hand they suppose that they are stationary, together with everything on board. In that same way, the motion of the earth can unquestionably produce the impression that the entire universe is rotating."[43]

[43]Copernicus' *On the Revolution*, Book 1, Chapter 8; Copernicus, 1978, p. 16.

This similarity is definitely intriguing. Did Copernicus know of the work of Āryabhaṭa I? Well, it is an issue that is not clearly resolved as yet. However, a strong possibility of Copernicus knowing the work of Āryabhaṭa I is provided in Section 5.5.

From the previous quotation from Āryabhaṭa I, at least the rotational (axial or spin) motion of the Earth is well-established. Does it lead to the heliocentric theory of solar system? Let us investigate further to answer this question.

According to Āryabhaṭa I, the Earth spins on its axis like a merry-go-round. However, we do not experience any fly-away feeling on the Earth as we do on a merry-go-round. The spin motion of Earth creates another issue of flying birds. How do they go back to their nest with the Earth spinning so fast, especially if they fly to the West? Assigning any motion to the Earth seems to be antithetical. It is a much bigger triumph to assign any kind of motion to the Earth than to add orbital motion to the already known spinning (rotational) motion of the Earth.

The moment one assigns spin motion to the Earth, it opens up a Pandora's box of other questions: Is there a motion of the sun? This question pops up as there is no longer a necessity of the sun's motion to explain day and night. The spinning motion of the Earth can take care of this. It is clear in Āryabhaṭa I's case that he considered some luminaries such as constellations to be stationary in the sky and attributed their apparent motion to the moving Earth.

This leads to the next question. If day and night are due to the Earth spinning in one place, why do we have different seasons? Why do we observe northward or southward motions of the sun? Why do we not observe a "fly-away" experience on the Earth? It suddenly becomes a complex problem to deal with if spin motion is assigned to the Earth. There are many unanswered questions that one has to deal with and explanations become more and more complex. No where did Āryabhaṭa I struggle in dealing with such questions in his book. His statements are fairly conclusive and straight forward. In summary, the axial rotation of the Earth complicates the simplicity of the geocentric system. On the contrary, in a heliocentric system, the axial rotation is a necessity.

In describing the spin motion of the Earth, Āryabhaṭa I makes another explicit statement, "The revolutions of the moon multiplied by 12 zodiac are signs [*rāśi*]. The signs multiplied by 30 are degrees [360°]. The degrees multiplied by 60 are minutes. . . . The Earth moves one minute in a *prāṇa*."[44]

Āryabhaṭa I provided the following definition of *prāṇa*, a unit of time: "One year consists of twelve months. A month consists of thirty days. A day consists of sixty *nāḍī*. A *nāḍī* consists of sixty *vināḍikā*. Sixty long syllables or six *prāṇas* make a sidereal *vināḍikā*. This is the division of time."[45]

Let us transcribe it in a modern set of standards. Assuming a day to be 24 hours long with a 360° rotation, *prāṇa* comes out to be four seconds. Therefore, Āryabhaṭa I's statement can be modified as follows: "The Earth rotates by an angle of one minute ($1'$) in 4 seconds." one minute in angle multiplied by 21,600 gives us 360°. Thus, 4 seconds multiplied by 21,600 should give us the time that is equal to one day (or 24 hours). This is the case when we multiply the two numbers and change the units to hours. Therefore, not only Āryabhaṭa I assigned spin motion to the sun, he also correctly provided the rate of the spin. Āryabhaṭa I makes explicit statement elsewhere: "The Earth rotates through [an angle of] one minute of arc in one *prāṇa*."[46]

Āryabhaṭa I provided the periods of revolution for different planets, the moon, and the sun in one *yuga*: "In a yuga the revolutions of the sun are 4,320,000, of the moon 57,753,336, of the Earth eastward 1,582,237,500, of Saturn 146,564, of Jupiter 364,224, of Mars 2,296,824, of Mercury and Venus the same as those of the sun."[47]. *Yuga* is an important concept in Hindu cosmology, as given in the *Vedas*, *Upaniṣads*, and *Purāṇas* (Section 5.2)

The mean motion of the planets is calculated and tabulated by Shukla and Sarma and an abridged form is provided in Table 5.3:[48]

As one can notice, all values are quite close to the modern accepted val-

[44] *Āryabhaṭīya, Dasgītika*, 4.
[45] *Āryabhaṭīya, Kālakriyā*, 2.
[46] *Āryabhaṭīya, Dasgītika*, 6.
[47] *Āryabhaṭīya, Dasgītika*, 3
[48] Shukla and Sarma, 1976, p. 7.

Table 5.3: Mean Motion of the Planets (sidereal period, in days)

| Planet | Āryabhaṭa I | Modern Value |
|---|---|---|
| Sun | 365.25868 | 365.25636 |
| Moon | 27.32167 | 27.32166 |
| Mars | 686.99974 | 686.9797 |
| Jupiter | 4332.27217 | 4332.5887 |
| Saturn | 10766.06465 | 10759.201 |

ues. Interestingly, the period of Mercury and Venus are considered equal to the sun by Āryabhaṭa I. Since Mercury and Venus are the only two planets inside the Earth's orbit, their orbital period, as observed from the Earth, comes close to that of the sun. Obviously, this observation is in a geocentric system. Was it because Āryabhaṭa I believed in the geocentric solar system or because it was a prevalent practice of the period to describe motion as viewed from the Earth?

In the very next verse Āryabhaṭa I erased any doubt about the two different motions assigned to Mercury and Venus (verse 4): " of the moon's apogee, 4,88,219; of [the *śighrocca* (conjunction)] of Mercury, 1,79,37,020; of (the *śighrocca*) of Venus, 70,22,388; of (the *śighrocca*) of the other planets, the same as those of the sun; of the moon's ascending node in the opposite direction (i.e., westward), 2,32,226. These revolutions commenced at the beginning of the sign Aries on Wednesday at sunrise at Lanka (when it was the commencement of the current yuga.)"[49]

The "moon's apogee" defines the point of the moon's orbit when it is farthest from the Earth. The *śīghrocca* of Mercury and Venus are the imaginary bodies that revolve around the Earth with the *heliocentric mean angular velocities* of Mercury and Venus, respectively, their direction from the Earth being always the same as those of the mean positions of Mercury and Venus from the sun. Thus, the *śīghrocca* of the various planets are equal to their periods around the sun. Thus, the sidereal period of Mercury and Venus comes

[49] *Āryabhaṭīya, Dasgītika*, 4.

out to 87.97 and 224.70 days, in the calculation of Shukla and Sarma.[50] This compares well with the modern values of 87.97 and 224.70, respectively for Mercury and Venus. Thus, Āryabhaṭa I provides one statement in the geocentric system while the other in the heliocentric system with the Earth as the point of observation. This has intrigued astronomers through the ages.

According to a detailed analysis given by B. L. van der Waerden,[51] the motion of Mercury and Venus as given by Āryabhaṭa I were in a heliocentric model.[52] Van der Waerden makes the following assertions to back up his conjecture that Āryabhaṭa I proposed a heliocentric model and not a geocentric model:[53]

1. In a geocentric system, there is no need to assume axial rotation for the Earth as most observations, though wrong, are easier to explain. However, in a heliocentric system, we are forced to think of axial rotation. Āryabhaṭa I clearly argued for the axial rotation of the Earth.

2. In the Midnight system of Āryabhaṭa I, the apogees of the sun and Venus are both at 80°, and their eccentricities are also equal. This fact can be explained by assuming that the system was originally derived from a heliocentric system.[54] "In genuine epicycle theory for Venus, the eccenter carrying the epicycle is independent of the sun's orbit. Its apogee and eccentricity are determined from observations of Venus, whereas the apogee and eccentricity of the sun are determined from eclipse observations. The probability that the apogee and eccentricity of Venus coincide with those of the sun is very small."[55]

The periods of revolution for the outer planets are essentially the same in the geocentric and heliocentric models. It is the planetary periods of

[50]Shukla and Sarma, 1976, p. 7.

[51]B. L. van der Waerden (1903 - 1996) was a prolific author with several books on algebra, geometry, and astronomy to his credit. He taught at the University of Leipzig in Germany, the University of Amsterdam in the Netherlands, and the University of Zurich in Switzerland.

[52]van der Waerden, The Heliocentric System in Greek, Persian, and Hindu Astronomy, in the book by King and Saliba, 1987, p. 534.

[53]van der Waerden, The Heliocentric System in Greek, Persian, and Hindu Astronomy, in the book by King and Saliba, 1987, p. 530 - 535.

[54]van der Waerden, in the book by King and Saliba, 1987, p. 532.

[55]van Waerden, in the book by King and Saliba, 1987, p. 531.

the inner planets, Mercury and Venus, that separates the two theories. In a geocentric theory, these two periods will essentially be the same as the solar period. However, in a heliocentric system, these periods are quite different. This is indeed the case in Āryabhaṭa I's model.

3. The revolutions of Mercury and Venus considered by Āryabhaṭa I are heliocentric revolutions, not geocentric. These periods provided by Āryabhaṭa I have "no importance whatsoever" in a geocentric system.[56]

Based on the descriptions given in *Āryabhaṭīya*, Van der Waerden concludes that "it is highly probable that the system of Āryabhaṭa [I] was derived from a heliocentric theory by setting the center of the Earth at rest."[57] The reason for this kind of torturous path in the work of Āryabhaṭa I is perhaps due to an overwhelming tendency among all early astronomers and their students, in the words of Van der Waerden, "to get away from the idea of a motion of the Earth."[58] In the history of astronomy, for convenience purposes, astronomers did transform the heliocentric theory into equivalent geocentric one. This was done by Tycho Brahe when he transformed Copernican heliocentric model into a geocentric one.[59]

Van der Waerden's conclusion that Āryabhaṭa I proposed a heliocentric model of the solar system has received independent support from other astronomers.[60] For example, Hugh Thurston came up with a similar conclusion in his independent analysis. "Not only did Āryabhaṭa believe that the Earth rotates, but there are glimmerings in his system (and other similar Indian systems) of a possible underlying theory in which the earth (and the planets) orbits the sun, rather than the sun orbiting the earth."[61] The evidence used by Thurston is in the periods of the outer planets and the inner planets. Āryabhaṭa basic planetary periods are relative to the sun which is not so significant for the outer planets. However, it is quite important for the inner planets (Mercury and Venus).

---

[56]Van Waerden in the book by King and Saliba, 1987, p. 534.
[57]Van der Waerden in the book by King and Saliba, 1987, p. 534.
[58]Van der Waerden in the book by King and Saliba, 1987, p. 530.
[59]Van der Waerden in the book by King and Saliba, 1987, p. 530.
[60]Billard, 1977; Thurston, 1994 and 2000.
[61]Thurston, 1994, p. 188.

Figure 5.2: Al-Bīrūnī's argument against Āryabhaṭa's assignment of the motion of the earth. (Designed with the help of David Valentino)

The motion that Āryabhaṭa I assigned to the Earth is not a mere speculation of modern astronomers. Āryabhaṭa I's thesis was well known in the Middle East even after six centuries. Al-Bīrūnī (973 - 1050 A.D.) criticized Hindu astronomers for assigning motion to the Earth. He referred to the work of Varāhamihira, a Hindu astronomer, to support his idea of the geocentric universe: "If that were the case, a bird would not return to its nest as soon as it had flown away from it towards the west,"[62] and "stones and trees would fall."[63] A similar argument was used by Aristotle (384 - 322 B.C.) to favor his theory of the geocentric universe. Such criticism are common in science. However, in the case of Al-Bīrūnī, this criticism has done more. His criticism has validated Āryabhaṭa I's work in India and the existence of this theory in the Middle East prior to the eleventh century.

After viewing various possibilities of the motion of the Earth, al-Bīrūnī favored the stationary Earth: "The most prominent of both modern and ancient astronomers have deeply studied the question of the moving of the

[62]Sachau, 1964, vol. 1, p. 276.
[63]Sachau, 1964, vol. 1, p. 277.

Earth, and tried to refute it. We, too, have composed a book on the subject called *Miftah-ilm-alhai'a* (*Key of astronomy*), in which we think that we have surpassed our predecessors, if not in the words, at all events in the matter."[64] This depicts that at least some six centuries after the heliocentric theory was proposed by Āryabhaṭa I, the Islamic philosophers from Arabia still could not grasp the idea of moving Earth.

To propose a heliocentric solar system to replace the geocentric theory required a philosophical triumph that discarded "common sense." Let us examine the results of the heliocentric system:

1. Earth rotates around its own axis with a time period of 24 hours. This allows day and night on Earth as the half globe facing the sun gets sunlight while the other half remains dark. By the same token, it implies that a person on the equator moves at approximately 1669 km/h (or approximately 1000 miles/hour). This is much faster than any machine made so far.

2. The Earth revolves around the sun at an incredibly fast speed of approximately 107 $\times 10^3$ km/h (67,000 miles/hour). Even spaceships in the outer atmosphere do not move with such high speed.

The above information does not seem to be plausible to our common sense. We feel dizzy when we sit even for a short time in a merry-go-round which rotates much slower than the Earth. The truth lies in the relative motion. On highways, when cars move with the same speed, if we focus on the car in front of us, it appears to be stationary. Something similar is observed when we sit on a chair in our garden and watch plants around us. They all look stationary to us although their actual speed is very high. However, their relative speed to us is zero. Therefore, we do not feel dizzy on the Earth watching other objects as we do on a merry-go-round.

## 5.2 Hindu Calendar

*Pañcāṅga* (*Pañcā* = five and *aṅga* = limb, meaning five-limbed) is the term used for the Hindu almanac. The five limbs are: day (*vāra*), date (*tithi*),

[64] Sachau, 1964, vol. 1, p. 277.

*nakṣatra* (asterism), *yuga* (period), and *kāraṇā* (half of *tithi*). *Pañcāṅga* is popular even today among the Hindus. Hindu priests use it for predicting eclipses, the performance time of various rituals, specific time of marriage, casting horoscopes, and for moving to a new house (*gṛah-praveśa*) or business. Hindu families use it to check the day of fasting, auspicious times for worshipping, and days of festivals. According to the scientific analysis of K. D. Abhyankar, a professor of astronomy from Osmania University in Hyderabad in India, the Hindu calendar dates back to 7000 B.C.[65] This conclusion is based on the occurrence of the winter solstice in different constellations in different years. As the Vedic documents list such astronomical events, it is possible to calculate the period of the event. Perhaps the resident of Mehrgarh in the Indus-Sarasvatī Civilization used this calendar in some primitive form as they were living in the region, as archaeological investigations have proven.[66]

Most cultures have calendars that are either based on the motion of the moon or the sun. The regular appearance of the new or the full moon forms a basis of most lunar calendars, like the Islamic calendar. Solar calendars are based on the cyclic motion of the sun in different zodiacs that is due to the orbital motion of the Earth around the sun.

The Hindu calendar is luni-solar in which the months are based on the motion of the moon while the sun guides the span of a year. A year is the time the Earth takes to complete one revolution around the sun, starting from *Meṣa* (Aries). This calendar is similar to the Jewish or Babylonian calendar which are also luni-solar.

In the Hindu Calendar, a month is divided into two equal parts, known as *pakṣa*, each of roughly fifteen days depicting the waxing and waning of the moon. The *pakṣa* starting from the new moon to the full moon is considered the bright-half (*Śukla-pakṣa*) while the second part starting from a full moon to a New moon, is known as the dark-half (*Kṛṣṇa-pakṣa*).[67] The new moon day, when the longitude of the sun and moon are equal, is called *amāvāsya*. The full moon night, when the sun and the moon are 180° out of phase, is

[65]Abhyankar, 1993.

[66]The Mehrgarh site is perhaps the oldest site so far from the region. It is believed to be around 8000-9000 years old.

[67]see *Arthaśāstra*, 108.

known as *pūrṇimā*. It gives a mean lunar year to be 354 days 8 hours 48 minutes and 34 seconds.

A day (*vāra*) begins at sunrise. The date (*tithi*) is indicative of the position of the moon relative to the sun. A month (*māsa*) starts when the longitude of the sun and the moon are the same, called *amāvāsya* (meaning dwelling together, implying conjunction of the sun and the moon, the new moon). This word *amāvāsya* is used in *Atharvaveda*[68] which signifies that the ancient Hindus knew the cause of the new moon during the Vedic period. At 180° difference in longitude, a full moon occurs, called *pūrṇimā* (the full moon).

During the ancient period, days were called by the appearance of the moon in the night sky. For example, full-moon night was called *pūrṇimā*, the new moon *amāvāsya*, and the days in between *ekādaśī* (eleventh day of the fortnight), *caturthī* (fourth day of the lunar fortnight), etc. As explained in this section, the ancient Hindus divided the lunar month into two parts (*pakṣa*) and numbered them. They did not divide into four equal parts defining a week; this happened much later in history, most likely after the Islamic invasion of India. It is possible that the idea of the seven-day week was borrowed from the Middle East. The ecliptic circle was divided into 27 parts, each consisting of 13°20′, called *nakṣatra*. A *kāraṇā* is half of a *tithī*. It is defined as the period in which the difference of the longitudes of the sun and the moon are increased by 6°.

To understand the features of the Hindu calendar, let us compare it with the Western calendar that is popular internationally. The Western calendar, also called the Gregorian calendar, was proposed by Pope Gregory XIII in 1582. It was a modified form of the calendar established by Julius Caesar. The Julian calendar was based on the Egyptian calendar of the period. The Catholic kingdoms adopted the Gregorian calendar soon after its inception. However, England resisted its use and adopted it in 1752, under some resistance from the Protestant majority.

The Western calendar is irregular and inconvenient to use because:

1. There exists no easy way to figure out the date of a particular day from

[68]defined in *Atharvaveda*, 7: 79.

simple observations.

2. Different months have different numbers of days. This creates difficulties in the business world where, at times, monetary transactions are made based on the day devoted to a particular task.

3. Because the span of a month is different for different months, performance records are difficult to compare.

4. There is a problem of the leap year. One has to remember the year to decide the number of days in the month of February. There is no possible way to figure it out using astronomical observations. One has to remember the empirically defined rules to figure this out.

Since people are familiar with the Western calendar since their childhood, they get used to it in the absence of any popular alternative.

In the lunar calendar, one year equals 354 days and, in the solar calendar, one year is roughly equal to 365 days. The difference of 11 days in a year can cause radically different seasons for the same month in two years that are about 15 - 17 years apart from each other. This is the case with the Islamic calendar.

Āryabhaṭa I explained the civil and sidereal days: "The revolutions of the sun are solar years. The conjunctions of the sun and the moon are lunar months. The conjunctions of the sun and Earth are [civil] days. The rotations of the Earth are sidereal days."[69]

This defines the sidereal day as the period from one star-rise to the next, civil days as one sun-rise to the next, and the lunar month, or synodic month, as from one new-moon to the next new-moon. The ancient Hindus, who knew both the lunar and solar calendars, realized that 62 solar months are equal to 64 lunar months. Therefore, they added one extra month after every 30 - 35 months.

The *Ṛgveda* described the moon as "the maker of months" (*māsa-kṛt*).[70] "True to his holy law, he knows the twelve moons with their progeny: He

[69] *Āryabhaṭīya*, *Kālakriyā*, 5.
[70] *Ṛgveda*, 1: 105: 18.

knows the moon of later birth,"[71]. Here "the twelve moons with their progeny" means the twelve months and "the moon of the later birth" means the 13th month, the supplementary or the intercalary month of the luni-solar calendar. This is a clear indication of the luni-solar calendar during the *Ṛgvedic*-period.

The *Atharvaveda* also mentions the 13th month in some years. "He [sun] who meets out the thirteenth month, constructed with days and nights, containing thirty members, . ."[72] The creation of 13th month or the intercalary month of thirty days is ascribed to the sun, the moon being the originator of the ordinary months of the year. This is a clear indication that the thirteenth month was added to keep up with the seasons since it is ascribed to the sun.

Al-Bīrūnī explained the Hindu luni-solar calendar in his book, *Alberuni's India*: "The months of the Hindus are lunar, their years solar; therefore their new year's day must in each solar year fall by so much earlier as the lunar year is shorter than the solar (roughly speaking, by eleven days). If his precession makes up one complete month, they act in the same way as the Jews, who make the year a leap year of thirteen months . . . The Hindus call the year in which a month is repeated in the common language *malamasa* [*malamāsa*]. *Mala* means the dirt that clings to the hand. As such dirt is thrown away out of the calculation, and the number of months of a year remains twelve. However, in the literature, the leap month is called *adhimāsa*."[73] Kauṭilya (ca. 300 B.C.), in his *Arthaśāstra*, mentions a separate intercalary month and calls it *malamāsa*.[74] This month was added approximately in the third year.[75] In modern days, this month is called *adhimāsa*.

In the Hindu calendar, most of the festivals have religious as well as seasonal importance. In societies where the lunar calendar is in practice, the seasonal festivals are not much celebrated. India is an agricultural country where approximately 65% of the population still lives in villages. During the *Holī* festival, a big fire is burned every year to symbolize the death of Holikā, the aunt of Lord Dhruva in Hindu mythology. People bring a sample of their

[71] *Ṛgveda*, 1: 25: 8.
[72] *Atharvaveda*, 13: 3: 8.
[73] Sachau, 1964, vol. 2, p. 20.
[74] *Arthśāstra*, 60; Shamasastry, 1960, p. 59.
[75] *Arthaśāstra*, 109; Shamasastry, 1960, p. 121.

harvest, and place it over a fire to roast the wheat or barley seeds which they tie to sugarcane. They share these seeds and sugarcane with friends and family members and decide whether the harvest of wheat, barley and cane sugar is ready or not. During *Daśaharā*, the quality of barley and wheat seeds is tested in a social gathering; people carry sprouted seeds and share them with their friends. Similarly, after the monsoon season from July to September, one needs to get ready for the winter in India. Cleaning spider webs, dusting rooms, painting walls, and decorating the houses are common chores before *Dīpavali* (Divali). It is common sense, but religious obligations enforce the need for an action.

Most Hindu festivals are defined either by the position of the moon or the sun. *Makara-Saṁkrānti* (the sun enters the sign of Makara (Capricorn) constellation in its northward journey), *Gaṇeśa-caturthī* (fourth day of the moon, starting from *amāvāsya*, the new moon), *Kṛṣṇa-Janmāṣṭamī* (eighth day of the moon) are all defined by the phase of the moon or the sun in a particular constellation. *Basant-Pañcamī* (fifth day of the new moon), *Rām-Navamī* (a day to honor Lord Rāma, falls on the ninth day of the new moon), *Guru-pūrṇimā* (a day to honor teachers, always fall on the full-moon), and *Nāga-Pañcamī* (a day to honor snakes, falls on the fifth day of the new moon) are some of these festivals that are defined by the moon.

The moon is seen in the sky on almost all nights unless it is close to the sun. The position of the sun can be fixed against a constellation only a little before sunrise or after sunset – the time when the sunlight is too weak to suppress the light of other stars. Hindu astronomy, unlike Western astronomy, mapped the sky with the phases of the moon rather than with the stars. It simplified their calculations – at full moon, the position of the sun can automatically be given by that of the moon if we know their diametrical positions. Similarly, the position of the sun can easily be determined with the different phases of the moon.

A change in the length of a day was observed and winter solstice (*Uttarāyana*, when the sun moves to the north) and summer solstice (*Dakṣiṇāyana*, when the sun begins to move to the south) were defined. The northward motion of the sun causes an increase of the length of a day and the corresponding decrease in the length of a night. The southward motion of the sun causes a reverse effect in the northern hemisphere. During the summer solstice, the length of the day is maximum, while night is longest for the winter solstice.

". . .the year consists of twelve months . . . and six seasons . . . sun goes south for six months and north for six months,"[76] suggests the *Taittirīya-Saṁhitā*.

Kauṭilya defined the seasons and the winter and summer solstices: "The seasons from *śiśira* (winter) and upward are the summer solstice (*Uttarāyana*, when the sun moves to the north) and those from *varṣā* (rainy season) and upward are the winter solstice (*Dakṣiṇāyna*, when the sun moves to the south)."[77]

In the Hindu calendar, a month spans from one new moon to the next new moon. The Hindu lunar month was equivalent to the moon's synodic revolution – the period during which the moon completes one cycle of *amāvāsya* (new moon) or *pūrṇimā* (full moon), equivalent to roughly 29.5 solar days.

The following are the months in the Hindu calendar:

1. *Caitra* (March-April)
2. *Vaiṣākha* (April-May)
3. *Jyaiṣṭha* (May-June)
4. *Āṣāḍha* (June-July)
5. *Śrāvaṇa* (July-August)
6. *Bhādrapad* (August-September)
7. *Āśvina* or *Kwār* (September-October)
8. *Kārttika* (October-November)
9. *Agarhayana* or *Aghan* (November-December)
10. *Pauṣa* (December-January)

[76] *Taittirīya-Saṁhitā*, 6: 5: 3: 1 - 4.
[77] *Arthśāstra*, 109; Shamasastry, 1960, p. 120.

11. *Māgha* (January-February)

12. *Phālguna* (February-March)

The names of the months are derived from the names of the *nakṣatra* (star or constellation) in which the sun dwells (or nearby). Xuan Zang (or Hiuen Tsang), a Chinese traveler who visited India during the seventh century, and al-Bīrūnī who traveled in India during the eleventh century also used the similar names for various months in India.[78]

Nothing was more natural for the sake of counting days, months, or seasons than to observe the twenty-seven places[79] which the moon occupied in her passage from any point to the sky back to the same point. The location of the moon and its shape provided the *tithī* (date) as well as the particular time in the night sky for astute observers. This procedure was considerably easier than determining the sun's position either from day to day or from month to month. As the stars are not visible during daytime and barely visible at sunrise and sunset, the motion of the sun in conjunction with certain stars was not an easily observable task. On the contrary, any Vedic shepherd was able to decide day and time easily with the observation of the moon.

The ancient Hindus formulated a theory of creation which was cyclic in nature. The universe followed a cycle of manifestated and non-manifestated existence. A new unit of time, *yuga*, was chosen to define the period of this cycle. The *yuga* system is based on astronomical considerations, and is frequently mentioned and explained in the *Purāṇic* literature.

In the *yuga* system of the Hindus, a *mahā-yuga* (*mahā* means big in the Sanskrit language) is divided into four *yugas*: *Satya* or *Kṛta*, *Tretā*, *Dvāpar*, and *Kali*. The periods of *Satya-yuga*, *Tretā-yuga*, *Dvāpara-yuga*, and *Kali-yuga* are in the ratio 4: 3: 2: 1. According to the Hindu scriptures, life on the Earth diminishes in goodness as we go from Satya- to *Kali-yuga*. At present, we live in the age of *Kali-yuga* that started in 3102 B.C. of the Julian Calendar. *Kali-yuga* began when the sun and the moon were at the zero point in Aśvini ($\beta$ and $\gamma$ Arietis).

---

[78]For Hiuen Tsang, see Beal, book II, p. 72; for al-Bīrūnī, see Sachau, vol. 1, p. 217.

[79]See section 5.3 for a description of the 27 places.

Table 5.4: *Yuga* System of *Garuḍa-Purāṇa*, Chapter 233

| Yuga | Period, in calendar years |
|---|---|
| *Kali-yuga* | 432,000 |
| *Dvāpara-yuga* | 864,000 |
| *Tretā-yuga* | 1,296,000 |
| *Satya-yuga* | 1,728,000 |
| *Mahā-yuga* | 4,320,000 |
| *Manvantara* = 70 *maha-yuga* | 302,400,000 |
| *Kalpa* = 14 *manvantara* | 4,233,600,000 |
| *day of Brahmā* = *Kalpa* | 4,233,600,000 |
| complete cycle of the creation | 8,467,200,000 |

A *Kali-yuga* is equal to 432,000 solar years. A *maha-yuga* is equal to 4 + 3 + 2 + 1 = 10 *Kali-yuga*, equivalent to 4,320,000 solar years. Seventy *maha-yuga* constitute a *manvantara* and fourteen *manvantara* constitute one *kalpa*. A *kalpa* is the duration of the day of Brahmā, the creator of the universe. The night is equally long and the creation dissolves into the unmanifested form during the night of Brahmā. After that, it starts again.[80]

According to the *Garuḍa-Purāṇa*, the cycle of creation takes about 8.5 billion years. Is it a coincidence that the ancients came up with a number that is of the same order of magnitude as the number suggested by modern scientists for the age of the universe? It is a difficult question to answer. In whatever circumstances, we must credit the authors of these ancient books that they could conjecture such a large number which is comparable to the modern accepted value. The oscillating universe theory provides a beginning and end to creation. However, the modern scientists do not suggest a number at this stage.

[80]Garuḍa-*Purāṇa*, Chapter 233.

## 5.3 Constellations

The word *nakṣatra* (stars or constellations) or related terms have appeared numerous times in the *Vedas*.[81] The celestial circle is divided into 27 parts of approximately 13°, each corresponding to a particular *nakṣatra*. The *nakṣatra* system was well-established during the Vedic period. Various ancient scriptures including the *Vedas*[82], the *Śatapatha-Brāhmaṇa*[83], the *Taittirīya-Brāhmaṇa*[84], and the *Vālmīki-Rāmāyaṇa*[85] mentioned the names of these constellations. The system of *nakṣatras* is based upon the sidereal revolution of the moon of about 27 days. Accordingly, a lunar zodiac of 27 *nakṣatras* near the ecliptic was made. Each night the moon enters a different lunar constellation (*nakṣatra*).

The *nakṣatra* system is a highly practical division of the zodiac, providing a different constellation for the moon to occupy on each night. Since it followes a sequence, one can predict the location of the moon with respect to the background stars.

The order of the *nakṣatras* is as follows:[86]

1. *Aśvinī* ($\beta$ & $\gamma$ Arietis)
2. *Bharaṇī* (35, 39 and 41 Arietis)
3. *Kṛttikā* ($\eta$ Tauri)
4. *Rohiṇī* (Aldebaran $\alpha$, $\gamma$, $\delta$, $\epsilon$, $\theta$, Tauri)
5. *Mrgaśiras* ($\lambda$, $\phi^1$, and $\phi^2$ Orionis)

---

[81] *Ṛgveda*, 1: 50: 2; 3: 54: 19; 6: 67: 6; 7: 81: 2; 7: 86: 1; 10: 88: 13; 10: 22: 10; 10: 85: 2; 10: 111: 7; 10: 156: 40. *Yajurveda*, 14: 19; 18: 18; 18: 40; 22: 28; 22: 29; 23: 4; 23: 43; 29: 2; 30: 10; 30: 21. *Atharvaveda*, 2: 2: 4; 3: 7: 7; 6: 86: 2; 6: 110: 3; 6: 128: 1 and 4; 9: 7: 15; 9: 8: 1; 10: 2: 22 and 23; 11: 6: 10; 13: 2: 17; 15: 6: 5; 19: 9: 9; 19: 19: 4; 20: 16: 11; 20: 47: 14.

[82] *Atharvaveda*,7: 13: 1; 19: 8: 2.

[83] *Śatapatha-Brāhmaṇa*, II: 1: 2: 18: 19.

[84] *Taittirīya-Brāhmaṇa*, II: 7: 18: 3; I: 5: 1: 1-5.

[85] *Bālkaṇḍa*, 71: 24; *Ayodhyākaṇḍa*, 2: 12; 4: 2; 4: 21; *Sundarakaṇḍa*, 17: 24; 19: 9; *Yuddhakaṇḍa*, 126: 54.

[86] Sarma, K.V. in Selin, Helaine (Editor), 1997, p. 519; Saha and Lairi, 1992, p. 184.

6. *Ārdrā* ($\alpha$-Orionis)
7. *Punarvasu* ($\alpha$ and $\beta$ Geminorum)
8. *Pusya* ($\gamma$, $\delta$, $\theta$ Cancri)
9. *Āśleṣā* ($\delta$, $\epsilon$, $\eta$, $\rho$, and $\sigma$ Hydrae)
10. *Māgha* ($\alpha$, $\gamma$, $\epsilon$, $\xi$, $eta$, and $\mu$ Leonis)
11. *Pūrva-phalgunī* ($\delta$ and $\theta$ Leonis)
12. *Uttara-phalgunī* ($\beta$ Leonis)
13. *Hasta* ($\alpha$, $\beta$, $\gamma$, $\delta$, and $\epsilon$ Corvi)
14. *Citrā* (Spica or a Virginis)
15. *Svātī* (Arcturus, or $\alpha$ Bootis)
16. *Viśākha* ($\alpha$, $\beta$, $\gamma$, and $\epsilon$ Librae)
17. *Anurādhā* ($\beta$, $\delta$, and $\pi$ Scorpionis)
18. *Jyeṣṭha* ($\alpha$, $\sigma$, $\tau$ Scorpionis)
19. *Mūlā* ($\epsilon$, $\zeta$, $\eta$, $\theta$, $\kappa$, $lambda$, $\mu$, and $\nu$ Scorpionis)
20. *Pūrva-āṣāḍha* ($\delta$, $\epsilon$ Sagittarii)
21. *Uttara-āṣāḍha* ($\alpha$ Sagittarii)
22. *Śravaṇa* or *Śroṇa* ($\alpha$, $\beta$, and $\gamma$ Aquilae)
23. *Dhaniṣṭhā* or *Śravisthā* ($\alpha$, $\beta$, $\gamma$, and $\delta$ Delphinis)
24. *Śatabhiṣaj* ($\gamma$ Aquarii, etc.)
25. *Purva-bhadrapadha* ($\alpha$, $\beta$ Pegasi)
26. *Uttara-bhadrapadha* ($\gamma$ Pegasi and $\alpha$ Andromedae)
27. *Revatī* ($\zeta$ Piscium)

A twenty-eighth *Abhijit* ($\alpha$, $\beta$ and $\zeta$ Lyrae) is also sometime included.[87] *Abhijit* lies between *Uttara-āṣāḍha* and *Śravaṇa.*

## 5.4 Hindu Cosmology

Cosmology is the study of the origin of the universe. Many religious books including the *Bible*, the *Vedas*, and the *Koran* provide their own account of the creation. For example, the first chapter of Genesis in the *Bible* provides an account of the creation that took seven days.

Till the beginning of the present century, it was a prevalent belief among Christians that the world was created around 4004 B.C. on one Friday afternoon. This was based on the genealogy presented in the book of *Genesis*, in the *Bible*, as concluded by Archbishop James Ussher (1581 - 1656 A.D.), who was also the Vice-Chancellor of the Trinity College in Dublin. In 1642, Dr. John Lightfoot (1602 - 1675 A.D.), then Vice Chancellor at the Cambridge University improved the calculation further and made it more precise to be 9 AM, October 23, 4004 B.C.[88] No medieval or renaissance scientist suggested it to be more than 7000 years old.[89] Today, it is difficult to accept these small numbers for the age of the universe. These explanations are not given much consideration even by devout Christians.

In 1940, George Gamow, a nuclear physicist, who defected from Russia and lived in Europe and the USA, proposed the Big Bang theory. His theory indicates that some 2 billion years ago, the density of matter in the universe was close to the density of the nucleus. The temperature of matter was nearly $10^{10}$ K, and the space was filled with intense radiation. In a sudden process, for some unknown reasons, matter exploded – something like a nuclear blast. After the explosion, neutrons evolved and decayed into protons and electrons; protons united with neutrons and formed nuclei. These nuclei combined with electrons to form atoms. These atoms aggregate to form compounds and matter.

---

[87] *Atharvaveda*, 19: 7: 1-5; *Taittirīya-Brāhmaṇa*, 1: 5: 1: 3; *Taittirīya-Saṁhitā*, 4: 4: 10: 1 - 3.

[88] White, 1897, p. 9.

[89] Crosby, 1997, p. 23.

After the explosion, the temperature of the matter dropped and the various fragments of the initial mass started to drift away from each other. Under the effect of gravitational force, the small fragments aggregated into larger masses, and evolved into the present form of the universe with galaxies, nebulas, stars, planets, satellites, comets, etc. This process of the expansion of the universe still continues.

One limitation of the Big Bang Theory is that it does not tell us if all of the fragments of the universe would keep on moving farther away from each other forever or if they would eventually return back to form a single clump under the gravitational force, as was the condition at the time of explosion. If the galaxies slow down their motion and return back under the gravitational force, it would lead us to an oscillating universe or a cyclic universe, as suggested by the ancient Hindus. The answer to this question lies in the precise measurement of the Hubble constant. This is not yet possible with sufficient accuracy to decide this issue.

Several questions related to the issue of creation are raised in the *Ṛgveda*: "What was the tree, what wood, in sooth, produced it, from which they fashioned forth the Earth and Heaven?"[90] Or, "What was the place whereon he took his station? What was it that supported him? How was it? Whence Visvakarman [God], seeing all, producing the Earth, with mighty power disclosed the heaven."[91]

In most Hindu accounts, the desire of God is the primary cause of creation (*sṛṣṭi*). Desire is more than just thought; it denotes the intellectual stir, the sense of active effort. It is the boundary between the nonexistent and the existent.

The *Ṛgveda* gives an account of creation and explains the state of the universe just before the blast: "Then was neither non-existent nor existent: there was no realm of air, no sky beyond it. What covered in, and where? And what gave shelter? Was water there, unfathomed depth of water? Death was not then, nor was there aught immortal: no sign was there, the day's

[90] *Ṛgveda*, 10: 31: 7
[91] *Ṛgveda*, 10: 81: 2.

and night's divider. That one thing, breathless, breathed by its own nature: apart from it was nothing whatsoever. Darkness there was: at first concealed in darkness this all was indiscriminate chaos. All that existed then was void and formless: by the great power of *warmth* was born that unit. Thereafter rose desire in the beginning, Desire, the primal seed and germ of Spirit. Sages who searched with their heart's thought discovered the existent's kinship in the non-existent."[92]

The *Srimad-Bhāgvatam* provided a description of the process of creation. According to it, the whole universe was in noumenal form and came to existence under the desire of the Creator (God). The present existence of the universe will continue for a period of time before the universe would again go back to its noumenal form of matter. This will form a cyclic (oscillating) process which will continue forever.[93] The concept of matter that cannot be experienced was not a part of science only 100 years ago. Today, the issues of dark matter and energy as forms of matter that we cannot experience are becoming popular.

The *Bhagavad-Gītā* described the creation of the universe as a transformation of noumenal matter into matter: "From the noumenal all the matter sprung at the coming of the day; at this coming of the night they dissolve in just that called the noumenal."[94]

In the belief of the ancient Hindus, the end of each creation comes with heat death, somewhat similar to the accelerated global warming which is causing concerns to modern scientists. This process of destruction is described in the *Viṣṇu-Purāṇa*. "The first, the waters swallow up the property of Earth, which is the rudiment of smell; and Earth, deprived of its property, proceeds to destruction. Devoid of the rudiment of odor, the Earth becomes one with water. The waters then being much augmented, roaring, and rushing along, fill up all space, whether agitated or still. When the universe is thus pervaded by the waves of the watery element, its rudimental flavor is licked up by the element of fire, and, in consequence of the destruction of the rudiments, the waters themselves are destroyed. Deprived of the essential

[92] *Ṛgveda*, 10: 129: 1 - 4.
[93] *Srimad-Bhāgvatam*, 3, 11 - 12.
[94] *Bhāgavad-Gītā*, 8: 8.

rudiment of flavor, they become one with fire, and the universe is therefore entirely filled with flame, which drinks up the water on every side, and gradually overspreads the whole of the world. While space is enveloped in flame, above, and all around, the element of wind seizes upon the rudimental property, or form, which is the cause of light; and that being withdrawn, all becomes of the nature of air."[95]

The *Ṛgveda*, while explaining the creation of universe, uses the analogy of a blast furnace of a blacksmith: "These heavenly bodies produced with blast and smelting, like a smith. Existence, in an earlier age of Gods, from non-existence sprang. Thereafter were the regions born. This sprang from the productive power."[96]

Al-Bīrūnī used the views of Varāhamihira, a fifth century Hindu philosopher, to explain the Hindu view of creation. "It has been said in the ancient books that the first primeval thing was darkness, which is not identical with the black color, but a kind of non-existence like the state of a sleeping person."[97] "Therefore, they [Hindu] do not, by the word creation, understand a formation of something out of nothing."[98] "By such a creation, not one piece of clay comes into existence which did not exist before, and by such a destruction not one piece of clay which exists ceases to exist. It is quite impossible that the Hindus should have the notion of a creation as long as they believe that matter existed from all eternity."[99]

Let us sum up the Hindus' theory of creation as provided above: There was a void in the beginning. This void was not the one that we perceive in a strict physical sense; this void was full of energy, in analogy, similar to the fields of modern physicists. The voluminous writings in Hindu literature do not describe the special creation as arising out of nothing. Almost all the scriptures of the Hindus, including the earliest *Ṛgveda*, advocate that the present form of the universe evolved from the noumenal form of matter.

"Void" or "nonexistence" is like the noumenal matter or dark matter

[95] *Viṣṇu-Purāṇa*, 6: 4.
[96] *Ṛgveda*, 10: 7: 2, 3.
[97] Sachau, 1964, vol. 1, p. 320.
[98] Sachau, 1964, 1, p. 321.
[99] Sachau, 1964, vol. 1, p. 323.

proposed by the modern scientists and philosophers; like the wavefunction of Schrödinger to explain the microscopic reality that cannot be experienced but, when squared, gives the probability of existence of the particle. The noumenal matter is beyond the senses' experience but gives rise to a manifested form of matter with a blast under the desire of God.[100] The language used to describe the process of creation in the *Vedas* is about four thousand years old. The account of the Hindu theory of creation by al-Bīrūnī is about 1,000 years old. Yet, there are striking similarities in the modern theory of creation and the Hindu theory of creation. In both theories, the present universe was created with a blast. Science is quite about the period before the Big Bang while the Hindu theory is providing some insights that explains the state of the universe before the blast. In addition, the ancient Hindus believed that the creation process is cyclic in nature, *i.e.* it goes through the cycle of creation and destruction. The destruction will start with an increase in heat that will give rise to an increase in the water levels of the oceans. Eventually, the heat will become so high that it will destroy all life forms.

Since everything animated in the universe should have a cause or the beginning, what was the beginning of the void or the beginning of the noumenal matter? Instead of drawing a picture of the beginning of the universe, the Hindus exposed the limit of human inquiry. The *Ṛgveda* says: "Who verily knows and who can here declare it, whence it was born and whence comes this creation? The Gods are later than this world's production. Who knows then whence it first came into being? He, the first origin of this creation, whether he formed it all or did not form it, Whose eye controls this world in highest heaven, He verily knows it, or perhaps *He knows not.*"[101] In this way, the *Ṛgveda* exposes the limits of human inquiries to define the first cause of creation.

---

[100]Kant defined the term, *noumenon*, which means a thing that cannot be perceived and can only be inferred from experience. It is a thing of purely intellectual intuition, like the interaction of electrical fields that gives rise to a force on a charge particle when placed near other charge particles. Such intuitions are an integral part of most religions as well science where realities are defined from inferred experiences. The analogy mentioned above is a meager effort to describe *nonexistence* of the ancient Hindus by sharing similar concepts in physics. However, we are dealing with two different domains of knowledge and the nonexistence of the ancient Hindus is not the dark matter or Schrödinger's wavefunction.

[101]*Ṛgveda*, X: 129: 6 - 7; for more information on Hindu cosmology, see Jain, 1975; Miller, 1985; Verma, 1996.

## 5.5 Diffusion of Hindu Astronomy to World Cultures

The work of Hindu astronomers certainly impacted astronomy in the Middle East and China. It indirectly impacted astronomy in Europe via the Middle East. Several noted European astronomers, including Copernicus and Kepler, read translations of the books from the Middle East that were based on Hindu astronomy.

### 5.5.1 Hindu Astronomy in the Middle East

If we had to choose a few popular books on the astronomy from the non-Western world, *Zīj al-Sindhind* of al-Khwārizmī (c. 800 - 850 A.D.) would be among these books. al-Khwārizmī, Abu Ja'far Muḥammad ibn Mūsā (ca. 800 - 847 A.D.) was a member of al-Ma'mun's *Bayt al-Hikma*. The motion of seven celestial bodies, the mean motions, and the positions of apogee and the nodes are described in his *Zīj al-Sindhind.*[102] As the title indicates, this book was based on Hindu astronomy. al-Khwārizmī's book agrees well with *Brahmsphuta-siddhanta*[103] (composed in 628 A.D.) by a Hindu astronomer Brahmgupta. al-Khwārizmī's tables were mostly created using Hindu procedures.[104]

"al-Khwārizmī's . . . treatise on astronomy was . . . a set of tables concerning the movements of the sun, the moon and the five known planets, introduced by an explanation of its practical use. Most of the parameters adopted are of Indian origin, and so are the methods of calculation described, including in particular use of the sine," concludes Régis Morelon in his analysis.[105] al-Khwārizmī's book has been translated into English by Neugebauer.[106] This book of al-Khwārizmī became one of the most documented books of astronomy in Europe during the medieval period. The

[102]Translated by Neugebauer, 1962.
[103]Salem and Kumar, 1991, p. 47.
[104]Goldstein, 1996; Kennedy and Ukashah, 1969.
[105]Régis Morelon, Eastern Arabic Astronomy Between the Eighth and the Eleventh Centuries in the book by Roshdi Rashed, 1996, vol. 1, p. 21. For more information on Ptolemy, see Pederson, 1974.
[106]Neugebauer, 1962.

famous Toledo Tables and the Alfonsine Tables were based on this book. In the title of his book, al-Khwārizmī has acknowledge Hindus' contribution to astronomy. Several medieval and modern historians have written about the connection of *Zīj al-Sindhind* to Hindu astronomy.

"Three Indian astronomical texts are cited by the first generation of Arab scientists: Aryabhatiya [*Āryabhaṭīya*], written by Aryabhata [Aryabhaṭa I] in 499 [A.D.] and referred to by Arab authors under the title *al-arjabhar*; *Khandakhadyaka* by Brahmgupta (598 - 668 A.D.), known in Arabic under the title *Zīj al-arkand*; and *Mahasidhanta* [*Mahāsiddhānta*], written towards the end of the seventh or at the beginning of the eighth century, which passed into Arabic under the title *Zīj al-Sindhind*"[107] A multitude of *Zīj*s were written in India and Afghanistan first, and in Persia and Baghdad later. A typical *Zīj* covered information on calendrical conversion; trigonometry; spherical astronomy; solar, lunar, and planetary mean motions; solar, lunar and planetary latitudes, parallax, solar and lunar eclipses, and geographical coordinates of various locations, particularly to locate *qibla*.[108]

"The first work of Arabic astronomy to have reached us in its entirety is that of Muḥammad b. Mūsā al-Khwārizmī (c. 800 - 850 A.D.) and is also called Indian Astronomical Table (*Zīj al-Sindhind*); it is in keeping with the preceding tradition but with the addition of elements from Ptolemaic astronomy. The Arabic text is lost and the work has been transmitted through a Latin translation made in the twelfth century by Adelard of Bath from a revision made in Andalusia by al-Majrīṭī (d. 1007 A.D.)," writes Régis Morelon, a professor in the University of Paris.[109]

Al-Khwārizmī even used metamorphosed Sanskrit terms in his astronomical calculations. For example, for the rules when finding the sizes of the sun, the moon and the Earth's shadow, al-Khwārizmī used the term *elbuht*, that comes from the Sanskrit word *bhukti* where the shadows on the Earth from

[107]Régis Morelon, General Survey of Arabic Astronomy in the book by Roshdi Rashed, 1996, vol. 1, p. 8

[108]King, in the book by Selin, 1997, p. 128; Mercier in Selin, 1997, p. 1057.

[109]Régis Morelon, Eastern Arabic Astronomy Between the Eighth and the Eleventh Centuries in the book by Roshdi Rashed, 1996, vol. 1, p. 21; see also Toomer, G. J., 1973, *Dictionary of Scientific Biographies*, vol. 7, p. 360.

the sun and the moon was observed at the same time daily.[110]

Ṣā'id al-Andalusī (1029 - 1070 A.D.) wrote that "a person originally from India [al-Hind, in the original] came to Caliph al-Manṣūr in A.H. 156 [773 A.D.] and presented him with the arithmetic known as *Sindhind* for calculating the motion of stars. It contains *ta'ādyal* [equations] that give the positions of stars with an accuracy of one-fourth of a degree. It also contains examples of celestial activities such as the eclipses and the rise of the zodiac and other information. . . . Al-Manṣūr ordered that the book be translated into Arabic so that it could be used by Arab astronomers as the foundation for understanding celestial motions. Muhammad ibn Ibrahim al-Fazārī accepted the charge and extracted from the book that astronomers called *al-Sindhind*. . . . This book was used by astronomers until the time of Caliph al-Ma'mūn, when it was abbreviated for him by Abu Ja'far Muḥammad ibn Mūsā al-Khwārizmī, who extracted from it his famous tables, which were commonly used in the Islamic world."[111]

Caliph al-Manṣūr was a ruler of Baghdad and his Abbasid dynasty was known for its respect of knowledge. He established *Bayt al-Hikma* (House of Wisdom) which became a model for other empires in Arabia and Europe. The House of Wisdom was a court or school where scholars worked in history, jurisprudence, astronomy, mathematics, and medicine, etc. These scholars were supported by Caliph al-Manṣūr and, in return, they helped the Caliph in his personal and kingdom affairs. It was a practice of the rulers of Abbasid dynasty to patronize scholars from foreign lands. With time, Baghdad became a center of knowledge. Scholars from the nearby regions visited Baghdad to acquire knowledge.

Abū Ma'sher, Jafar ibn Moḥammad ibn Amar al-Balkhī (787 - 886 A.D.), a Persian astronomer who mostly lived in Baghdad, also mentioned Kanaka's role in Baghdad. He labeled Kanaka as the foremost astronomer among all the Indians of all times.[112] We do not know much about Kanaka. Most of the information about him has come from the manuscripts written later in Arabia, the Mediterranean region and Europe. Obviously, Kanaka made his

---

[110] Neugebauer, 1962, p. 57; Goldstein, 1996.

[111] Salem and Kumar, 1991, p. 46 - 47.

[112] Pingree, 1968, p.16.

impact outside India.

In Persia, under the Sasanids (226 - 651 A.D.), observational astronomy was practiced under the influence of Indian and Greek astronomy. We know from al-Hashīmī (fl. ninth century) that Shāh Anūshirwān compared the work of Ārybahaṭa I's *Arkand* with Ptolemy's *Almagest*. He found Āryabhaṭa I's work better than Ptolemy's. Thus, the king asked his astronomers to compile a *Zīj* on Āryabhaṭa I's system. This is how the "Royal tables" (*Zīj al-Shāh*) were compiled.[113] Al-Manṣur decided the auspicious time for the foundation of the capital Baghdad using a Pahlavi version of *Zīj al-Shāh*.[114]

Like the Greenwich observatory in England has become a standard location to define the time and longitude of various locations in the world, al-Khwārizmī used Arin (Ujjain), the Greenwich of the ancient and medieval worlds, as the central place of the Earth.[115] This is an important piece of information. Almost any point on the Earth can be chosen as a standard for this. Al-Khwārizmī could have chosen Baghdad as the prime meridian, his place of residence. However perhaps due to the prevalent practice in Arabia and al-Khwārizmī's dependence on the Hindu astronomical tables, he preferred to choose Ujjain. It is the city that was also chosen by Āryabhaṭa I and Brahmgupta, from whom al-Khwārizmī derived his work. al-Khwārizmī defined one sidereal year equal to 365.15302230 days. This is exactly the same value used by Brahmgupta.[116] In Al-Khwārizmī's book, the "era of flood" was the era of *Kaliyuga* (February 17, 3102 B.C.). Al-Khwārizmī's *elwazat*, a procedure to calculate the mean positions of the planets, was similar to the *ahargaṇa* method of Hindu astronomy.[117]

---

[113]van der Waerden, The Heliocentric System in Greek, Persian, and Hindu Astronomy, in the book by King and Saliba, 1987; Régis Morelon, General Survey of Arabic Astronomy in the book by Roshdi Rashed, 1996, vol. 1, p. 8.

[114]David King, in the book by Selin, 1997, p. 126; F. Jamil Ragep, in the book by Selin, 1997, p. 395.

[115]Neugebauer, 1962, p. 10, 11.

[116]Neugebauer, 1962, p. 131.

[117]Sen, 1970.

### 5.5.2 Hindu Astronomy in Europe

Hindu astronomy was not only known in the Islamic region in the Middle East, it became known in Europe. Even in the eleventh century, the European scholars in Spain knew that al-Khwārizmī's work was an extension of Hindu astronomy. Ṣā'id al-Andalusī, a prominent scholar from Spain, provided ample information about the work of Āryabhaṭa I, Brahmgupta, etc. He believed that the Hindus' work on astronomy formed a basis for Arab astronomy,[118] as mentioned in the previous subsection.

Ṣā'id al-Andalusī was not the only person stating this fact. It was also noted by several other scientists from Spain. Rabbi Abraham Ibn Ezra (1096 - 1167 A.D.) provided a similar story of this transfer of knowledge from India to Arabia.[119] Ibn Izra was based in Spain and wrote about Hindu mathematician and astronomer, Kanaka, who shared his knowledge and allowed the Arabs to know about Hindu astronomy: "The scholar, whose name was Kanaka, was brought to the king, and he taught the Arabs the basis of numbers, i.e., the nine numerals. Then, from this scholar with the Jew as Arabic-Indian interpreter, a scholar named Jacob b. Sharah translated a book containing the tables of seven planets [five planets, the sun, and the moon]. . ."[120]

Abū Ishaq Ibrāhim ibn Yahya al-Naqqash, better known as al-Zarqalī (ca.1029 - 1087 A.D.), a Spanish astronomer who worked under Ṣā'id al-Andalusī, compiled the famous Toledan Tables that are based on *the Sindhind* system.[121] "One of the first Latin authors to use tables of Arabic origin was Raymond of Marseilles. In 1141 A.D., he composed a work on the motions of the planets, consisting of tables preceded by canons and an introduction in which he claims to draw on al-Zarqāllu [Zarqalī]. In fact, his tables are an adaptation of those of al-Zarqāllu [Zarqalī] to the Christian calendar and the longitude of Marseille.[122]

---

[118] Salem and Kumar, 1991

[119] Goldstein, 1967, p. 1478.

[120] Goldstein, 1996.

[121] Sarton, 1927, vol. I, p. 759.

[122] Henri Hugonnard-Roche, Influence of Arabic Astronomy in the Medieval West, in the book by Roshdi Rashed,1996, vol. 1, p. 287; also see, Toomer, 1968.

"The Toledan tables are a composite collection, including the parts taken from the tables of al-Zarqāllu [Zarqalī] alongside extracts from al-Khwārizmī (notably the planetary latitudes), elements from al-Battānī (in particular, the tables of planetary equations), and yet other parts derived from the *Almagest* or the *Handy Tables* of Ptolemy."[123] The *Toledan tables* were translated by Gerard of Cremona[124] and played an important role in the growth of European astronomy.

Al-Zarqalī is not the only person in Europe who studied and wrote astronomy books on the Hindu system of *Sindhind*. As mentioned in the previous chapter, al-Majrīṭī (d. 1007 A.D.) modified al-Khwārizmī's text of *Sindhind*. It was the Latin translation of this text that has survived and now available in English. A student of al-Majrīṭī, Ibn al-Samḥ (d979 - 1035 A.D.) composed a *zij* that was based on the *Sindhind* system.[125] Another student of al-Majrīṭī, Ibn al-Saffār, also authored a brief astronomical table that was based on the *Sindhind* system."[126] The third scholar from Spain, Ibn al-Ādāmi also compiled the astronomical tables known as *Kitab Naẓm al-'Iqd* (*Book on the Organization of the Necklace*) that was completed after his death by his student al-Qāssim bin Moḥammad ibn Hashim al Madā'inī, better known as al-'Alawī, in 950 A.D.. "This book contains all that was known about astronomy and the calculation of the motions of the stars according to the system of the *Sindhind*, including certain aspects of the trepidant motions of the celestial bodies, which were never mentioned before," writes Ṣā'id al-Andalusī.[127]

It is the Latin version of al-Khwārizmī's text that was used by Adelard of Bath, a British scientist and Arabist of the early twelfth century, to teach Henry Plantagenet, the future King Henry II of England.[128] Thus, the work of the Brahmins on the banks of Ganges became the subject matter of learning to the Royalty in England.

---

[123]Henri Hugonnard-Roche, Influence of Arabic Astronomy in the Medieval West, in the book by Roshdi Rashed, 1996, vol. 1, p. 289.

[124]Henri Hugonnard-Roche, Influence of Arabic Astronomy in the Medieval West, in the book by Roshdi Rashed, 1996, vol. 1, p. 292.

[125]Salem and Kumar, 1991, p. 64.

[126]Salem and Kumar, 1991, p. 65.

[127]Salem and Kumar, 1991, p. 53.

[128]Burnett, 1997, p. 31.

Alfonso X, King of Castile[129] (1252 - 1284 A.D.) gathered a group of Muslims, Jewish, and Christian scholars in the tradition of *Bayt al Hikma* or *House of Wisdom* of Baghdad. These scholars translated the earlier works in Arabic that were based on Hindu astronomy and compiled them into a single book in the Castilian language, called Alfonsine Tables. This book is composed of canons and tables. The book reached Paris in the early 14$^{th}$ century and were translated into Latin. Soon the book spread throughout Europe and used as computing tool for European astronomers.[130] Thus, medieval astronomy in Europe was a derived and improved mixture of Hindu, Persian, and Arabic contributions.

It is established that a bound volume of the Alfonsine Tables was owned by Copernicus and he did copy planetary latitudes from these tables.[131] The extent to which Copernicus was in debt of the Hindu astronomy is a matter of interpretation and debate. However, there is no dispute that he did use Hindu numerals and mathematics in his calculations. His dependence on Hindu astronomy is an open question which only the future scholarships will ascertain.

Johannes Kepler gave his table of parallax of the moon which is essentially identical with the theory of parallax given in *Khaṇḍa-Khādyaka* of Brahmgupta.[132] The theory of Brahmgupta was proposed about one millennium before Kepler. Although there are similarities between these two theories, Kepler's similarities with Brahmgupta's method could be purely anecdotal; there is no documentary evidence so far to relate the two. A possible connection could be al-Khwārizmī's book on Hindu astronomy, *Zīj al-Sindhind*. This book contained the work of Brahmgupta and was quite popular in Europe when Kepler was learning astronomy in his youth.

[129]Castile is presently a part of Spain. As a side note, soaps made from olive oil and sodium hydroxide or any hard soap from fat or oil are called Castile soaps. It is the legacy of this region.

[130]Chabás, 2002; King and Saliba in the book by Gingerich, 1993.

[131]Swerdlow and Neugebauer, 1984, p. 4 and Chabás, 2002.

[132]Neugebauer, 1962, p. 124. This is the conclusion of Neugebauer (1899 - 1990) who was trained in Gröningen, Germany, and taught at Brown University, USA.

### 5.5.3 Hindu Astronomy in China

The exchanges between India and China have a long record that goes back to the ancient period due to the close geographical location of these two countries. Indian thoughts influenced Chinese intellectuals especially with the introduction of Buddhism into China, which changed the social fabric of Chinese society. Such exchanges between India and China were not restricted only to social, cultural, and religious domains; they extended even to science. Between 67 A.D. and 518 A.D., some 1,432 works in 3,741 fasciculi were translated from Sanskrit to Chinese and were catalogued in 20 disciplines.[133]

By 433 B.C., Chinese astronomers recorded a system of 28 hsiu (lunar mansions or moon stations) marked by a prominent star or constellation. The moon travelled past and lodged in each of these mansions. This system probably originated from the Hindu system of 28 *nakṣatra* (star or constellation).[134] The *Vedas* provide a complete list of these *nakṣatra*. One difference between these two systems is that whereas the Hindus named their *nakṣatra* after their gods, the Chinese honored their emperors, queens, princes, and even bureaucrats by assigning their names to stars.

*The Navagraha-siddhānta* (*Nine Planet Rule*) was translated into *Kaiyuan Zhanjing* in 718 A.D. by Indian astronomer Gautama Siddhārtha (Qutan Xida, his Chinese name) and is still preserved in a collection from the Tang period. This book has the Indian calendar, known as *navagṛaha* (nine houses; used for the five planets, the sun, the moon, and the two nodal points, Rāhu and Ketu). This calendar was known as *Jiuzhi li* (Nine Upholder Calendar) in China.[135] This calendar is based on Varāhamihira's *Pañca Siddhānta* that was written during the sixth century. It contains the astronomical tables along with the methods to calculate an eclipse. It has a table of sines and the Hindu numeral zero in the form of a dot (*bindu*).[136] The Chinese, in following the Hindu tradition, also used nine planets in their astronomical work that included the sun, the moon, Mercury, Venus, Saturn, Jupiter, Mars, and the ascending and descending nodes known as Rāhu and Ketu in the Hindu

---

[133] Gupta, 1981; Mukherjee, 1928, p. 32, taken from Gupta, 1981.
[134] Sarma, 2000.
[135] Yoke, 1985, p. 162
[136] Yoke, 1985, p. 162.

work.[137]

During the Tang Dynasty (618 - 907 A.D.), three schools of Indian astronomical systems were based in China to guide the emperors. These schools are: Siddhārtha school, Kumāra school, and Kaśyapa school.[138] At least two experts from the Siddhārtha school served as Director of the astronomical bureau during Tang dynasty. It is important to mention here that these directors were crucial for emperors in their day-to-day activities as they had to find auspicious times for rituals and government actions, consult the director on astrological matters, and even take advice on dealing with people. This *Navagṛaha* calendar was later officially adopted in Korea.[139]

The *Chiu Chih Calendar* in China is derived from Varāhamihira's work. *Chiu Chih* mean "nine forces" and refers to nine planets of the Hindu astronomy.[140] In 759 A.D., an Indian Buddhist astrological work, the *Hsiu Yao Ching* (*Hsiu and Planet Sūtra*), was translated by Pu-Khung (Amoghavajra) and years later Yang Ching-Fêng, his pupil, added a commentary. According to him, "those who wish to know the positions of the five planets adopt Indian calendrical methods. . . So we have the three clans of Indian calendar experts, Chiayeh (Kaśyapa), Chhüthan (Gautama), and Chümolo (Kumāra), all of whom hold office at the Bureau of Astronomy."[141] It is established that the Indian scholars were employed in the astronomical observatory in China.[142] In Chinese as well as Hindu astronomy, the pole star (Mount Meru) played an important role and the universe revolved around it, whereas the Europeans laid more importance on the movement of the sun.

Rāhu (*Chiao chhu*) and Ketu (*chiao chung*) were frequently mentioned in the Chinese texts written during and after the Tang Dynasty (618 - 907 A.D.).[143] The book, *Qi Yao Rang Zai Jue* (*Formulae for Warding off Calamities According to the Seven Luminaries*), was compiled by an Indian Buddhist monk, Jin Ju Zha, in China. This book has detailed ephemerides of Rāhu

[137]Bagchi, 1950, p. 169 and Yoke in Selin, 1997, p. 78.
[138]Yoke in Selin, 1997, p. 110.
[139]Yoke, 1985, p. 162
[140]Needham and Ling, 1959, vol. 3, p. 175.
[141]Needham and Ling, 1959, vol. 3, p. 202; Yoke in Selin, 1997, p. 78.
[142]Needham and Ling, 1959, vol. 3, p. 37.
[143]Needham and Ling, 1959, vol. 3, p. 416.

and Ketu. According to this book: "The luminary E Luo Shi Rāhu is also known by the following names: The Yellow Standard (*Huang Fan*), The Head of the God of Eclipse (*Shi Shen Tou*), Superposition (*Fu*), and The Head of the sun (*Tai Yang Shou*). It always moves invisibly and is never seen when it meets the sun or the moon, an eclipse occurs; if the meeting is at a new moon or a full moon, then an eclipse necessarily occurs; when it is opposite to the sun or the moon, there will also be an eclipse."[144] In summary, the text supports the following points:

1. Rāhu and Ketu are invisible luminaries.

2. The motion of Rāhu and Ketu have a bearing on the occurrence of eclipses.

3. The theory depicted in the tales of Rāhu and Ketu is different than the ancient theory of eclipses in China.

4. Rāhu and Ketu execute a uniform motion against the background of fixed stars and its speed does not vary.

Obviously, the Chinese planets, Rāhu and Ketu, are taken from the Hindu tradition, and demonstrate the spread of Hindu astronomy to China.

[144]Wei-xing, 1995.

# Chapter 6

# Physics

Physics deals with matter and energy and their interactions. Representative disciplines of such studies include mechanics, acoustics, optics, heat, electricity, magnetism, radiation, atomic structure and nuclear phenomenon. The interaction of matter and energy, as it manifests itself in sound, light, and motion, excited the natural curiosity of ancient thinkers.

Length, time, and mass are considered as the three most important physical quantities in physics, called the fundamental quantities. Most other physical quantities are generally expressed in their terms of mass, length, and time. For example, speed is measured in miles per hour (or kilometer per hour) and involves a measurement of space (distance) and time. This means that a car moving with 65 miles/hour moves 65 miles (one measurement) in one hour (second measurement). These fundamental physical quantities were defined by most cultures; their standards were different from culture to culture. However, these standards served similar purposes.

## 6.1 Space (*Ākāśa*)

Space is a three dimensional matrix into which all objects are placed and move without producing any interaction between the object and the space. In the classical sense, space allows a physical ordering of objects without any reference to time. Objects appear to be near or distant due to this physical ordering. Though space and time collectively describe events in the physical universe, in the Newtonian (classical) world, they are independent of each other, and are considered as separate fundamental quantities. In the Newtonian world, space provides a sense of "orientation" to observers. In the relativistic world, where objects move with a velocity that is comparable to the velocity of light ($3 \times 10^8$ m/s), space and time do not have their independent status; they are integrated and form a new space-time (spatio-temporal) reality. In the present context, only the classical picture of space and time as two independent realities are considered.

All objects exist in space, and space is not made up of something else. Space is a great receptor of all things; if an object moves from one place to another, the space does not move with it. The motion is only in the object.

In Greek thought, space was considered an all-pervading "ether," or as some sort of container. Since objects can move in space from one place to another, space was considered to be different from material objects; it was considered as a subtle jelly-like medium within which material objects are located.[1]

*Ākāśa* is one of the terms used to describe "space" in the Sanskrit language. According to the *Chāndogya-Upaniṣad*, space was the first entity in the creation of the universe. To a question, "To what does the world go back?", the *Chāndogya-Upaniṣad* suggested: "To space, all things are created from space and they dissolve into space. Space alone is greater than any manifestation; space is the final goal."[2]

The *Chāndogya-Upaniṣad* identified space as an entity that contains the sun, the moon, and other material objects. "In Space are both the sun and the moon, lightening, the stars, and fire. Through space one calls out; through Space one hears; through Space one answers. In Space one enjoys himself. In space one is born. Reverence Space."[3] Thus, even the heavenly objects were a part of something larger and more basic than them; this was space that encompassed humans, sound, the sun, the moon, and the stars. It is space that provides individuality to objects as they are recognized based

[1] Algra, 1994.
[2] *Chāndogya-Upaniṣad*, 1: 9: 1.
[3] *Chāndogya-Upaniṣad*, 7: 13: 1 - 2.

on the space they occupy: "Space is the accomplisher of name and form (individuality)."[4]

The *Vaiśeṣika-Sūtra* of Kaṇāda explained the attribute of space: "That which gives rise to such cognition and usage as, this is remote from this, is the mark of space."[5] Thus, distance is an attribute of space. Kaṇāda considered the various directions (east, west, north, south) as the attributes of space.[6]

A standard of length is imperative for any measurement of space. The *Mārkaṇdaya-purāṇa* provided the following standards of length: "A minute atom, a *para-sukṣma*, the mote in a sunbeam, the dust of the earth, and the point of a hair, and a young louse, and a louse, and the body of a barley-corn; men say each of those things is eight times the size of the preceding thing. Eight barley-corns equal an "aṅgula" (finger-breadth), six finger-breaths are a "pada" (step), and twice that is known as bitasti; and two spans make a cubit measured with the fingers closed in at the root of the thumb; four cubits make a bow, a "daṇḍa" (stick), and equal to two "nadikayaga"; two thousand bows make a "gavyūti"; and four times that are declared by the wise to be a "yojna"; this is the utmost measure for the purpose of calculation."[7]

The *Baudhayāna-Śulbasūtra* listed the standard of length, as given in Table 6.1.[8] *Āpastambā-Śulbasūtra*[9], *Kātyāyana-Śulbasūtra*[10] and *Mānava-Śulbasūtra*[11] provide their own standards of length that are similar to each other.

We no longer know the magnitude of these standards. However, one can make a rough estimate. We know that the standard of length is "foot" in the official American system of units. This is the approximate distance covered

[4] *Chāndogya-Upaniṣad*, 7: 14: 1.

[5] *Vaiśeṣika-Sūtra*, 2: 2: 10.

[6] *Vaiśeṣika-Sūtra*, 2: 2: 11 - 13. For this book, I have used a Hindi edition by Pundit Shriram Sharma Ācarya.

[7] *Mārkaṇdeya-purāṇa*, 49: 37.

[8] *Baudhayāna-Śulbasūtra*, 1: 3

[9] *Āpastambā-Śulbasūtra*, 15: 4.

[10] *Kātyāyana-Śulbasūtra*, 2: 1; 5: 9.

[11] *Mānava-Śulbasūtra*, 2: 1; 4: 4.

Table 6.1: A Standard of Length from the *Baudhayāna-Śulbasūtra*

| | |
|---|---|
| 1 aṅgula | 14 aṇu or 34 tila |
| 1 small pāda, meaning step | 10 aṅgula |
| 1 pradeśa, meaning region | 12 aṅgula |
| 1 īṣā | 188 aṅgula |
| 1 akṣa | 104 aṅgula |
| 1 yuga | 86 aṅgula |
| 1 jānu | 32 aṅgula |
| 1 Śāmya | 36 aṅgula |
| 1 bāhu | 36 aṅgula |
| 1 prakrama | 2 pāda |
| 1 aratni | 2 pradeśa |
| 1 aratni | 24 aṅgula |
| 1 puruṣa | 5 aratni |
| 1 puruṣa | 120 aṅgula |
| 1 vyāma | 5 aratni |
| 1 vyāyāma | 4 aratni |

by a person in one step. In the Indian system, this equals to "pāda," literally meaning a human step. If we consider "pāda" to be equal to one foot and calculate the size of an atom using *Mārkaṇdaya-purāṇa*, it comes out to be $2.9 \times 10^{-9}$ m which is strikingly closer to the actual value (of the order of $10^{-10}$ m). Again, if it is a mere coincidence, it is still remarkable.

Kauṭilaya, the teacher and Prime Minister of Chandragupta Maurya (Candragupta, reigned, 322 - 298 B.C.) defined several standards of length in his book, *Arthaśāstra*[12] that are compiled in Table 6.2. If we use a similar procedure and assume "pāda" (meaning step) to be one foot. Then, the smallest particle (*paramāṇu*) is about $6.7 \times 10^{-7}$m.

It is difficult to know the exact length of *yojna* since a slight variation in the size of a minute atom can make a big difference in the standard of *yojna*. Let us define the literal meaning of several standards used: *aṅgula* means finger-breadth, pada is a step (like a foot), *daṇḍa* is a stick, *dhanu* is a bow without knowing the exact magnitude. *Yojna* is a fairly standard unit of length that is commonly used in the *Vedas*,[13] all the way to Āryabhaṭa I who provided the dimensions of various heavenly objects in *yojna*.

## 6.2 Time

Time is easy to measure. Most people carry a watch and some of these watches can be bought for just a few dollars. To define the nature of time is another matter. The nature of time is a philosophical issue that is dealt with by philosophers as well as physicists. Despite consistent efforts for at least two millenniums, it is still an open question. A practical standard for time is an easy task to define; it is a challenge to set an abstract definition of time that can exemplify its nature.[14]

St. Augustine (354 - 430 A.D.) is known to have thought about the nature of time and came up with the following answer: "If no one asks of me,

[12] *Arthaśāstra*, Chapters 106 - 107; Book 2, Chapter 20; Shamasastry, 1960, p. 117.

[13] *Ṛgveda*, 1: 123: 8; 2: 16: 3; 10: 78: 7; *Atharvaveda*, 4: 26: 1.

[14] For the history of time and clocks, read Balslev, 1983; Landes, 1983; Panikkar, 1984; and Prasad, 1992. Balsev, Panikkar and Prasad have compiled a good review on the philosophical developments in the concept of time by the Hindus.

Table 6.2: A Standard of Length from Kauṭilaya's *Arthaśāstra*

| | |
|---|---|
| 8 paramāṇu, atoms | 1 particle thrown off by the wheel of a chariot |
| 8 particles | 1 likṣā (egg of louse or young louse) |
| 8 likṣā | 1 yūka (louse of medium size) |
| 8 yūka | 1 yava (barley of medium size) |
| 8 yava | 1 aṅgula (finger breadth) |
| 4 *aṅgula* | 1 dhanurgṛaha |
| 8 aṅgula | 1 dhanurmuṣaṭi |
| 12 aṅgula | 1 vitasti |
| 14 aṅgula | 1 śamya or pāda |
| 2 vitasti | 1 aratni or prajāpatya hasta (arm-length) |
| 42 aṅgula | 1 kiṣku (forearm) |
| 54 aṅgula | 1 hasta used in measuring timber forests. |
| 84 aṅgula | 1 vyama |
| 4 aratini | 1 rajju (rope) |
| 10 daṇḍa | 1 rajju |
| 2 rajju | 1 paarideśa |
| 3 rajju | 1 nivartana |
| 66.5 nivartana | 1 goruta |
| 4 goruta | 1 yojna. |

I know: If I wish to explain to him who asks, I know not."[15] This bizarre answer demonstrate the complexity of knowing something at a superficial level and defining it at a scholarly level. Time is something that occupies no space and does not have weight. It can be measured and yet we cannot see it or touch it. We can use time, save time, even waste time. However, we cannot destroy it or create it.

Time is interrelated and interwoven with many other philosophical questions of existence, space, and causality. Time is defined in terms of events. In one philosophical approach, time is an ordering of events from the past to the present. In general terms, time can be considered to be a dimension in which we order events from the past to the present and into the future. Time is not an independent substantial entity with its own characteristics; it exists when superimposed on changes (events). Time is a fundamental quantity in the Newtonian world. It is connected with space in the relativistic world. A detailed philosophical study of the nature of time, which has intrigued scholars for millenniums, is beyond the scope of this book.

Time is the basis of all actions (events); since there is no action without a change and change separates two states (events), there is no change without time. Time and action (event) are related with each other and are born from each other. Since events are discrete in nature, this makes time discrete in nature. What happens when events move faster and the time between the two events becomes smaller and smaller? Do we have continuous events and, therefore, continuous time? Philosophers have considered the question as to whether time is continuous or discrete for ages, without coming to a definite conclusion.

Time deals with the order and the duration of events. If there is a sequence of events at a definite time interval, careful measurements of such a body constitutes a clock. In simple words, time divides two events from one another.

Aristotle considered time to be eternal because "there could not be a before and an after if time ceased to exist."[16] Time, just like space, is a con-

[15] *Confessions*, 11: 14; Parsons, 1964, p. 18
[16] *Metaphysics*, 12: 6: 1071b.

tinuous quantity in his mind.[17] For Aristotle, time is a measure of motion; as a body fills space, it is the movement that "fills" time. Plato, in relating time to events, wrote that there were no days or nights, months and years, before the Heavens came into being.[18] This is similar to the Hindu doctrine of the creation of the universe as explained in the *Ṛgveda*.[19] (Chapter 5) Thus, creation and its various manifestations, as they change and evolve every instant, become a source of time.

Any temporal existence is evolved with time and vice-versa. It is time that is the origin of the first cause, as the *Atharvaveda* tells us: "Time created living things and first of all, Prājapati."[20] In Hindu mythology, "Prājapati" is the first God created who, in turn, created the present universe. In this view, it is time that unfolds the spatio-temporal universe and divides the past from the present and the future. "In time, texts are produced – what is, and what is yet to come."[21] "Time is the supreme power in the universe."[22]

The *Maitreyī-Upaniṣad* relates time to the *mūrta* (existent) world. "There are two forms of Brahmā: Time and Timeless. That which is before the sun is Timeless and which begins with the sun is Time."[23] This quotation clearly states that the sun was created with the creation of universe and that time existed only after the creation. Prior to the creation, time did not exist, in the absence of any event.

In the above quotation, "anything before the sun" defines the period before the physical manifestation of the world, the situation before the Big Bang. Matter was in noumenal form, and no events were possible to experience this matter. Therefore, time could not have existed in the absence of events. Only after the creation, matter became accessible to experience. Thus, it was the "timeless" situation before the universe was evolved.

The *Vaiśeṣika-Sūtra* considered time as an entity that exists only in the

[17] Aristotle, *Categories*, 5a, 8 - 14.
[18] Plato, *Timaeous*, 37e.
[19] *Ṛgveda*, 10: 129: 1 - 4.
[20] *Atharvaveda*, 19: 53: 10.
[21] *Atharvaveda*, 19: 54: 3.
[22] *Atharvaveda*, 19: 54: 6.
[23] *Maitreyī-Upaniṣad*, 6: 15.

manifested world (non-eternal), like in the *Maitreyī-Upaniṣad* mentioned above. "The name Time is applicable to a cause, in as much as it does not exist in eternal substances and exists in noneternal substances."[24] Therefore, time can only be noticed in a dynamic world (temporal world) where events are a distinguishing factor. In a void, after this manifested world dissolves into "darkness" or "non- existence," and time cannot be experienced.[25]

A similar view of time is provided in the *Yoga-Vasiṣṭha.*[26] "Time is not destroyed. Just as man after a day's activity rests in sleep, as if in ignorance, even so Time after the cosmic dissolution sleeps or rests with the creation-potential hidden in it. No one really knows what this time is."[27] This explanation of time basically supports the idea that time cannot exist without events. Also, as mentioned in Section 5.4, in the cyclic creation of the universe, there is a stage where matter cannot be experienced. It is during this stage where events cannot be defined after the cosmic dissolution and time ceases to exist.

The *Yoga-Vasiṣṭha* suggests that time is indestructible and cannot be analyzed since we cannot define its smallest unit: "Time allows a glimpse of itself through its partial manifestation as the year, the age, and the epoch; but its essential nature is hidden . . . Time cannot be analyzed; for however much it is divided, it still survives indestructible."[28]

The *Patañjali-Yoga-Sūtra* defines events in the pursuit to define time: "The difference between that which is past and that which has not yet come, according to their attributes, depends on the difference of phase of their properties."[29] Time cannot exist without attributes; there has to be a change in the system to comprehend time. This provides an operational approach to the question of the nature of time.

"Every object has its characteristics which are already quiescent, those

[24] *Vaiśeṣika-Sūtra*, 2: 2: 9.
[25] see *Ṛgveda*, 10: 129: 1 - 4; see Chapter 5.
[26] Venkatesananda, 1984, p. 17.
[27] Dasgupta, 1932, p. 230 - 233.
[28] Venkatesananda, 1984, p. 16.
[29] *Patañjali-Yoga-Sūtra*, 4: 12.

which are active, and those which are not yet definable, wrote Patañjali.[30] "Through perfectly concentrated meditation on the three stages of development comes a knowledge of past and future. The series of transformations is divided into moments. When the series is completed, time gives place to duration."[31] Patañjali's definition of time is in tune with the *Vaiśeṣika-Sūtras*, *Atharvaveda*, and *Maitreyī-Upaniṣad*.

The *Yoga-Vasiṣṭha* also introduces the idea of time-relativity, only on the psychological level.[32] The relativity of time, as introduced in *Yoga-Vasiṣṭha*, is due to a different state of mind, and not due to time relativity as defined by Albert Einstein. In an observation of events, the mind plays an important role; if an event is not observed, the duration of events (time) cannot be measured.

To explain the subtlety of time and its quantized nature, the *Viṣṇudharmotra-Purāṇa* mentioned: "if one pierces 1,000 lotus petals (put on top of each other) with a needle, the foolish man thinks that they are pierced simultaneously, but in reality they were pierced one after the other and the subtle difference between the instants in which the successive petals have been pierced represents the subtlety of time."[33]

Āryabhaṭa I explains the nature of time and defines a method to measure it: "Time, which has no beginning and no end, is measured by (the movements of) the planets and the asterisms on the sphere."[34]

To measure time, the apparent motion of the sun defined the events; sunrise and sunset were two easily observed events. A scale based on the average-solar day was established by the ancient Hindus. The duration of an average day was divided into several segments that became a standard of time. To measure the magnitude of time, a simple method is defined in the *Srimad-Bhāgvatam*.[35] (Table 6.3)

---

[30] *Patañjali-Yoga-Sūtra*, 3: 14 - 16.
[31] *Patañjali-Yoga-Sūtra*, 4: 33.
[32] *Yoga-Vasiṣṭha*, 3: 20, 29.
[33] *Viṣṇudharmotra-Purāṇa*, 1: 72: 4-6.
[34] *Āryabhaṭīya*, *Kalākriya*, 11.
[35] *Srimad-Bhāgvatam*, 3: 11: 6 - 11.

Table 6.3: A Standard of Time from *Srimad-Bhāgvatam*

| | |
|---|---|
| 3 trasareṇu | 1 truti |
| 100 truti | 1 bedha |
| 3 bedha | 1 lāva |
| 3 lava | 1 nimeṣa |
| 3 nimeṣa | 1 kṣaṇa |
| 5 kṣaṇa | 1 kāṣṭha |
| 15 kāṣṭha | 1 laghu |
| 15 laghu | 1 nāḍikā or daṇḍa |
| 2 nāḍikā | 1 muhūrta |
| 6 or 7 daṇḍa | 1 prahara (one-fourth of a day or night) |

The duration of "nāḍikā" is measured as follows: "Take a copper vessel measuring six "pala"; make a hole into the copper vessel by a pin made of gold of which the length shall be four fingers and measure four "maśa" (unit of mass). Put the vessel on water. The time taken to make the vessel filled with the water and sink constitutes one nāḍikā." If we consider the average length of a day to be 12 hours then the smallest unit of time, "trasareṇu", comes out to be about $1.7\times10^{-4}$s.[36]

Within the accuracy of a solar clock, the method provided in the *Srimad-Bhāgvatam* for the measurement of time is fairly good. All factors that can influence a measurement are provided – the dimension and weight of the copper vessel, the size of the hole, and the density of the liquid.

Table 6.4 defines a standard of time provided in Kauṭilaya's *Arthaśāstra*.[37] One *nālika* is defined as the time during which one *adhaka* (a measurement of mass) of water passes out of a pot through an aperture of the same diameter as that of a wire of 4 *aṅgula* and made of 4 *maśa* (measurement of mass) of gold. The length and mass of gold defines the diameter of the wire. Here, truṭi comes out to be 0.06 s, if we assume 1 day to be 12 hours.

---

[36] *Srimad-Bhāgvatam*, 3: 11: 6 - 11.
[37] *Arthaśāstra*, 107; Shamasastry, 1960, p. 119.

Table 6.4: A Standard of Time from Kauṭilaya's *Arthaśāstra*

| | |
|---|---|
| 2 truṭi | 1 lava |
| 2 lava | 1 nimeṣa |
| 5 nimeṣa | 1 kāṣṭhā |
| 30 kāṣṭhā | 1 kalā |
| 40 kalā | 1 nālika |
| 2 nālika | 1 muhūrta |
| 15 muhūrta | 1 day or night |

Al-Bīrūnī explained that the smallest unit of time used in India was *aṇu* that is equal to $(24 \times 60 \times 60)/(88,473,600) = 9.766 \times 10^{-4}$ s.[38] He did not like such a small unit for time, just as much he did not like the Hindus to define large numbers. "The Hindus are foolishly painstaking in inventing the most minute segment of time, but their efforts have not resulted in a universally adopted and uniform system."[39] I am reasonably sure that al-Bīrūnī would not criticize the ancient Hindus in view of the modern developments in science. We deal with nano-second ($10^{-9}$s) or femto-second ($10^{-15}$s) in the scientific work these days. History shows that the Hindus were simply ahead of their time, and the other cultures failed to appreciate their innovations in the sciences.

## 6.3 Matter and Mass

The term "matter" and its cognates have played active parts in philosophical debates throughout our intellectual history. The natural philosophers studied material objects and related them to energy and fields, while the mathematicians distinguished them by their geometrical shapes, in order to attribute physical properties to them.

[38]Sachau, 1964, vol. 1, p. 337.

[39]Sachau, 1964, vol. 1, p. 334. Again, al-Bīrūnī has used the word "foolishly" that is not common in scientific writings these days. However, such usages were practiced in scholarly writings of the ancient and medieval periods.

In order to define and distinguish matter, Kaṇāda classified matter (*padārtha*) into six categories.[40]

1. Substance (*dravya*)
2. Quality (*guṇa*)
3. Action (*karma*)
4. Generality (*sāmānya*)
5. Individuality (*viśeṣa*)
6. Inherence (*samavāya*)

Kaṇāda defined nine properties to distinguish substances:

1. Earth (*pṛthivī*)
2. Water (*āpaḥ*)
3. Fire (*tejas*)
4. Air (*vāyu*)
5. Space (*ākāśa*)
6. Time (*kāla*)
7. Space or direction (*dik*)
8. Self (*ātman*)
9. Mind (*mana*)

Obviously, Kaṇāda's substance were more than just "matter," as viewed by modern science.

Table 6.5 defines seventeen attributes of *dravya* (substance).[41] According to Kaṇāda, matter possesses action, attributes, and the combinative cause.[42]

[40] *Vaiśeṣika-Sūtra*, 1: 1: 4.
[41] *Vaiśeṣika-Sūtra*, 1: 1: 6.
[42] *Vaiśeṣika-Sūtra*, 1: 1: 15; 10: 2: 1; 10: 2: 2.

Table 6.5: Attributes of matter from *Vaiśeṣika-Sūtra* (1: 1: 6)

| | | | |
|---|---|---|---|
| 1. | color (*rūpa*) | 10. | priority (*paratva*) |
| 2. | taste (*rasa*) | 11. | posteriority (*aparatva*) |
| 3. | smell (*gandha*) | 12. | intellect (*buddhi*) |
| 4. | touch (*sparśa*) | 13. | pleasure (*sukha*) |
| 5. | number (*saṅkhyā*) | 14. | pain (*duḥkha*) |
| 6. | measure (*parimāṇa*) | 15. | desire (*icchā*) |
| 7. | individuality (*pṛthaktā*) | 16. | aversion (*dveṣa*) |
| 8. | conjunction (*saṁyoga*) | 17. | volition (*prayatnaḥ*) |
| 9. | disjunction (*vibhāga*) | | |

Attributes are different in different matter, and are the defining principles for the differences in various types of matter. For example, identical sizes of iron and cotton would have different weights. Apples can have differences in their colors and yet be called apples. It is the assembly of attributes that defines matter.

Kaṇāda defines attributes for different types of matter: Earth has the attributes of color, taste, and touch;[43] water has the attributes of color, taste, touch, fluidity, and viscidity;[44] fire with color, and touch;[45] air has the attribute of touch;[46] while space has no such properties.[47]

In analyzing air on his classification scheme, Kaṇāda concluded that "air is a substance since it has action and attributes."[48] "Air is matter, but its non-perception, in spite of being substance, is due to the non-existence of color in it."[49] Thus, Kaṇāda defines a case of reality beyond appearance. Not all that exists in the world is visible to human eyes. In his attempt

[43] *Vaiśeṣika-Sūtra*, 2: 1: 1.
[44] *Vaiśeṣika-Sūtra*, 2: 1: 2.
[45] *Vaiśeṣika-Sūtra*, 2; 1: 3.
[46] *Vaiśeṣika-Sūtra*, 2; 1: 4.
[47] *Vaiśeṣika-Sūtra*, 2; 1: 5.
[48] *Vaiśeṣika-Sūtra*, 2: 1: 12.
[49] *Vaiśeṣika-Sūtra*, 4: 1: 7.

to clarify a difference between space and air, Kaṇāda used the attribute of touch which is absent in space: "Air has the property of touch while space does not have such a property."[50] To support that air is matter, Kaṇāda argued that a breeze can easily move the leaves of grass.[51] Thus, air, though invisible, can exert force and move things.

In the classical picture, matter and mass are practically identical; mass is a directly observable quality used to identify matter. The modern concept of mass has no sensory or any direct measurement to reveal it; it is a construct that is related to or is a form of energy. Einstein's mass-energy relationship connects them as: $E = mc^2$, where $E$, $m$, and $c$ are energy, mass, and the velocity of light, respectively.

In the classical sense, mass may be described as the inherent property of matter to resist a change in the state of motion (acceleration), as indicated in Newton's first law of motion, which is considered to be appropriate for our purpose in this book.

The practical necessities of commerce led to a measurement of mass. Mass, in the ancient Hindu literature, is not considered a measure of inertia, as indicated by Newton, but more appropriately as weight. Weight was not perceived by them as a force but an inherent quality that a substance possesses.

Kauṭilaya's *Arthaśāstra* provided standards of mass, as given in Table 6.6. Kauṭilaya suggested that the standards of mass should be made of iron or stones, available in Magdha and Mekala, both parts of Central India. A replacement could be of objects that do not change their physical conditions. These standards should not contract or expand when wet, and a change of temperature should not affect them.[52]

Though mass was described as weight, a measurement was done using a beam-balance that measures mass since the effect of gravity cancels out. In this device, the fulcrum is in the middle. Two identical pans hang on

[50] *Vaiśeṣika-Sūtra*, 2: 1: 4, 5.
[51] *Vaiśeṣika-Sūtra*, 5: 1: 14.
[52] *Arthaśāstra*, 103; Book 2, Chapter 19; Shamashastri, 1960, p. 113.

Table 6.6: A Standard of Mass from *Arthaśāstra* (Chapter 103.

| | |
|---|---|
| 10 seeds of maśa (Phraseolus Radiatus) or 5 seeds of *gunga berry* (arbus precatorius) | 1 Suvarṇa-maśa |
| 16 suvarṇa-maśa | 1 suvarṇa or karśa |
| 4 karśa | 1 pala |
| 88 white mustard seeds | 1 silver-maśa |
| 16 silver-maśa or 20 saibya seeds | 1 dharaṇa |
| 20 grains of rice | 1 dharaṇa of diamond |

both sides of the fulcrum at equal distances. Standard masses are placed on one side and the unknown mass on the other. The beam is balanced by selecting appropriate standard masses. Kauṭilaya defines balances with different lever arms and scale-pans for a range of masses.[53] He explained different kinds of balances for different accuracy: "public-balance" (*vyāvahārikā*), "servant-balance" (*bhājini*) and "harem-balance" (*antaḥpura-bhājini*).[54] Superintendents were assigned to stamp mass-standards for public use to avoid cheating, and money was charged for such services.[55] Traders were not allowed to use their own standards, and were fined if found guilty of doing so.[56]

### 6.3.1 Conservation of matter

Conservation of matter is a fundamental law in science that has been known to the Western world for the last two centuries. John Dalton (1766 - 1844 A.D.) proposed the law of conservation of mass in 1803 A.D.. Einstein's mass-energy relation unites mass and energy; mass can be transformed into energy and vice-versa. They become two different manifestations of the same reality. Therefore, the law of conservation of matter is now transformed into

[53] *Arthaśāstra*, 103; Book 2, Chapter 19; Shamasastri, 1960, p. 114.
[54] *Arthaśāstra*, 104; Book 2, Chapter 19; Shamasastri, 1960, p. 114.
[55] *Arthaśāstra*, 105; Book 2, Chapter 19; Shamasastri, 1960, p. 116.
[56] *Arthaśāstra*, 105; Book 2, Chapter 19; Shamasastri, 1960, p. 116.

a broader law of conservation of energy to incorporate mass-energy transformations.

Conservation of matter, in the ancient Hindu literature, is an extension of the theory of reincarnation. The ancient Hindus noted that life is a continual process and birth and death are just two different stages of life. In an extension of birth and death to the inorganic world, they believed that matter, like the soul, cannot be destroyed or created; only a transformation from one form to another is possible. The possibility of creation from "nothing" is rejected by the Hindus (Chapter 5).

The *Vaiśeṣika-Sūtra* tells us that a substance is produced from other substances.[57] Mixing or separation of different substances creates new substances and it is impossible to create matter from something that does not exist.[58] In his attempt to explain the properties of air, Kaṇāda suggests the law of conservation of matter that is strikingly similar to the definition accepted by the scientists about hundred years ago. "Matter is conserved. Since air and fire are made up of atoms, these are also conserved."[59] There is also an explicit statement that "molecules and materials are eternal," implying that they are conserved.[60] The Sanskrit term *nitya* is used to define that atoms and matter are conserved. The most common translation of *nitya* is "eternal," defining something that will exist forever. This essentially means that atoms are eternal or conserved. Interestingly, the shape of atom is defined as spherical.[61] Obviously, the ancient seers could not see atom through a microscope as we do today. How could they derive this knowledge? This is obviously a mystery.

## 6.4 Atom (*Paramāṇu*)

The English language has only twenty-six letters. With various combinations of these letters, an innumerable number of words are fabricated that allows

[57] *Vaiśeṣika-Sūtra*, 1: 1: 23 and 27.
[58] *Vaiśeṣika-Sūtra*, 1: 1: 25 and 28.
[59] *Vaiśeṣika-Sūtra*, 7: 1: 4; 7: 1: 19 - 20.
[60] *Vaiśeṣika-Sūtra*, 7: 1: 8.
[61] *Vaiśeṣika-Sūtra*, 7: 1: 20.

the English literature to be massive and rich. These twenty-six letters are the building blocks of the English literature. In analogy, what are the building blocks of matter?

This is an perennial question that the natural philosophers thought of during the ancient period. The same question still intrigues modern scientists. It is much easier to find connections between a stone and a mountain. These connections lead us to understand that a large number of stones can turn into a mountain and vice versa. Such connections are found in many cultures. The triumph of Kaṇāda was to connect a mountain to a stone and then beyond. This "beyond" led him to explain a stone as an aggregate of something that is so minute that it is beyond the senses and invisible. Thus, he conceded a reality beyond what the sense-perception could give him – a remarkable feat not much duplicated in other cultures.

Thales considered this question and resolved that water to be the building block of all matters. Water was considered an element that created the whole universe. Empedocles, Plato, and Aristotle improved this theory further and suggested air, water, earth, and fire to be the building blocks of matter. This four-element theory was similar to the periodic table of today, that has more than hundred elements. A fifth element, called the aether, was also considered as a building block for heavenly materials such as the planets. These elements represented geometrical atomism rather than material atomism. A major breakthrough came when Robert Boyle (1627 - 1691 A.D.) wrote his *The Sceptical Chymist* in 1661 A.D. and gave the first modern definition of chemical elements. John Dalton (1766 - 1844 A.D.), in 1808 A.D., improved element theory further and suggested atoms as the building blocks of matter. Was he the first one? A study in the history of science takes us to Greece and India where Democritus and Kaṇāda lived, respectively.

There is not a single book that can be assigned to Democritus (c.460 - 370 B.C.) and yet he is generally credited for his theory of atomism in most science texts. Our knowledge about him comes from the later writers who wrote about him. One possible explanation for his lost books is that his theories were not favored by Plato and Aristotle. Plato was against Democritus' atomism and even considered burning all the books of Democritus in his collection.[62] Aristotle, on the other hand, simply rejected atomism using reasoning. Democritus' work gained prominence after John Dalton proposed his atomic hypothesis in the nineteenth century.

Democritus was born in a wealthy family from Abdera, a region between Greece and Turkey. His father helped the Persian King Xerxes in his expedition to control the region. After the death of his father, Democritus inherited a lot of money and decided to spend it on travels to Egypt, Babylon, Persia, and perhaps India.[63] Democritus was not the only Greek philosopher who possibly visited India. Several other philosophers including Plotinus, Aplollonius, and Pythagoras are known to have visited India or known to have made efforts to do so. Pythagoras' possible journey to India is documented in Appendix A.

Karl Marx (1818 - 1883), the father of communism, wrote a thesis on the differences of the philosophies of Democritus and Epicurus in 1841. Due to Marx's influence, Democritus was extensively studied in the former Soviet Union. According to Karl Marx, Democritus came in contact with Indian gymnosophists which is also the view of Diogenes Lucretius (First Century B.C.; IX: 35) whom Marx studied extensively for his thesis. Marx writes, "Demetrius in the *Homonymois* [*Men of the Same Name*] and Antisthenes in the *Diadochais* [*Successions of Philosophers*] report that he travelled to Egypt to the priests in order to learn geometry, and to the Chaldeans in Persia, and that he reached the Red Sea. Some maintain that he also met the gymnosophists in India and set foot in Ethiopia."[64]

In his pursuit to understand the building blocks of matter, Democritus proposed that matter is composed of small particles, which sometimes cannot even be seen with our eyes, and called them atoms. According to Democritus, observable changes are based upon a change in the aggregation

---

[62]Horne, 1963

[63]Asimov, 1982, p. 20; Horne, 1963; Marx, 1841, Part 1, Chapter 3; Sarton, 1993, p. 252. Asimov and Sarton are legends in the history of science. George Sarton (1884 - 1956) was a professor at Harvard University and was the founder to two leading journals in the history of science: ISIS and Osiris. However, these travels are hardly ever mentioned in introductory textbooks.

[64]Marx, 1841, The Difference Between the Democritean and Epicurean Philosophy of Nature, 1841, Doctoral Thesis, Part 1, Chapter 3. online version.

of atoms. During such aggregations, however, the atoms remain intrinsically unchanged. The only change is in the perception whereby one can see them. Atoms do not change their nature, size, or shape; they remain indivisible. For Democritus, the indivisibility of the atom was an absolute indivisibility. Also, it was taken for granted that atoms were indestructible so that matter must be conserved. Matter could not be created out of nothing, nor completely destroyed during any action. This was a law in science until Einstein developed the concept of the conservation of mass-energy, using the relation $E = mc^2$.

Aristotle (384 - 322 B.C.) did not believe that matter is made up of small particles, especially air and fire. He argued that if air and fire are made up of particles, they should fall like pebbles or any other heavenly body. He refuted the atomic doctrines and came up with his four-elements theory, which prevailed for some two thousand years till the period of Dalton.

Roger Bacon (1214 - 1292 A.D.) ridiculed Democritus' idea and rejected that "the world is composed of an infinite number of material particles called atoms."[65] In Roger Bacon's view this idea was an error that caused more hindrance to the following natural philosophers than any other idea in history.[66]

The greatest achievement of Greek atomism is the general view of nature that the multitude of phenomena are based on some unifying concept. Greek atomism was an outcome of metaphysical reality or an ethical system. The atomic theory is an outcome of empty space theory where particles can exist. If there is no separation between particles, it would lead to a continuum. Thus, for a particle to exist in space, the separation of particles and the existence of space are the absolute necessities. These particles can move, like air particles, and form aggregates in the void and become visible as a lump. This explanation can easily explain the compressibility of cotton or the density of a substance.

The Hindu concept of matter shares some similarities with the Greek

[65]Burke, 1928, vol. 1, p. 173. Roger Bacon was responsible for the introduction of gunpowder in the West.

[66]Bacon, 1928, vol. 1, p. 173.

theories. In *Śvetāsvetara-Upaniṣad*, *pṛthivī* (earth), *āpaḥ* (water), *agni* (fire), *vāyu* (air), and *ākāśa* (space) are considered to be the five elements (*pañca-mahābhūta*; *pañca* = five, *mahābhūta* = gross elements) of matter plus four other elements.[67] These five elements are considered the basic constituents of the manifested world.[68]

Kaṇāda, the founder of the *Vaiśeṣika* system of philosophy, mainly occupied himself with the study of matter and the nature of atoms and molecules, using *dravya* to indicate *pañcabhūta*, the five elements of matter.[69] In the *Vaiśeṣika-Sūtra*, there are nine categories of substance: earth, water, fire, air, space (ākāśa), time, place, self, and mind (*manas*).[70] However, the physical elements are only five.[71] These are the elements that create, nurture, and sustain all life forms (organic); they are also manifested in all inorganic materials. These elements have distinctive sense-capacities: for example, earth possesses color, taste, smell, and touch; water has color, taste, and touch. Water is also distinguished as fluid and viscid. Fire has the sense capacity of color and touch; air has touch despite its invisibility; space has no perceived sense.[72]

Kaṇāda's atomic theory was formulated sometime between the sixth century B.C. and tenth century B.C..[73] Kaṇāda was from Pabhosa, Allahabad, and was perhaps the first atomist in India.[74] His theory was quite popular in ancient and medieval India. The *Vāyu-Purāṇa*, *Padma-Purāṇa*, *Mahābhārata*, and *Srimad-Bhāgvatam* list Kaṇāda's work and bear infallible testimony to the antiquity and popularity of the *Vaiśeṣika-Sūtra*.

In Kaṇāda's theory, the combination of molecules (*aṇu*) is possible and

[67]*Śvetāsvetara-Upaniṣad*, 1: 2.

[68]Preisendanz, 2010. This article has a detailed discussion on the issue and lists the scriptures where this term is used.

[69]*Vaiśeṣika-Sūtra*, 1: 1: 5.

[70]*Vaiśeṣika-Sūtra*, 1: 1: 4.

[71]*Vaiśeṣika-Sūtra*, 8: 2: 4.

[72]*Vaiśeṣika-Sūtra*, 2: 1: 1 - 11. For a detailed information on the history of *Vaiśeṣika* system in Indian philosophy, read Karin Preisendanz, 2011.

[73]Sinha, 1911, p. VI; Sarsvati, 1986, p. 302; Subbarayappa, 1967. A recent paper by Karin Preisendanz assigns the second century A.D. to the commentary of Candrananda on *Vaiśeṣika-Sūtra*.

[74]Sarasvati, 1986, p. 295.

the non-perception of atoms disappears when they amass together to become bigger (*mahat*).[75] This indirectly defines limits to the resolving power of human eyes. The smallest state of matter is *paramaṇu* (atom) which cannot be seen. These atoms aggregate to form the world we see. The shape of atoms is spherical.[76] With regard to such explanation, Bernard Pullman in his book, *The Atom in the History of Human Thought*, suggests that "Hindus would have been far more advanced than the Greeks in this respect."[77]

The attributes such as color, taste, smell, and touch of earth, water, fire, and air are non-eternal on account of their substrata.[78] These attributes disappear with the disappearance of their substance. Some attributes of the substrata are prone to change in its combinative (chemical action; *pākajah*) causes under the action of heat.[79] Thus, odor, flavor, and touch are attributes of atom in the *Vaiśeṣika-Sūtra* whereas Democritus denied such connections. The Greek atoms possessed only minimal properties: size, shape, and weight.[80]

The roots of Indian atomism were epistemological and based on empiricism and observation while those of Greek atomism were ontological or *a priori*.[81] Kaṇāda gained knowledge of atoms through empiricism, intuition, logic, observation, and experimentation.

The invisibility of an object, as we divide it into small parts, can simply be observed from the experience of wet clothes hung in air. The water in clothes can be seen and experienced. As the sunlight and air dry the cloth, the moisture escapes from the cloth and becomes invisible. The same water drop that was visible to the human eye disappears with extended exposure

---

[75] *Vaiśeṣika-Sūtra*, 4: 1: 6; 7: 1: 9 - 11.

[76] *Vaiśeṣika-Sūtra*, 7: 1: 20. As a physicist, I have mixed feelings about this issue. On one side we have a clear statement from Kaṇāda. On the other side, we know that it was not at all possible for the ancient scholars to see an atom. How can you define a shape for something that you cannot even see? It is perplexing to me. I have added this controversial statement to promote more scholarship on this issue. May be, it is a result of the imagination of the translator.

[77] Pullman, 1998, p. 84.

[78] *Vaiśeṣika-Sūtra*, 7: 1: 2.

[79] *Vaiśeṣika-Sūtra*, 7: 1: 6.

[80] Horne, 1960.

[81] Horne, 1960

to air and sunlight. Thus, smallness of an object causes this invisibility, as suggested by Kaṇāda. Continual division of matter does not lead it to disappear from existence, as it becomes invisible. It is for this reason "an absolute non-existence of all things is not possible because an atom remains in the end."[82] The *Vāyu-Purāṇa* suggested a similar ultimate division of matter, "A *paramāṇu* (atom) is very subtle. It cannot be seen by the eye. It can be imagined. What cannot be (ultimately) split in the world should be known as atom."[83]

The *Srimad-Bhāgvatam* suggested a theory of the atom that is similar to Kaṇāda's: "That which is the ultimate division of matter, that which has not gone through any change, that which is separated from others, and that which helps the perception of objects, that which remains after all is gone, - all those go under the name of *parāṃanu* (atom).[84] "Two atoms make one *aṇu* (molecule) and three *aṇu* make one *trasareṇu.* This *trasareṇu* is discovered in the line of solar light that enters into a room through a window and due to its extreme lightness such *trasareṇus* couresth the way to sky."[85]

Kaṇāda's theory was analyzed by Cyril Bailey (1871 - 1957) in his book, *The Greek Atomists and Epicurus.* In his view, "It is interesting to realize that at an early date Indian philosophers had arrived at an atomic explanation of the universe. The doctrines of this school were expounded in the Vaicesika Sutra [*Vaiśeṣika-Sūtras*] and interpreted by the aphorisms of Kanada [Kaṇāda]. While, like the Greek Atomists, they reached atomism through the denial of the possibility of infinite division and the assertion that indivisible particles must ultimately be reached in order to secure reality and permanence in the world, there are very considerable differences between the Indian doctrines and that of the Greeks."[86] On the relation of atoms and matter in Kaṇāda's system and its similarity to Greek atomism, Bailey continues, "Kanada [Kaṇāda] works out the idea of their combinations in a detailed system, which reminds us at once of the Pythagoreans and in some respects of modern science, holding that two atoms combined in a binary compound and three of these binaries in a triad which would be of a size to

[82] *Nyāya-sūtra*, 4: 2: 16.
[83] *Vāyu-Purāṇa*, 2: 39: 117.
[84] *Srimad-Bhāgvatam*, 3: 11: 1.
[85] *Srimad-Bhāgvatam*, 3: 11: 5.
[86] Bailey, 1928, p. 64.

be perceptible to the senses."[87]

There are marked differences between Kaṇāda and Democritus. Atoms, as defined by Kaṇāda, are indestructible, indivisible, and have the minutest dimension. In the case of Democritus, atoms of all elements (substances) are made up of just one substance; they differ only in regards to form, dimension, position, etc. These atoms were indivisible, like Kaṇāda suggested, and had infinite shapes and, therefore, infinite types of substances.[88]

On the possibility that an exchange between India and Greece concerning Kaṇāda's atomic theory led to Democritus theory of atomism, Garbe wrote in his book *The Philosophy of Ancient India*: "It is a question requiring the most careful treatment to determine whether the doctrines of the Greek philosophers . . . were really first derived from the Indian world of thought, or whether they were first constructed independently of each other in both India and Greece, their resemblance caused by the natural sameness of human thought. For my part, I confess I am inclined toward the first opinion without intending to pass an apodictic decision . . . The *historical possibility* of the Grecian world of thought being influenced by India through the medium of Persia must unquestionably be granted, and with it the possibility of the above-mentioned ideas being transferred from India to Greece. The connexions [connections] between the Ionic inhabitants of Asia Minor and those of the countries to the east of it were so various and numerous during the time in question that abundant occasion must have offered itself for the exchange of ideas between the Greeks and the Indians, then living in Persia."[89] Mabilleau in the nineteenth century suggested that Hindu atomism played a role in shaping Greek atomism.[90] Other historians argue that both theories are independent and unrelated.[91] In a recent scholarship of Thomas McEvilley, a parallelism between *Vaiśeṣika* system and Aristotelian metaphysics is pointed out. McEvilley has explored the possibility of Aristotelien influence on *Vaiśeṣika-Sūtras* and denied it. "When, after 326 B.C., Greek philosophy entered India, it found long-established traditions of philosophy already present which were not about to be discarded wholecloth in favor

[87]Bailey, 1928, p. 65.
[88]McDonell, 1991, p. 11 - 12.
[89]Garbe, 1897, p. 37 - 38.
[90]Mabilleau, 1895; taken from Pullman, 1998, p. 77.
[91]Partington, 1939.

of imported brands."[92] According to McEvilley, "[i]f Aristotle came into the Vaiśeṣika zone it could only have been in a messy organic weld producing not a clone but a new child."[93] This is a question that requires further investigations before any judgement can be made. At this stage, the historical records and modern scholarship are inconclusive on this issue of exchange between India and Greece. There are indications of a possible travel of Democritus to India for learning purposes.

## 6.5 Motion

The relative motion of two objects was explained by Āryabhaṭa I in the following words: "Divide the distance between the two bodies moving in the opposite directions by the sum of their speeds, and the distance between the two bodies moving in the same direction by the difference of their speeds, the two quotients will give the time elapsed since the two bodies met or to elapse before they will meet."[94] This implies that, if two objects are moving in opposite directions along a line, the distance between them at an instant divided by the sum of their velocities gives the time it takes for them to meet from that instant. If the objects have already met and moving away from each other, then the above procedure gives the time since the object were at one place. In the second case, if the two objects were moving in the same direction along a line, the distance between them at an instant divided by the difference of their velocities gives time it will take before they meet. If the slow moving object is trailing, in that case, the above procedure gives the time when they were at one place.

Āryabhaṭa I, as mentioned in Section 5.1, also explained the relativity of motion with an example of a boat moving in a river and the person sitting in the boat observes stationary objects on the bank of the river moving backwards. Similarly, we observe that stars apparently move, while it is the earth that is in motion. This illusion of the relative motion is known to

---

[92]McEvilley, 2002, p. 536.

[93]McEvilley, 2002, p. 536. McEvilley has beautifully compared parallels in Greek and Indian philosophy, including the one on Aristotle and *Vaiśeṣika-Sūtras*. However, there is no historical support for this comparison at this stage.

[94]*Āryabhaṭīya*, *Gaṇita-pada*, 31.

have proposed by Nicolas Oresme (1320 - 1382), a French scholar. He used a similar example of two boats, one stationary and another moving. Observers on different boats cannot distinguish which boat is moving by simply looking at the boat while a person at the shore can easily figure it out. Similarly, he suggested that a person in the heavens will observe diurnal movement in the earth while a person on the earth will observe oppositely directed motion in the heavens. As mentioned earlier, even Copernicus made a similar argument using similar language. This argument is remarkably similar to the argument of Āryabhaṭa I.

## 6.6 Gravitation and Ocean Tides

Aristotle believed that objects fall because they look for their right place in the universe. Since the earth can give them stability, they fall in the direction of earth. Heavy objects fall faster to look for a right place for themselves. Weight was a direct reflection of the divinity of matter.

Newton (1642 - 1726 A.D.) wrote in *Principia Mathematica* that each particle in the universe attracts every other particle of the universe. The force of attraction is proportional to the product of their masses and is indirectly proportional to the square of the distance between them. Mathematically,

$$F = \frac{M_1 M_2}{r^2}$$

Where $M_1$, $M_2$ are their masses and $r$ is the distance between the particles. Whether it is the force of attraction between apple and earth, the force of attraction between the Earth and the moon, or the planets and the sun, all forces can be calculated with the help of this formula. Using this formula, the escape velocity of a space shuttle from earth's gravitational force can be calculated. Newton could not provide an explanation for the gravitational formula. He empirically suggested it, and provided an explanation for the planetary motions. With the work of Newton, the Aristotelian scientific enquiry of "why it happened" was replaced by the Newtonian "how it happened."

The idea of gravitation is credited to Newton due to the mathematical formulation that he came up with. The mathematical form was a powerful intuition that only a genius like Newton could provide. However, the idea of gravitation was known long before Newton.

The *Vaiśeṣika-Sūtra* suggested that objects fall downward when dropped as a result of gravity. "The falling of water, in absence of conjunction, is due to gravity,"[95] and "flowing results from fluidity."[96] Kaṇāda used *gurutva* word for heaviness or gravity; *gurutvākarṣaṇ* (implying attraction of the earth) is the word that is used for the gravitational force today.

The above quotation differentiates fluid nature and gravitational force. To elaborate the gravitational properties, Kaṇāda wrote: "In absence of conjunction, falling results from gravity."[97] In absence of any conjunction, due to gravity, only downward motion is possible. "In absence of propulsive energy generated by action, falling results from gravity."[98] "Owing to the absence of a particular conjunction, there arise no upward or sideward motion."[99]

Kaṇāda also explained why water should fall due to gravitational effect, even though it goes up in the form of rain-clouds during the evaporation process. "The sun's rays cause the ascent of water, their conjunction with air."[100] Kaṇāda explains the upward motion of the objects, "Throwing upward is the joint product of gravity, volition, and conjunction."[101] An example is the upward motion of water in a tree.[102]

Al-Bīrūnī (973 - 1050 A.D.) wrote: "The difference of the times which has been remarked is one of the results of the roundity of the earth, and of its occupying the centre [center] of the globe. . . the existence of men on earth is accounted for by the attraction of everything heavy toward its centre [center], i.e. the middle of the world."[103] He continues, ". . . we say that the earth on all its sides is the same; all people on earth stand upright, and

[95] *Vaiśeṣika-Sūtra*, 5: 2: 3.
[96] *Vaiśeṣika-Sūtra*, 5: 2: 4.
[97] *Vaiśeṣika-Sūtra*, 5: 1: 7.
[98] *Vaiśeṣika-Sūtra*, 5: 1: 18.
[99] *Vaiśeṣika-Sūtra*, 5: 1: 8.
[100] *Vaiśeṣika-Sūtra*, 5: 2: 5.
[101] *Vaiśeṣika-Sūtra*, 1: 1: 29.
[102] *Vaiśeṣika-Sūtra*, 5: 2: 7.
[103] Sachau, 1964, 1, p. 270.

all heavy things fall down to earth by a law of nature, for it is the nature of the earth to attract and to keep things. . . ."[104]

Al-Bīrūnī provides an analogy of the kadamba flower (Fig. 5.1) to describe the earth where creatures live around the earth like spikes on the flower. The force of gravitation can allow them to live where nothing is up or down. He ascribed this knowledge to Varāhmihir (505 - 587 A.D.).[105] As mention in Chapter 5, Āryabhaṭa I also knew this because he used the same analogy. He said that there was no "up" or "down" side of the earth. What this means is that, from its location in the great vastness of space, among the immense number of stars and other celestial bodies, it is impossible to designate an "up" or "down" side, or direction, with the earth. However, because the force of gravitational attraction is, at every point on the earth, downward toward the center of the earth, gravity thus creates, along the *local vertical*, a local sense of "up" and "down." As we explained earlier, although gravity was known to the Hindus much before Newton, the mathematical expression for the force of gravity is the sole work of Newton.

Johannes Kepler (1571 - 1630 A.D.) is generally credited with suggesting the correlation of ocean tides (high tides and low tides) with the phases of the moon. Since the moon and the sun are so far away from the Earth, it was difficult for scientists of that period to comprehend the significance of Kepler's theory of ocean tides. Newton's action-at-a-distance description of the force of gravity had not yet been established.

Galileo knew of the work of Kepler but was against the idea that the moon could influence ocean tides. Galileo rejected the possibility that the heavenly bodies (moon, sun, etc.) could control terrestrial events. Due to the greater influence of Galileo, Kepler's theory of ocean tides was overlooked by his fellow scientists. Newton revived Kepler's theory of the ocean tides and explained it with his law of universal gravitation.

Ocean tides result from the differential gravitational effects of the sun and the moon on ocean water. Vālmīki, in his famous epic *Rāmāyaṇa*, connects high ocean tides to the full moon: "the roaring of the heaving ocean during

[104] Sachau, 1964, vol. 1, p. 272.
[105] Sachau, 1964, vol. 1, p. 272.

the fullness of the moon."[106]

A simple explanation could be that the high tides are apparently due to more water in the ocean. Oceans can be like a pot of water which will spill when full. The spill of water in a pot is commonly observed. It is much easier to extend this analogy to explain the swelling of ocean which cannot easily be verified experimentally. However, this explanation was contradicted in the *Viṣṇu-Purāṇa*. According to this book, the ocean tides can be explained in terms of the phases of the moon, "In all the oceans the water remains at all times in the same quantity, and never increases or diminishes; but like the water in a pot, which, expands with heating, so the water of the ocean expand and contract with the phase of the moon. The waters, although really neither more or less, expand or contract as the moon increases or wanes in the light and dark fortnights."[107]

The *Matsya-Purāṇa* also gives a similar picture, "When the moon rises in the East, the sea begins to swell. The sea becomes less when the moon wanes. When the sea swells, it does so with its own waters, and when it subsides, its swelling is lost in its own water . . . the store of water remains the same. The sea rises and falls, according to the phases of the moon."[108]

This is an excellent analogy that indicates that there is an expansion in water during high tides in the ocean and the quantity of water in the ocean does not increase or decrease with a change in the phase of the moon. Additionally, this also explains that the Hindus knew the thermal expansion of matter.[109]

Al-Bīrūnī claims that "the educated Hindus determine the daily phases of the tides by the rising and setting of the moon, the monthly phases by the increase and waning of the moon; but the physical cause of both phenomena is not understood by them."[110] His statement that the Hindus were not aware of the cause of ocean tides is only partly correct from a modern

[106] Vālmīki's *Rāmāyaṇa*, 2: 6: 27.
[107] *Viṣṇu-Purāṇa*, 2: 4.
[108] *Matsya-Purāṇa*, 123: 30 - 34.
[109] The expansion of matter with an increase in temperature has been known to modern scientists only from the last 2-3 hundred years.
[110] Sachau, 1964, 2, p. 105.

perspective. Hindus had an explanation for the ocean tides; however, it was not based on Newton's law of gravitation.

# Chapter 7

# Chemistry

In India, chemistry, like medicine, evolved primarily as a science of rejuvenation. In Hindu literature, Soma – the elixir of life – was produced with a knowledge of chemistry, and is referred to widely in the *Ṛgveda* and *Atharvaveda*.[1] Suśruta defined chemistry (*Rasāyana-tantra*) as the science "for the prolongation of human life, and the invigoration of memory and the vital organs of man. It deals with the recipes that enable a man to retain his manhood or youthful vigor up to a good old age, and which generally serve to make the human system immune to disease and decay."[2] Many types of pestles and mortars and distillation devices that allowed extraction of oils or other products were designed.

Caraka and Suśruta mentioned calcination, distillation, and sublimation processes in the chemical transformation or purification of metals. Several other authors have provided detailed accounts of these metals and their uses in ancient India.[3] Caraka mentioned gold, copper, lead, tin, iron, zinc, and mercury used in drugs and prescribed various ointments made from copper sulphate, iron sulphate, and sulfur, for external application in various skin diseases.[4] The oxides of copper, iron, lead, tin and zinc were also used for medicinal purposes. All these metals are native to the Indian Peninsula. Thin sheets of gold, silver, and iron were treated with salts and alkali in the

---

[1] *Ṛgveda*, 10: 57: 3 - 4; *Ṛgveda*, 9: 62: 1; *Ṛgveda*, 9: 2: 1; *Ṛgveda*, 8: 2: 1; *Atharvaveda*, 19: 24: 3.

[2] *Suśruta-Saṁhitā*, *Sūtrasthanam*, 1: 10.

[3] Biswas and Biswas, 1996 and 2001; Ray, 1948; and Bhagvat, 1933.

[4] Ray, 1956, p. 61 - 62.

preparation of drugs by Caraka.[5] The *Chāndogya-Upaniṣad* tells us of the alloys (reaction or joining) of gold with salt, silver with gold, tin with silver, and copper with lead.[6] Bleaching, dyeing, and tanning were also common procedures.

Poisons were prepared and the drugs to counter poison were also developed. The *Mahābhārata* and *Rāmāyaṇa* and Kauṭilaya's *Arthaśāstra* mention chemical weapons (*astra*) that were used in wars. For example, Kauṭilaya provided a recipe of poison gas that was deadly: "The smoke caused by burning the powder of *śatakardama*, *uchchidiṅga* (crab), *karavira* (*nerium odorum*), *kaṭutumbi* (a kind of bitter gourd), and fish, together with chaff of the grains of *madana* and *kodrave* (*paspalam scrobiculatum*), or with the chaff of the seeds of *hastikarṇa* (castor oil tree) and *palāśa* (*butea frondosa*) destroys animal life as far as it is carried off by the wind."[7]

Several deadly poisons were also concocted using complicated procedures: "The smoke caused by burning the powder made of the mixture of the dung and urine of pigeons, frogs, flesh-eating animals, elephants, men, and boars, the chaff and powder of barley mixed with *kāsīsa* (green sulphate of iron), rice, the seeds of cotton, *kuṭaja* (*nerium antidysentericum*), and *kośātaki* (*luffa pentandra*), cow's urine, the root of *bhāṇdi* (*hydroeotyle asiatica*), the powder of *nimba* (*nimba meria*) *śigru* (*hyperanthera morunga*), *phaṇirjaka* (a kind of basil or *tulsī* plant), *kshibapiluka* (ripe *coreya arborea*), and *bhāṅga* (a common intoxicating drug), the skin of a snake and fish, and the powder of the nails and tusk of an elephant, all mixed with the chaff of *madana* and *kodrava* (*paspalam scrobiculatum*) or with the chaff of the seeds of *hastikaraṇa* (castor oil tree) and *palāśa* (*butea frondosa*) causes instantaneous death where ever the smoke is carried off by the wind."[8] One can notice that the concoction is fairly complicated to produce and the ingredients are not so common in most recipes these days. The dung of pigeons, frog, cow's urine, and skin or a snake are not common items in most chemical preparations. However, this is how most drugs/poisons were made during the ancient period. The *Ebers Papyrus* of Egypt has similar recipes for various diseases. For example, crocodile dung was used to make contraceptives in

---

[5] Ray, 1956, p. 62.

[6] *Chāndogya-Upaniṣad*, 4: 17: 7.

[7] *Arthaśāstra*, 411; Book 14, Chapter 1; Shamasastri, 1960. p. 442.

[8] *Arthaśāstra*, 411; Book 14, Chapter 1; Shamasastri, 1960. p. 442-443.

ancient Egypt. Kauṭilaya provided formulae for poisons that could cause blindness, gonorrhea, madness, fever, and deafness. He suggested poisons that could wipe out whole populations once mixed in the water supply.

## 7.1 Sanskrit Names in Modern Chemistry

The names of chemical elements and compounds provide the cultural legacy to modern science. For example, the Arabic words such as alembic, alcohol, and alkali in chemistry indicate the exchanges between Europe and the Middle East. Though it is a subject for discussion, the word *Bitumen* is derived from the Sanskrit word *jatu*, meaning pitch, and *jatukrit*, meaning pitch creating. This Sanskrit word was incorporated into Latin language as *gwitu*, *gwitumen*, or later as *bitūmen*, meaning pertaining to pitch.[9] The Latin Dictionaries also agree with this assertion.

Similarly, several chemical elements are thought to have had their names derived from the Sanskrit language. Gold owes its name to its color yellow. It was called *geolo* in the Anglo-Saxon name which perhaps came from the Sanskrit word *jval* [*jvālā*][10], meaning a shining flame. The term *jvālāmukhī* (*jvālā* means "flame", *mukhī* means "mouth") is used for volcanoes due to the yellow color lava that comes from it. Gold is called *Aurum* in Latin, implying a yellow color. According to Ringnes, it is derived from the Sanskrit word *harit* meaning yellow.[11] The online Oxford dictionary mentions that the origin of this word has "Indo-European root share." Similarly, the word silver is perhaps Sanskrit in origin. It is called *argunas*, meaning bright, or *rajat*, meaning silver in the Sanskrit language. It was called *argentum* in Latin, meaning shining. Table 7.1 lists elements whose names are considered to be derived from the Sanskrit language.[12] This table is mostly compiled from the articles of David W. Balle and Vivi Ringes. The works of Balle and Ringes do not provide any basis for this information. However, both articles were published in the prestigious *Journal of Chemical Education*. This is an official publication of the American Chemical Society where articles go

---

9 Speight, 1999, p. 2.
10 Ringnes, 1989.
11 Ringnes, 1989.
12 Balle, 1985; Ringnes, 1989.

Table 7.1: Some Elements and Their Epistemology

| Name | Symbol | Epistemology |
|---|---|---|
| Gold | Au | *jval* (*jvālā*, Sanskrit) to *geolo* (Anglo-Saxon), meaning shine. *Harit Haranyam* (Sanskrit), or *Aurum* (Latin), meaning yellow. |
| Silver | Ag | *Argunas* (Sanskrit) to *Argentum* (Latin), meaning shining. |
| sulfur | S | *Sulveri*or *śulbāri* in Sanskrit, meaning the enemy of copper because it destroys metallic properties. |
| Lead | Pb | *Bhlei* to *Bahumala* in Sanskrit, to *Blei* or *Bly* in German, implying dirty. |

through rigorous review process.

There may be some flaws in this interpretation. More scholarly work is needed to find the extent to which the modern chemistry owes credit to the ancient Hindus. Table 7.1 is not the final word on this issue; it is rather anecdotal. However, as the *Journal of Chemical Education* is a prestigious journal, I expect that there was some basis on the part of the editorial board for publishing these articles. The purpose of this section is clearly to promote more scholarship on this topic and not to provide the final word.

## 7.2 Mining and Metallurgy

Most ancient metals such as iron, copper, gold, silver and lead were naturally available in various parts of India.[13] The remains of Mohanjo-daro and Harappa contain metallic sheets and pottery that are inscribed. Bronze, copper, iron, lead, gold, and silver were used to make axes, daggers, knives, spears, arrow heads, swords, drills, metal mirrors, eating and cooking utensils, and storage utensils. The composition of various elements in these

[13] Agarwal, 2000; Biswas and Biswas, 1996.

metals indicate their local origin in India.[14] Out of the 324 objects from these archaeological sites that have been analyzed, about 184 objects are of pure copper[15] and four archaeological copper processing kilns are reported at Harappa, Lothal, and Mohenjo-daro's sites.[16] The smelting operation is defined in *Ṛgveda*[17] where ore was dumped into fire and bellows were used to fan the fire.

Mining in India was highly organized from the earliest period. Kauṭilya realized the importance of mining to the state economy. He suggested that "mines are the source of treasury; from treasury comes the power of government."[18] It is believed that India was the first to introduce brass and zinc metal. Rajasthan had a zinc ore mine from the second millennium BC period, Taxila housed a fourth-century B.C. brass vase that used zinc metal, and Nāgārjuna described the extraction process of zinc.[19] Various cities in India have archaeological sites that clearly demonstrate mining operations in the pre-Christian era,[20] and these sites are scattered all over India.[21] It is noteworthy that kilns were used to cast liquid metal in a desired mold. It is also noteworthy that the molding of gold, silver, copper, and iron requires furnaces that must reach high temperatures. (Table 7.2)

The *Vedas* mention the five ancient metals which were used for various purposes. The words *Harita*, meaning yellow, was used for gold.[22] Similarly, *candra*[23] (meaning the moon) and *hiraṇya*[24] were also used for gold. Because of the color of a flame or volcano, *jvālā* could also be a word for gold, as mentioned in the previous section. The word *rajata*, meaning white, was used

---

[14]Ray, 1948.

[15]Lahiri, 1995.

[16]Agrawal, 2000, p. 40.

[17]*Ṛgveda*, 9: 112.

[18]*Arthaśāstra*, 85; Book 2, Chapter 12; Shamasastri, 1960. p. 89.

[19]Biswas, 2006.

[20]Willies *et al*, 1984.

[21]Biswas and Biswas, 1996; Ray, 1948.

[22]*Atharvaveda*, 5: 28: 1, 5, and 11: 3: 8.

[23]*Ṛgveda*, 2: 2: 4; 3: 31: 5; *Atharvaveda*, 12: 2: 53.

[24]*Ṛgveda*, 1: 22: 5; 1: 33: 8; 1: 43: 5; 1: 122: 14; 1: 162: 16; 2: 33: 9; 3: 34: 9; 4: 10: 6; 4: 17: 11; 5: 54: 11; 8: 7: 25; ; *Atharvaveda*, 1: 9: 2; 2: 36: 7; 5: 28: 6; 6: 38: 2.

Table 7.2: Melting Temperature of Metals

| Metal | Melting Point (°C) |
|---|---|
| Tin | 232 |
| Lead | 328 |
| Silver | 960 |
| Gold | 1063 |
| Copper | 1083 |
| Iron | 1540 |

for silver.[25] It was called *argentum* in Latin. *Śyāma*[26] (meaning black) and *kārsnāyasa*[27] (meaning black metal) were the words used for iron. *Trapu* was the word used for tin. *Trap* means "to be ashamed or shy" indicating the low melting point (232 °C).[28] *Ayas* is the word for iron as well as for other metals. The word *śīśā* was used for lead. *Atharvaveda* tells us that lead is a poison if consumed in large quantities. However, in small quantities, it is good in the cure of tuberculosis.[29]

The *Chāndogya-Upaniṣad* mentions metallurgical processes for gold, silver, tin, lead, and brass: "So, as one mend (binds) gold with borax-salt, silver with gold, tin with silver, lead with tin, copper with lead, wood with copper and leather with wood . . ."[30]

Ktesias, a Greek traveler who lived in Persia during the 5th century B.C. mentioned of the vast amount of gold mined from "high-towering mountains"[31] and the use of four-legged birds for mining and protection of treasure in India.[32] He also briefly mentioned a congealing processing to get quality

[25] *Ṛgveda*, 8: 25: 22; *Atharvaveda*, 5: 28: 1; 13: 4: 51; *Chāndogya-Upaniṣad*, 4: 17: 7.
[26] *Atharvaveda*, 11: 3: 7; 9: 5: 4.
[27] *Chāndogya-Upaniṣad*, 6: 17: 7.
[28] Biswas and Biswas, 1996, vol. II, p. 19.
[29] *Atharvaveda*, 1: 162: 2 and 4; 12: 2: 1.
[30] *Chāndogya-Upaniṣad*, 4: 17: 7.
[31] McCrindle, 1973, p. 16, 17.
[32] McCrindle, 1973, p. 44. The identity of the bird is not known. Other Greek writers such as Megasthenes and Arrian have also mentioned this bird.

gold.[33] Several ancient Greek historians, such as Herodotus, Ktesias, Arrian, and Megasthenes, mentioned the use of metals in India.[34]

The Painted Grey Ware of the Gangetic Valley were manufactured around 600 B.C.[35] Two glass bangles found at Hastinapur, near New Delhi, are from the 1100 - 800 B.C. period. The bangles are made from soda-lime-silicate glass with ornamental coloring from iron. Similar bangles of green color were also found elsewhere in northern India.[36] Glass bangles are the ornamental jewelry that are worn by Hindu females even today. Beads of green glass with cuprous oxide coloring were found in Taxila belonging to the 700 - 600 B.C. period.[37]

Aristotle, in his book, *On Marvelous Things Heard*, wrote that Indian-copper is known to be good and is indistinguishable from gold.[38] This mention was made perhaps to acknowledge the high quality of the Indian bronze made from copper that was "indistinguishable from gold" due to the mixing of different metals or minerals.

Kauṭilya (3rd century B.C., also known as Cāṇakya) described various metallic ores and mining operations along with the processes of distillation, refinement, and quality control. "Possessed of the knowledge of the science dealing with copper and other minerals (*sulbadhātu-śāstra*), experienced in the art of distillation and condensation of mercury (*rasapāka*) and of testing gems, aided by experts in mineralogy and equipped with mining laborers and necessary instruments, the superintendent of mines shall examine mines which, on account of their containing mineral excrement, crucibles, charcoal, and ashes, may appear to have been once exploited or which may be newly discovered on plains or mountain slopes possessing mineral ores, the richness of which can be ascertained by weight, depth of color, piercing smell, and taste."[39]

---

[33] McCrindle, 1973, p. 68.
[34] Bigwood, 1995.
[35] Dikshit, 1969, p. 3.
[36] Dikshit, 1969, p. 3.
[37] Dikshit, 1969, p. 4.
[38] Aristotle, *On Marvelous Things Heard*, 49, part of Aristotle's *Minor Work*.
[39] *Arthaśāstra*, 82; Book 2, Chapter 12; Shamasastri, 1960. p. 83.

Kauṭilaya provided clues for finding good locations for mining: observe soil, stones, or water that appears to be mixed with metal and the color is bright or if the object is heavy or with strong smell, this indicates the possibility of a mine nearby.[40] He provides techniques to locate glass, gold, silver, iron, and copper mines.[41] Mining operations were leased to private investors at a fixed rate.[42] According to him, the head of mining operations must know metallurgy, chemistry, and the refinery processes.[43] This tells us that metallurgy was an established discipline during the third century B.C. in India. Kauṭilaya defined the ethical rules of conduct for goldsmiths. Severe penalties were prescribed for goldsmiths who fraudulently adulterated gold. Any mixing of inexpensive impurities of tin, copper, and brass during the melting process of gold and silver was considered a crime and the criminals were prosecuted. Thieves of mineral products were punished with a financial penalty that was eight times the value of the stolen products.[44]

Kauṭilaya described gold, silver, arsenic, copper, lead, tin, and iron ores and their purification processes.[45] Coins, as currency, were used in business transactions during his period. Compositions of different coins are defined in Kauṭilaya's book,[46] and rules of punishment are defined for people making counterfeit coins.[47] Kauṭilaya also explained the methods to catch manufacturers of counterfeit coins[48] and defined a penalty for the coin-examiner to accept a counterfeit coin into the treasury.[49]

Sixteen gold standards based on the percentage of gold content in a specimen, that is similar to the current "carat" system, were defined by Kauṭilaya. Copper was mixed with gold to form an alloy of different carat standards and sixteen varieties were defined.[50]

---

[40] *Arthaśāstra*, 82; Book 2, Chapter 12; Shamasastri, 1960. p. 84.
[41] *Arthaśāstra*, 82; Book 2, Chapter 12; Shamasastri, 1960. p. 84.
[42] *Arthaśāstra*, 83; Book 2, Chapter 12; Shamasastri, 1960. p. 86.
[43] *Arthaśāstra*, 81; Book 2, Chapter 12; Shamasastri, 1960. p. 83.
[44] *Arthaśāstra*, 83; Book 2, Chapter 12; Shamasastri, 1960. p. 86.
[45] *Arthaśāstra*, 93; Book 2, Chapter 14; Shamasastri, 1960. p. 98.
[46] *Arthaśāstra*, 84; Book 2, Chapter 2; Shamasastri, 1960. p. 86 - 87.
[47] *Arthaśāstra*, 59; Book 2, Chapter 5; Shamasastri, 1960. p. 57, 58.
[48] *Arthaśāstra*, 212; Boook 4, Chapter 4; Shamasastri, 1960. p. 239.
[49] *Arthaśāstra*, 203; Book 4, Chapter 1; Shamasastri, 1960. p. 230.
[50] *Arthaśāstra*, 86; Book 2, Chapter 13; Shamasastri, 1960. p. 90.

Several gold alloys were defined by Kauṭilaya that were blue, red, white, yellow, and parrot (green) in colors. This was achieved by the processing of gold with rock salt, lead, copper, silver, mercury, etc. The processes of chemical reactions were well-defined with exact proportions of each element or compound.[51] For example, pure gold was mixed with rocksalt and heated to the melting point of gold and mixed with other substances to make alloys in blue, red, white, yellow, parrot (green), and pigeon (grey) colors. Gold-plating and other metal-plating procedures are defined using amalgams, heating processes, rock salt, mica, wax, etc.[52] Kauṭilaya also defined color, weight, characteristics, hammering, cutting, scratching, and rubbing as tools to test precious stones.[53]

### 7.2.1 Delhi's Iron Pillar

The Iron Pillar near Qutab-Minar in New Delhi is a testimonial to the metal forging skills of the ancient Hindus (Figure 7.1). The pillar was constructed and erected by King Candra on a hill in the town of Viṣṇupadagiri, as indicated in the three-stanza and six-line long Sanskrit inscription on the pillar. This pillar marked his renunciation of kingly duties and entrance to an aesthetic life.

The inscription is dated between 400 - 450 A.D. and the inscribed letters have minimal corrosion despite 1600 years of weathering in the open air..[54] Air, heat, cold and heavy rains of Northern India have not caused significant rusting of the pillar. Expert observers agree that the pillar is indisputably a long standing permanent record of the excellent metallurgical skills and the engineering skills of the ancient Hindus.

King Candra is most likely King Candragupta Vikramāditya II (375 - 413 A.D.). He was also called Candra, in short, as inscribed in the gold coins of the period.[55] Viṣṇupadagiri is perhaps the present town of Udaygiri in the

[51] *Arthaśāstra*, 88; Book 2, Chapter 13; Shamasastri, 1960. p. 92.
[52] *Arthaśāstra*, 92; Book 2, Chapter 14; Shamasastri, 1960. p. 97.
[53] *Arthaśāstra*, 92; Book 2, Chapter 14; Shamasastri, 1960. p. 97.
[54] Balasubramaniam, 2001.
[55] Balasubramaniam, 2001.

Vidisha-Sanchi region in India.[56] The current site of the pillar was chosen by Tomar King Anaṅgpāla who erected it on the site of a temple. Between 1192 - 1199 A.D., after the defeat of Pṛthvīrāj Cauhān by Qutb-ud-din Aibak, a mosque was erected on the premise.[57] In the eighteenth century, a cannon was fired at the pillar to break the pillar and it failed.[58] This canon ball was most likely used by Nadir Shah in 1739 A.D. when he came to Delhi, his army killed about 30,000 people in just one day, looted the region and left the area with an enormous amount of wealth.

The pillar is about 7.16 m long (23 feet, 6 inches) with a 42.4 cm (16.4 inches) diameter near the bottom and about 30.1 cm (11.8 inches) at the top. It weighs over six tons. The pillar is a solid body with a mechanical yield strength of 23.5 tons per square inch and ultimate tensile strength of 23.9 tons per square inch. It is made of wrought iron and no other pillar from the early medieval period of that size has been found anywhere else in the world.[59]

The composition of the wrought iron is as follows:[60] carbon 0.15%, silicon 0.05%, sulfur 0.005%, manganese 0.05%, copper 0.03%, nickel 0.05%, nitrogen 0.02%, phosphorous 0.25%, and 99.4% pure iron.

This composition provides strong evidence of an iron refining process in the making of this pillar. Such a good quality of pure iron is not available in Indian mines. The pillar has a coating of a thin protective layer of $Fe_3O_4$ using salts and quenching. The excavation of the buried portion revealed that the base of the pillar was covered by a sheet of lead, about 3 mm in thickness.

The lack of rusting is partly due to the dry climate of Delhi where the relative humidity is less than 70% for most of the year, except for the three months of July, August, and September. However, iron beams of the Sun Temple in coastal Orissa and the Iron Pillar of Dhar (Madya Pradesh), both in relative higher humidity regions, do not show much rusting either. This is perhaps due to a higher phosphorus content of the iron.

[56] Balasubramaniam, 2001.

[57] Balasubramaniam, 2001.

[58] Balasubramaniam, 2001; Balasubramaniam, Prabhakar and Shanker, 2009.

[59] Lal, B. B., On the Metallurgy of the Meharauli Iron Pillar, in the book by Joshi and Gupta, 1989, p. 25 and 26; Balasubramaniam, 2001.

[60] Biswas and Biswas, 1996, vol. 1, p. 394.

Figure 7.1: The Iron Pillar of New Delhi.

Once the surface of iron is rusted, it becomes porous and allows phosphorus to react with other chemical compounds which reduces to phosphoric acid. This acid interacts with iron to form dihydrogen phosphate. The chemical reactions are as follows:[61]

$$2H_3PO_4 + Fe \rightarrow Fe(H_2PO_4)_2 + H_2$$

$$2H_3PO_4 + FeO \rightarrow Fe(H_2PO_4)_2 + H_2O$$

This further dissociates into two forms:

$$3Fe(H_2PO_4)_2 \rightarrow Fe_3(PO_4)_2 + 4H_3PO_4$$

$$Fe(H_2PO_4)_2 \rightarrow FeHPO_4 + H_3PO_4$$

Both of these phosphates are amorphous and insoluble in water. These amorphous phosphates reorganize themselves into crystalline ferric phosphate, and a large reduction in the porosity of the surface results. This large reduction in porosity reduces any further rusting of the iron.[62] Thus, it is the phosphorus content of the pillar that does the trick, making the pillar to be rust free.

As mentioned earlier, the Iron Pillar of Delhi is not an isolated artifact. The Sun Temple of Koṇārka in Orissa was built during the ninth century. There are some 29 iron beams of various dimensions used in its construction. The largest beam is 10.5 meter (35 feet) in length and about 17 cm (about 7 inches) in width with a square cross-section and is about 2669 kg (6000 lbs.) in weight.[63] Iron nails are used to connect stone pieces. The composition of the iron beams and nails is similar to those of the famed Iron Pillar of Delhi. Similarly, the iron pillar of Dhar is about thousand years old and has little or no rusting.[64] The pillar is 7.3 tons in mass and 42 feet long. Dhar is an old capital of Malwa which is near the well known city of Indore. The pillar was

---

[61]Balasubramaniam, 2001.
[62]Balasubramaniam, 2001.
[63]Ray, 1956, p. 212.
[64]Balasubramaniam and Kumar, 2003.

possibly constructed by King Bhoja (1010 - 1053 A.D.) and currently lies in three parts in front of Lāṭ mosque in Dhar. King Bhoja was well versed in metallurgy and wrote a book on metallurgical processes and metal weapons, called *Yuktikalpataru*. Bahadur Shah, a Muslim king captured the region and decided to move the pillar to Gujrat. In the process of taking it out from the ground, the pillar fell down and was broken into two pieces, as we know from the memoirs of Emperor Jahāngir. The pillar has been weathering the monsoons of India for over a thousand years and has little rusting. The surface of the pillar is coated with a thin optically dull layer and on top of it is another thick layer of optically bright material. Professor R. Balasubramaniam of the Indian Institute of Technology, Kanpur has compiled a nice summary of the history of the pillar and its chemical constitution. [65]

## 7.3 *Wootz* or Damascus Steel

Steel is an alloy of iron containing 0.10 to 1.5% carbon in the form of cementite ($Fe_3C$).[66] The properties of steel vary greatly with a minor change in carbon content, along with other impurities. Metals such as manganese, silicon, chromium, molybdenum, vanadium, or nickel are purposely mixed in the process depending on the desired outcome. For example, stainless steel has approximately 12% or more chromium content.

The word steel comes from the Old High German (German language, around 11th century A.D.) word *stahal*. This implies high resistance of steel and is related to the Sanskrit word *stakati*, meaning "it resists"[67] or "strike against." The name could be due to the steel's popular use in sword-making dur to its hardness and strength.

Steel was prepared and used in India for various purposes from the ancient period.[68] Ktesias, who was at the court of Persia during the 5th century B.C., mentioned two high quality Indian steel swords that were presented to him. One sword was presented by the King of Persia and the other by King's

[65] Balasubramaniam, 2002.
[66] Bhardwaj, 1979, p. 159.
[67] Le Coze, 2003.
[68] Prakash and Igaki, 1984.

mother, Parysatis.[69] Nearchus (fl. 360 - 300 B.C.) mentioned that the Indians generally carried a broad three cubit long sword with them.[70]

King Poros, after losing the battle with Alexander the Great, gave 100 *talents* (6000 pounds)[71] of steel as a precious gift to Alexander, as we know from the accounts of Quintus Curtius (9: 8: 1), a Roman historian during the first century A.D. who wrote a biography of Alexander the Great. Steel was called *ferrum candidum*, meaning white iron, by Curtius.[72] It would be worthwhile to recall the history. Poros, the Indian king, lost the war with Alexander and was about to be killed. After a debate with Poros, Alexander realized the futility of wars, and released Poros and gave his territory back to him. After his release, Poros must have felt that steel was a precious gift to give to Alexander, something Alexander needed and did not have – more precious than gold, gems, or spices. After all, Porous received the gift of life from Alexander.

Steel was called *wootz* in India and was traded in the form of castings (cakes) of the size of ice-hockey pucks.[73] Persians made swords from *wootz* and these swords were later erroneously known as Damascus swords.[74] As is the case with the Arabic numerals, the name Damascus steel came about when the Europeans encountered steel in the Middle East around 1192 A.D. during the War of the Crusades.[75] While noticing the remarkable hardness and strength, they became interested in knowing the secrets of making ultra-high carbon steel. The European did not know at that time that the manufacturing process had originated in India.

Archaeological sites in the Periyar district of Tamil Nadu, which date back to about 250 B.C., provide indications of crucibles used to mix iron and carbon for the steel-making process.[76] Varāhmihir, during the sixth century

---

[69] *Fragments*, 1: 4; McCrindle, 1973, p. 9.
[70] Bigwood, 1995.
[71] Casson, 1989, p. 114.
[72] Casson, 1989, p. 114.
[73] Sherby and Wadsworth, 2001
[74] Sherby and Wadsworth, 2001
[75] Joshi, Narayan, R., Tough Steel of Ancient India, p. 293, in the book by Sharma and Ghose, 1998.
[76] Agrawal, 2000, 197; Prakash and Igaki, 1984.

A.D., wrote a chapter on swords (*khaḍgalakṣanam*) in his *Bṛhat-Saṁhitā* (50: 23 - 26), and provided a recipe for the hardening of steel. His processes used chemical techniques as well as heating and quenching. Swords treated with these processes did not "break on stones" or become "blunt on other instruments," as reported by Varāhmihir.[77]

Steel production was very much limited due to the high consumption of fuel and the required high temperature for melting. The situation prevailed until 1850 A.D., when high temperature furnace technology improved. Thus, the use of steel was largely for making blades for knives and swords. India, Syria, Iraq, and Iran were centers for the making the blades during the early medieval period. The high quality steel was imported from India.[78] Similarly, the so-called Damascus steel was also imported from India.[79]

Damascus steels are ultrahigh carbon steels that are high in strength. Even in its softest condition, Damascus steel is one-and-a-half times stronger than severely worked wrought iron. With proper processing, these steels can be made to a strength that is about five times greater than that of the strongest wrought iron.[80] Damascus steel has an attractive swirling surface pattern that is an outcome of the cooling process. The patterns result from the alignment of the $Fe_3C$ particles that form on the surface during the cooling process. The most common use of this type of steel was in the making of swords and daggers.[81] Thus, Damascus swords became famous for their hardness and could absorb blows in combat without breaking. It did not fail a Middle Age warrior during combat. This use of steel in sword-making contributed to the winning of many battles during the period.

Making *wootz* steel is complex as even minute impurities play an important role. Temperature and the cooling period also affect the quality of steel. Giambatlista della Porte in 1589 A.D. wrote on the importance of temperature in treating *wootz*, and suggested to avoid "too much heat." Similarly, Joseph Maxon, in 1677 A.D., cautioned against forging the alloy above a

[77]Biswas and Biswas, 1996, vol. 1, p. 276.
[78]Pacey, 1990, p. 80.
[79]Pacey, 1990, p. 80; Sherby and Wadsworth, 1985.
[80]Sherby and Woodworth, 2001
[81]For more information, please read, Sache, 1994; Figiel, 1991.

blood-red heat.[82]

Oleg D. Sherby and Jeffrey Wadsworth, researchers from Stanford University, found that the steel was produced with a slow equilibrium cooling. When iron and carbon (1.3 - 1.9%) are heated to 1200°C, they reach a molten state and the slow cooling allows the carbon to diffuse through the iron to form white cementite patterns, called the Damascus steel. The pattern results from alignment of the $Fe_3C$ (cementite) particles during the cooling process. With polishing, the $Fe_3C$ particles appear white in the near black steel matrix. The carbide particles serve the role of strengthening without making the metal brittle.[83]

The blade was hardened by heating it to 727°C which allows a change in the crystal structure. Iron molecules that were distributed as body-centered ferrite begin to form a face-centered lattice. The blade was then quenched in water.[84] If the heating was done above 800°C before quenching, it made the metal brittle.

*The Periplus Maris Erythraei* text, written around 40 - 70 A.D.,[85] provides information on the importation of steel to the West. Steel was an item that was subjected to a custom duty under Marcus Aurelius (121 - 180 A.D.) and his son Commodus (161 - 192 A.D.) in Rome.[86] In the *Digest of the Roman Law*, (39, 15, 5 - 7), Indian iron (*Ferrum Indicum*) was in the list of objects that were subject to import duty,[87] as defined by Emperor Justinian (482 - 565 A.D.). Similarly, iron was taxed and traded in Alexandria, Egypt in the early first century A.D. and was called *koos*. This word is similar to *ukku* or *urku*, the Telgu word for iron.[88] Steel is called *wuz* in the Gujrati language, and *Wooku* (or *ukku*) in the Kannada language, a prominent language in South India.[89] Perhaps this explains the origin of *wootz*, the Syrian word for steel. This word was new to the Syriac literature and appeared late

---

[82]Smith, 1982.
[83]Sherby and Wadsworth, 1985.
[84]Sherby and Wadsworth, 1985.
[85]Casson, 1989, p. 7.
[86]Casson, 1989, p. 114.
[87]Schoff, 1915.
[88]Toussaint, 2000.
[89]Agarwal, 2000, p. 198; Biswas and Biswas, 1996, vol.1, p. 121, Le Coze, 2003.

in chronology.

On the eastward journey, wootz steel was an import item in China during the sixth and seventh century and was called "bin iron".[90] This steel was imported as a raw material and not a finished artifact. Several books, such as *Wei Shu*, written during the sixth century, and *Zhou Shu*, written during the seventh century, mentioned this term as an imported item and labeled it as fine steel. In 1259, an embassy was sent to the Middle East. Liu Yu, a member of this embassy, mentioned "bin iron" and indicated its source as Yindu (similar to Hindu).[91]

The Arabs called steel *Hundwániy*, meaning Indian.[92] This word perhaps evolved into *andanic* or *ondanique* for swords and mirrors, used by the medieval writers. This also led to the words *alhinde* [or al-Hind, meaning India] and *alinde* [for steel mirror] in Spain.[93] The best steel in Persia was called *foulade Hind*, meaning steel of India. Another kind of steel, *jawābae Hindu*, meaning a Hindu answer, was also popular because it could cut a steel sword.[94]

## 7.4 *Agni-astra* and Gunpowder

Human history is full of wars. Most wars were fought either in the name of religion, race, or culture. In many cases, they were fought to satisfy the desire of one person to subjugate others. Scientists have always worked to create more effective weapons to defeat the enemy. Swords, bows and arrows, boiling oil, melted sugar, naphtha (in Greek fire), and concave mirrors are some selected examples that were used by the ancient warriors. Armor was designed to protect a knight from injury. How do you pierce strong armor with a sword?

It was considered better to hit the enemy at a distance to avoid any per-

[90]Wagner, 2007.
[91]Wagner, 2007.
[92]Schoff, 1915.
[93]Schoff, 1915.
[94]Royle, 1837, p. 47.

sonal injury. Projectiles with deadly powers were bound to be invented. A major breakthrough came with the invention of gunpowder that made it possible to hit the enemy from a far distance with deadly power. It allowed for the discharge of a missile that could penetrate the armor of Middle Age warriors and made them susceptible to destruction. Fortresses were not immune either; they were susceptible to projectiles with fire and could not provide refuge for a king or for an army. This technology was mastered by the European nations and, as a result, they were able to colonize most of the world during the early part of the twentieth century.

Roger Bacon is generally credited with the invention of gunpowder. He has given a recipe for gunpowder in his book, *Opus Majus.* Bacon wrote, "From the force of salt called saltpeter so horrible a sound is produced at the bursting of so small a thing, namely, a small piece of parchment, that we perceive it exceeds the roar of sharp thunder, and the flash exceeds the greatest brilliancy of the lightening accompanying the thunder."[95] He was interested in alchemy, which was labeled as an Islamic art during the period. Like many other chemists of the period, Roger Bacon came under criticism by the church and was accused of dealing with forbidden magic. He was incarcerated by his Franciscan Order. He was released by Pope Clement IV for a while. However, he was again imprisoned by Pope Nicholas III.

Was he the first person to invent gunpowder? The answer is a clear "no". A popular theory is that Roger Bacon learned this idea from his friend, William Rubruck. In 1237, Mongol armies captured parts of Russia, Poland, and Hungary. After a brief occupation, the armies retreated from Hungary when Ogeidei Khan died. King Louis IX of France sent Franciscan William Rubruck to meet the Mongol Prince Sartak in the Mongol capital of Karakorum in 1253. He came back in 1257 and wrote an account of his travels. William Rubruck experimented with rockets and gunpowder in Cologne. His close friend, Roger Bacon, another Franciscan, perhaps came in contact with this information and wrote the first account of gunpowder in 1266.[96] Another theory is that Roger Bacon learned the recipe for gunpowder during his travels in Spain. The recipe of gunpowder was known to the Moors.[97]

[95]Bacon, vol. 2, p. 629 - 630.
[96]Pacey, 1990, p. 45.
[97]Oppert, 1967, p. 46

An Arabic treatise on gunpowder that was written in 1249 is still preserved in the Library of the Royal Escurial.[98]

The Chinese experimented with gunpowder and came up with various uses. The Chinese, though they initially used gunpowder for fireworks, used the gunpowder as explosives, and made rockets and guns. This was all done before the thirteenth century.[99] In the middle of the thirteenth century, gunpowder was used in the war between the Sung dynasty and the Mongols, to open city gates and blow walls.

In India, *Dhanur-vidyā* is the term that translates into "the knowledge of the bow." The *Yajurveda* and *Agni-Purāṇa* have significant hymns dealing with the science of the bow. It was expected of every prince in ancient India to master this art. Although the literal meaning restricted the art to using bows and arrows, this term was also used as a subject and comprised all forms of arms, including firearms. *Agney-astra* (or *agni-astra*)is the term commonly used in the *Vedas* to denote fire weapons. The *Rāmāyaṇa* of Vālmīki and the *Mahābhārata* mention devices that can be assumed to have material comparable to modern gunpowder.[100] A popular book on the art of warfare was written by Śukra with title, *Nīti-Śāstra* or *Śukranīti*, at the beginning of the Christian era and was first translated by Gustav Salomon Oppert (1836 - 1908) in 1880.[101] Since Śukra is a popular name, it is not certain if we are dealing with the same book and the same person, or not. The title of the book is made up of two terms: Śukra is the name of the author and *nīti* means rule. Thus, the title signifies the rule of Śukra. Oppert also wrote a book, *On the Weapons, Army Organisation and Political Maxims of the Ancient Hindus*, that primarily dealt with the history of firearms in

[98]Oppert, 1967, p. 46. Perhaps Oppert is alluding to a library that is about 27 miles NW of Madrid.

[99]Pacey, p. 47

[100]Kokatnur, 1948 and Oppert, 1967. Oppert's work has been published several times by different publishers with added authors. However, the contents have remained the same.

[101]It is doubtful that the book was written during the first century of the Christian era. However, there are no such studies on this manuscript to my knowledge. This book is mentioned in *Mahābhārata* and *Pañcatantra*. However, it is not in the context of gunpowder. There may have been another book with the same title. Some suggest that the book was written around the ninth century, or even the sixteenth century. However, all speculations are tentative at this stage.

India.[102]

Gunpowder has been known in India as well as China for a long period of time, so that "Who invented it first?" is a difficult question to answer. In the popular literature, after the pioneer work of Joseph Needham was published, gunpowder is generally accepted as the Chinese invention. Joseph Needham mostly taught at Gonville and Caius College in Cambridge. He collaborated with a few Chinese scholars, received support from the British and Chinese governments, and compiled a monument work, *Science and Civilisation in China,* in 24 volumes over a period of about five decades. There is nothing comparable to Needham's work that is done on India. As a result, about a century ago, most books also included India in their list for the invention of gunpowder. However, even with the absence of comparative studies on this topic, due to the massive literature produced on the existence of gunpowder in China during the last fifty years, the credit mostly goes to China.

We cannot deal with the issue of priority between India and China in this brief section in the absence of comparative scholarships on this issue. The purpose of this section is not to deny China its credit; the purpose is to promote scholarship on the existence of this knowledge in India. For this reason, I have decided to share the past knowledge which is yet to be proven wrong. It is established that the Hindus did not use gunpowder in cannons or guns prior to the Chinese or Europeans. However, it is likely that the Hindus knew of the explosive property of saltpeter and may even had used it as the so-called bombs at the tip of their arrows.

Apollonius of Tyana was a Pythagorean philosopher who visited India during the first century A.D.. He explained the reason as to why Alexander the Great, Hercules, and Dyonysus decided to return from their invasion of India. Apollonius, citing the case of Hercules and Dyonysus, explained that the Indian sages, "instead of taking the field against them, lay quiet and passive, as it seemed to the enemy; but as soon as the latter approached they were driven off by rockets of fire and thunderbolts which were hurled obliquely from above and fell upon them."[103] This is clearly an indication of

---

[102]Oppert lived in Madras and worked as a professor of Sanskrit in Presidency College there. His main job was to translate Telgu and Sanskrit documents for the British government. This book contains a chapter on explosives and their delivery devices.

[103]Philostratus, 1960, vol. 1, p. 205.

some kind of projectiles that were used as firearms. According to Apollonius, Alexander did write about these projectiles in his letters to Aristotle.[104] A passage from Quintus Curtius seems to indicate the existence of missiles, with attached explosives, that were used against the army of Alexander the Great.[105]

Following is an excerpt from the fourth book, seventh section of the *Śukranīti* that provides a recipe for gunpowder:[106] "According as a new weapon and missile varies in its size, whether it is small or large, in its shape or blade, experts name it differently. The tubular weapon should be known as being of two kinds, divided into large and small. The tube is five spans long, its breech has a perpendicular gun and horizontal hole, at the breech and muzzle is always fixed a sesambead for aligning the sights. The breech has at the vent a mechanism which, carrying stone and powder, makes fire by striking. Its breech is well wooded at the side, in the middle is a hole an *aṅgula* broad. After the gunpowder[107] is placed inside, it is firmly pressed down with a ramrod. This is the small gun which ought to be carried by foot-soldiers. In proportion as its outside (bark) is hard, its hole is broad, its ball is long and broad; the ball reaches far. . . . Five weights (pala) of saltpeter, one weight of sulfur, one weight of charcoal . . . and is prepared in such a manner that the smoke does not escape. If all this is taken after having been cleansed, is then powdered and mixed together, one should squeeze it with the juice of *calatropis gigantea*, *Euphorbia neriifolia* and *allium sativum* and dry in the sun; having ground this like sugar, it will certainly become gunpowder."

Śukra tells us that the gunpowder is made in many ways and of white or other colors, depending on the mixing of other substances to make it toxic. "One should clean the tube first and then put gunpowder, carry it down with the ramrod to the bottom of the tube till it is tight. then put a good ball, and place gunpowder at the vent, and by setting fire to the gunpowder at the vent discharge the ball towards its mark."[108]

---

[104]Oppert, 1967, p. 54.
[105]McCrindle, 1896, p. 189 and 205.
[106]Oppert, 1967, p. 105 - 107.
[107]*agni-cūrṇa* is the Sanskrit word for gunpowder, meaning fire mixture.
[108]Oppert, 1967, 108.

For about 75% saltpeter, 12 % sulfur, and 13 % carbon, the following is the reaction:

$$2\ KNO_3 + 3\ C + S = K_2S + 3\ CO_2 + N_2 + \text{about 500 kJ of energy}$$

Similarly, ammonium nitrate, a common fertilizer, with 10 to 20% charcoal, can explode in the reaction:

$$2\ NH_4NO_3 + C = CO_2 + 4\ H_2O + 2\ N_2 + \text{energy}$$

As mentioned above, the main ingredient of gunpowder is saltpeter. Its chemical name is potassium nitrate ($KNO_3$). It is scarce and mostly found naturally in India, Egypt, and America. It is found as a natural deposit in the sanitation pipes in tropical climatic conditions since the urine of livestock have potassium nitrate. Today ammonium nitrate, a similar derivative of potassium nitrate, is commonly used as a fertilizer. This makes fertilizer a deadly weapon.

Potassium nitrate was commonly used as a fertilizer and was used in medicine as tonic. In ayurveda, it is suggested for people suffering from indigestion, stomach ache and constipation.[109] In India, it was long known as *khāri-miṭṭī*, found near the Kutch region and was used for making fireworks from the earliest periods. Alum, known as *sphaṭikārī* in Sanskrit and popularly called *phiṭakarī* in Hindi, was also in use from the earliest period. Similarly, sulfur, another ingredient of gunpowder is native to India. As mentioned earlier, the word "sulfur" is Sanskrit in origin. Charcoal, the third ingredient, is commonly available in most lands.

Saltpeter is defined by Suśruta[110] as *yavakṣāra*. Elsewhere, it is defined as "inflammatory as fire" and helpful in curing piles and enlarged spleen and anti-spermatopoietic.[111] Since salt is commonly used in cooking, and usually placed in the close vicinity of fire for convenience, its explosive properties were bound to be discovered due to an accident. This view is also supported

[109] Oppert, 1967, p. 60
[110] *Suśruta-Saṁhitā, Sūtrasthāna*, 42: 19.
[111] *Suśruta-Saṁhitā, Sūtrasthāna*, 46: 323.

by William Wellington Greener.[112] Greener summarized that "there seems little doubt that the composition of gunpowder has been known in the East from times of dimmest antiquity. The Chinese and the Hindus contemporary with Moses are thought to have known of even the more recondite properties of the compound."[113] In a similar tone, John Timbs (1801 - 1875 A.D.), an English antiquarian and prolific writer, in his quest to define inventions and discoveries of the past, concluded that "[g]unpowder has been known in China as well as in Hindoostan [land of the Hindus] far beyond all periods of investigation.[114]

Johann Beckmann[115] (1739 - 1811), a professor in the University of Göttingen wrote a book, in 1780, *Beiträge zur Geschichte der Erfindungen.* This book was first translated into English in 1797 with the title, *History of Inventions, Discoveries, and Origins.* Beckman strongly supported the opinion that gunpowder was invented in India: "In a word, I am more than ever inclined to accede to the opinion of those who believe that gunpowder was invented in India, and brought by the Saracens from Africa to the Europeans, who however improved the preparation of it, and found out different ways of employing it in war, as well as small arms and cannon. In no country could saltpeter, and the various uses of it, be more easily discovered than in India, where the soil is so rich in nitrous particles that nothing is necessary but to lixiviate it in order to obtain saltpeter; and where this substance is so abundant, that almost all the gunpowder used in different wars with which the sovereigns of Europe have tormented mankind was made from Indian saltpeter."[116] In 1837, J. F. Royle, professor of materia medica and therapeutics at King's College, London suggested that the invention of gunpowder and fireworks

---

[112]Greener, William Wellington, The Gun and Its Development, Cassell, 1907, p. 14 (first published in 1881). Greener was a prominent gun-maker and studied the history of gunpowder and guns extensively. His book became a classic with numerous editions. His company, W. W. Greener Company, is still in existence.

[113]Greener, 1907, p. 13.

[114]Timbs, 1860, p. 43.

[115]Beckmann is credited for coining the term "technology." He is definitely among the first to teach technology as a subject and write books on it.

[116]Beckmann, Johann, 1846, vol. 2, p. 505 - 506. In a footnote, Beckmann mentioned a paper, written by M. Langles in 1798, that claimed the knowledge of gunpowder came to Arabia from India. In 690 A.D., it was employed at the battle in Mecca. Greener (1907, p. 14) has also mentioned the use of gunpowder at the siege of Mecca. I have changed the spelling saltpetre, used by Beckmann, to saltpeter.

"has often been assigned" to the ancient Hindus.[117]

On the use of gunpowder and the fact that saltpeter and other ingredients were present naturally in India, Beckmann writes, "But if it be true that this salt was known at a much earlier period in India, it is not improbable that both gunpowder and aquafortis were used by the Indians and the Arabians before they were employed by the Europeans, especially as the former were the first teachers of chemistry to the latter. In my opinion, what I have already related proves this in regard to gunpowder; and what I shall here add will afford an equal proof in regard to aqufortis."[118]

Beckmann describes the saltpeter trade that existed during his time. "But experience has shown that all the means and coercive measures hitherto employed have rendered the European saltpeter much dearer than that obtained by commerce from Bengal. This will be readily comprehended, when it is known that earth richly impregnated with salpetre [saltpeter] abounds in India, and that it may be extracted by lixiviation with any addition, and brought to crystallize in that warm climate without the aid of fire; that the price of labor there is exceeding low; that this salt is brought from India instead of ballast by all the commercial nations of Europe, where the competition of the sellers prevents the price from ever being extravagantly high, while the preparation of it in Europe, in consequence of the still increasing price of labour [labor], fuel and ashes, is always becoming dearer."[119]

The Chinese text, *Chin Shih Pu Wu Chiu Shu Chüeh* (Explanation of the Inventory of Metals and Minerals according to the Numbers Five and Nine), tells us that a Sogdian Buddhist monk[120] Chih Fa-Lin visited China from Samarkand. He brought Sanskrit books with him for translation in 664 A.D.. This was a standard practice for the Buddhist monks visiting China to take Sanskrit books with them and translated them into Chinese for the general Chinese population. These books were in great demand in China. Chih Fa-Lin knew the explosive properties of saltpeter. When he reached the

[117]Royle, 1837, p. 46.

[118]Beckmann, 1846, p. 506; aquafortis is another word for the nitric acid.

[119]Beckmann, 1846, p. 511.

[120]Sogdian means a person from Sogdiana, a province in ancient Persia which is presently in the region of Samarkand (longitude, 67° E; latitude, 39 - 42° N). This region is above the Hindukush region and was a cultural center for Buddhism during the period.

Ling-Shih district in Fên-chou, he noticed that the land had natural deposits of saltpeter and suggested its use for fireworks or firearms: "This place must be full of saltpeter, why isn't it collected and put to use?" However, he collected saltpeter and found it not suitable for fireworks. Later, they went to Tsê-chou and collected saltpeter. This time it was of good quality and they tested it by burning and it produced "copius purple flames." The Sogdian monk said: "This is a marvelous substance which can produce changes in the five metals, and when the various minerals are brought into contact with it they completely transmuted into liquid form."[121] This is perhaps the obvious use of saltpeter as nitric acid and as a flux in smelting.

Let us analyze the above description. Chih Fa-Lin was a Buddhist monk from ancient Persia, presently Samarkand, who took a large number of Sanskrit books to China for translation and also introduced gunpowder to China. Well, from this statement it is clear that gunpowder was introduced to China by a Buddhist monk and not a Chinese invention. It was Buddhist knowledge, that was known to the people who were a part of the greater Indian empire, spreading the Indian knowledge abroad. The ingredients for gunpowder were plentifully available in India. All that was needed was a spark that could cause an explosion. As mentioned before, prior to Needham's work, India was considered, by experts on gunpowder or experts in the history of science, to be the country where gunpowder was invented. The old scholarship that was established prior to Needham's work was not refuted. However, with the popularity of Needham's work, China started getting the credit with the invention of gunpowder.

## 7.5 Fermentation

The *Ṛgveda* mentions fermentation of barley, a key ingredient in beer making these days: "A mixture of a thick juice of soma with barley powder."[122] In another place, there is a mention of fermented alcoholic drink which took about 15 days of processing. "Fifteenth day old highly intoxicating soma,"[123] which probably refers to the broth fermented in the vat for 15 days.

[121] Needham, 1981, p. 29

[122] *Ṛgveda* IX: 68: 4.

[123] *Ṛgveda* X: 27: 2.

The *Yajurveda-Sukla* tells us about various kinds of alcoholic drinks:[124] "Âtithya's sign is Mâsara, the Gharm's symbol *Nagnahu.* Three nights with *Surâ* poured, this is the symbol of the *Upasads.* Emblem of purchased Soma is Parisrut, foaming drink effused." Elsewhere in the same book, the fermentation process is mentioned. "Like shuttle through the loom the steady ferment mixes the red juice with the foaming spirit."[125] Here *Nagnahu* is root of a plant that is used as yeast, Parisrut is a kind of beer, and *Surâ* is another word for liquor.

The *Chāndogya-Upaniṣad* tells us that drinking liquor on a regular basis is as bad as stealing gold, killing a Brahmin, or having an affair with your teacher's wife.[126] Elsewhere, Aśvapati Kaikeya tells his visitors that "no one drinks" in his kingdom.[127]

Caraka mentions some 84 different kinds of alcoholic liquors.[128] For the fermentation, Caraka mentions nine sources of sugar: sugarcane juice, guḍa (jaggery), molasses, honey, coconut water, sweet palmyra sap and *mahua* flowers. Also some sweet fruits such as grape, date, mango, banana, apricot, jackfruit, rose-apple, *jāmun*, pomegranate, *kādamba*, *bilva*, etc. are used in the fermentation. Also, rice and barley were used from the grain category for fermentation. By the time of Kauṭilaya, the superintendent of liquor was designated who supervised the manufacturing of liquor and controlled liquor traffic in and out of village boundaries.[129] Liquor shops were required to have beds and seats and the rooms were filled with the scents of flowers. Some recipes for liquor making are also provided: "Medaka is manufactured with one droṇa (measure of capacity, volume) of water, half an āḍhaka (unit of mass) of rice, and three prasthas (mass) of *kiṇva* (ferment)."[130]

Some ferments were used for medicinal purposes. Kauṭilaya suggested

---

[124] *Yajurveda-Sukla*, 19: 13 -15.

[125] *Yajurveda-Sukla*, 19: 83.

[126] "A man who steals gold, drinks liquor, and kill a Brahmin; A man who fornicates with his teacher's wife – these four will fall." *Chāndogya-Upaniṣad*, 5: 10: 9.

[127] *Chāndogya-Upaniṣad*, 5: 11: 5.

[128] Achaya, 1991.

[129] *Arthaśāstra*, 119; Shamasastri, 1960. p. 131.

[130] *Arthaśāstra*, 120; Book 3, Chapter 25; Shamasastri, 1960. p. 133.

that patients should learn the preparation of these *ariṣṭa* (fermented and distilled liquor and medicine) from physicians. He has even provided several recipes: "one hundred pala of *kapittha* (*Feronia Elephantum*), 500 pala (mass) of sugar, and one prastha (mass) of honey form *āsava*."[131] Some hard liquors with rice and lentils were also suggested: "one droṇa of either boiled or unboiled paste of *māsa* (*Phraseolus Radiatus*), three parts more of rice, and one karṣa of *morata* (*Alangium salviifolium*) and the like form *kiṇva* (ferment)."[132] Additives were mixed in order to improve the taste or appearance of liquors. These additives were used sweeteners, spices, and astringents.[133]

## 7.6 *Rasāyan-śāstra* as Alchemy

During the medieval period, the practice of chemistry was in the hands of craftsmen who used it in pottery making, soldering, tempering material, etc. The other place for the growth of chemistry was in the hands of mystics who used it to impress people by demonstrating the mystical powers of their religion. These mystics also used chemistry or pharmaceutical chemistry in the healing of people. Many kings hired mystics to enhance their longevity. These healers are today commonly called alchemists and their art is known as alchemy.[134]

The word alchemy has evolved from the Arabic word *al-kīmyā*, indicating transmutation. Alchemy is related to chemistry in the same way as astrology is related to astronomy. The present developments in astronomy started with the ancient belief that the heavenly bodies influence our lives. For this reason the ancient natural philosophers studied the motion of the planets. Similarly, the search for the elixir of life, the potable gold, and transmutation of elements from one form to another made a few advances in alchemy which in turn grew into the present form of chemistry, biochemistry, and medical chemistry.

In chemistry, the transmutation of one compound to another is common.

[131] *Arthaśāstra*, 120; Book 2, Chapter 25; Shamasastri, 1960. p. 132.
[132] *Arthaśāstra*, 120; Shamasastri, 1960. p. 133.
[133] For more information, read Achaya, 1991; Prakash, 1961; and Singh *et al*, 2010.
[134] Dubs, 1947; Sheppard, 1972.

Chemists conduct such transmutations in their laboratory experiments on a regular basis. However, alchemists even tried transmutation of the elements which cannot be accomplished even today using simple chemical techniques. Of course, they were unsuccessful. However, there were many anecdotal reports of such transmutations, with no real evidence provided. Today, only high-energy accelerators can accomplish such tasks, albeit with a limited success rate.

The origin of alchemy cannot be assigned to any particular culture or civilization. The need to prolong life was important to all cultures. Therefore, alchemy was bound to prosper in all cultures. Modern alchemy appeared in the Roman world around the middle of the twelfth century. This happened when Robert of Ketton (ca. 1110 - 1160 A.D.) translated an Arabic book into Latin, *De Compositione alchemiae of Morienus.*[135] Roger Bacon (1214 - 1294 A.D.), in his *The Opus Majus,* mentioned the use of alchemy to manufacture gold from lead and manufacture of silver from copper.[136] This knowledge was known to Aristotle as shared by Roger Bacon. Aristotle shared with Alexander the Great this "greatest secret, for not only would it procure an advantage for the state and for everyone his desire because of the sufficiency of gold, but what is infinitely more, it would prolong life."[137] The direct use of alchemy in prolonging life prompted Albert Magnus (ca. 1193 - 1280 A.D.), also known as Albert the Great, to compare alchemists to physicians – as alchemists first clean and purify old metal just as doctors employ emetics and diaphoretics to purge his patients.[138] He provided eight suggestions to all alchemists:[139]

1. Be reserved and silent.
2. Work in a remote private house.
3. Choose your working hours prudently.
4. Be patient, watchful, and tenacious.
5. Work on a fixed plan.

---

[135]Newman, 1989.
[136]Bacon, 1928, p. 626 and 627.
[137]Bacon, 1928, p. 627.
[138]Magnus, 1967, p. 178 and 179.
[139]Doberer, 1948.

6. Use only glass or glazed earthenware crucibles.

7. You must be rich enough to pay for your experiments.

8. Have nothing to do with princes and nobles.

These eight suggestions can be understood from the fact that alchemy entered into Europe as an Islamic art. Most universities in the West did not offer courses in alchemy until the early seventeenth century.[140] This explains the reason that several chemical terms are Arabic in origin (for example, alkali, alcohol, alembic, etc.). A backlash from Christianity was almost inevitable in view of the conflict between these two religions. Pope John XXII issued a decree in 1317 A.D. forbidding the use of alchemy. Excerpts of the decree are as follows: "Alchemists are here prohibited and those who practise [practice] them or procure their being done are punished. They must forfeit to the public treasury for the benefit of the poor as much genuine gold and silver as they have manufactured of the false or unadulterated metal. If they have not sufficient means for this, the penalty may be changed to another at the discretion of the judge, and they shall be deprived of any benefices that they hold and be declared incapable of holding others."[141] St. Thomas Aquinas (1225 - 1274 A.D.) even mentioned alchemists in the same breath as demons.[142] In 1323 A.D., the situation worsened and Pope John XXII issued another decree stating that all clergies who practiced alchemy would be excommunicated. King Henry IV issued a statute to forbid alchemy and its practice as a felony.[143] This statute was repealed in 1689 due to Robert Boyle's influence in the British Parliament.[144]

As alchemy entered into Europe as an Islamic art, some scholars consider alchemy to be of Islamic origin. On the Islamic origin of Indian alchemy, Nasr, a Persian scholar, writes: "The early historians of alchemy such as von Lipmann believed that alchemy was brought to India by Muslims. Later research has unveiled references to alchemy ante-dating the rise of Islam, and it is now conceded by most scholars that Indian alchemy has a much more ancient history than imagined until now and that it is in fact closely related

[140]Newman, 1989.
[141]Stillman, 1960, p. 274; the full text is provided by Walsh, 1912, p. 125 - 126.
[142]Newman, 1989.
[143]Geoghegan, 1957.
[144]Geoghegan, 1957; More, 1941.

to certain forms of Yoga, especially Tantrism, with which it became closely associated later to the extent that the power to perform alchemical transmutation came to be considered as one of the siddhis of yogis."[145]

*Rasāyana* is the rejuvenation therapy in ayurveda which was practiced to prolong life. This therapy is used to replenish the *rasa* and other *dhātu* of the body.[146] Caraka explains the purpose of *rasāyana* in the following words: "Long life, excellent memory and intelligence, freedom from disease, a healthy glow, good complexion, a deep powerful voice, strong bodily and sensory powers, and beauty can be obtained from *rasāyana*."[147]

Alchemists, with the help of chemicals, tried to prolong life and cure diseases. This allowed doctors to learn a way to heal themselves and to not look for teleological aids. This simply contradicted those who consider diseases as a curse of God. As Carl Jung wrote: "The alchemist carries the idea of the limitation a stage further and reaches the conclusion we mentioned earlier, namely that complete assimilation to the Redeemer would enable him, the assimilated, to continue the work of redemption in the depths of his own psyche. This conclusion is unconscious, and consequently the alchemist never feels impelled to assume that Christ is doing the work in him."[148] Jung continues, "He [alchemist] would then have had to recognize to only himself as the analogue of Christ, but Christ as a symbol of the self. This tremendous conclusion failed to dawn on the medieval mind. What seems like a monstrous idea to European Christendom would have been self-evident to the spirit of *Upaniṣads*."[149]

The ancient Hindus practiced alchemy, as in the other parts of the world, and used metal, minerals, gems and jewels to treat obstinate incurable diseases. Al-Bīrūnī provides the accounts of alchemical practices of the Hindus.[150] "They [the Hindus] have a science similar to alchemy which is quite peculiar to them. They call it *Rasāyana*, a word composed with *rasa*, i.e. gold. It means an art which is restricted to certain operations, drugs, and

[145] Mahdihassan, 1979, p. III.
[146] *Caraka-Saṁhitā*, *Cikitsāstānam*, 1: 5
[147] *Caraka-Saṁhitā*, *Cikitsāstānam*, 1: 6 - 7.
[148] Jung, *Collected Works*, vol. 12, p. 339.
[149] Jung, *Collected Works*, vol. 12, p. 341.
[150] Sachau, vol. 1, p. 187 - 193.

compound medicines, most of which are taken from plants. Its principles restore the health of those who are ill beyond hope, and give back youth to fading old age, so that people become again what they were in the age near puberty; white hair becomes black again, the keenness of the senses is restored as well as the capacity for juvenile agility, and even for cohabitation, and the life of people in this world is even extended to a long period."[151]

Alchemy evolved as an offshoot of herbalism; ayurvedic medicines were prepared with herbs and metals.The dependency of human life on plants was recognized during the *Ṛgvedic* period. Plant life provided food, shelter and medicine. A large number of hymns were composed in the *Vedas* praising plants. The ancient Hindus tried to make an "elixir" to stop, delay, or reverse the process of aging. In the *Vedas*, the drink "soma" is frequently mentioned as a tonic to rejuvenate a person and also to slow down the aging process. The method of preparation of "soma" is not presently known. The *Atharvaveda* suggests: "Invest this Soma for long life, invest him for great hearing power."[152]

C. J. Thompson, in his book *The Lure and Romance of Alchemy*, writes: "In India the earliest allusions to alchemical ideas appear in the Atharva Veda (*Atharvaveda*), where mention is made of the gold which is born from fire."[153] The *Ṛgveda* mentions *surā* as an intoxicating drink that was used along with soma. Today, *surā* is a term used in India for alcoholic drinks. The *Arthaśāstra* of Kauṭilaya gives recipes for fermented alcoholic drinks made from rice (similar to Sake, popular in Japan), sugarcane (similar to rum), grapes (similar to wine), and various spices. Suśruta wrote a complete chapter on the elixirs for rejuvenation of humans and suggested recipes for people to become immune to disease and decay.[154]

*Cyavanaprāśa* is a very popular tonic for rejuvenation in India. The name has stemmed from the name of Cayavana *ṛṣi* (seers) who rejuvenated his body with the help of Aśvins, two Vedic doctors. This story is mentioned at several places in the *Ṛgveda*.[155] People in northern India take *Cyavanaprāśa* as a

[151]Sachau, vol. 1, p. 188 - 189.
[152]*Atharvaveda* 29: 24: 3.
[153]Thompson, 1932, p. 54.
[154]*Suśruta-Saṁhitā, Cikitsāstānam*, 27: 1.
[155]*Ṛgveda*, 5: 74: 5 and 6; *Ṛgveda*, 7: 71: 5.

precautionary measure particularly in the winter season against the common cold and cold-related symptoms. The use of *Cyavanaprāśa* is fairly ancient as Caraka has referred to it for rejuvenation.[156]

Caraka, a famous physician from ancient India, provided several recipes for making iron tonics. The process of making these elixirs is known as the *killing of metal* in the Sanskrit language, and the final product is called *bhasma*. To accomplish the killing of iron, Caraka recommended the use of fine thin plates of iron with *āmlā* (a fruit) extract and honey for one year in an underground pot; it reduces iron to a ferrous compound. Also, Caraka mentioned the conversion of yellow gold into a red colloidal form using plant extract, before being used as a tonic. Many of these reactions were done to make the metals nontoxic. Suśruta devoted one whole chapter on the use of alkalis (*kṣāra*) in various diseases and noticed its corroding, digestive, mucous destroying, and virile potency destroying properties.[157]

---

[156] *Caraka-Saṁhitā, Cikitsāstānam,* 1: 74.
[157] *Suśruta-Saṁhitā, Sūtrasthānam,* Chapter 11.

# Chapter 8

# Biology

In the *Caraka-Saṁhitā*, an early Hindu treatise on medicine, biology was considered the most important of all sciences. "The science relating to life is regarded by the philosophers as the most meritorious of all the sciences because it teaches mankind what constitutes their good in both the worlds [spiritual and physical]."[1] Good health helps people to fulfil their four purposes in life: *arth* (prosperity), *kāma* (sensuality), *dharma* (duty), and *mokṣa* (Salvation).[2]

The ancient Hindus systematically studied various life forms and noticed inter-dependencies and commonness in them. It was due to these commonness between plants and animals that led Suśruta to suggest that new medical practitioners dissect plants before they performed any dissection on animals or humans. About 739 plants and over 250 animals are mentioned in the ancient literature of the Hindus.[3] The 24th chapter of *Yajurveda* has a large variety of birds, animals, and snakes mentioned. Plants and animals are characterized with respect to their utility, habitat, or some other special features: Four-footed, reptiles, claws, born from an embryonic sac, born from an egg, born from sprouts (plants), or domesticated.[4] For example, the botanical world was divided into grasses, creepers, shrubs, herbs, and trees.[5] The

[1] *Caraka-Saṁhitā*, *Sūtrasthānam*, 1: 43.
[2] *Caraka-Saṁhitā*, *Sūtrasthānam*, 1: 15.
[3] Kapil, 1970.
[4] Smith, 1991; This is a fairly extensive article.
[5] Kapil, 1970; Smith, 1991.

*Ṛgveda* mentions the heart, lung, stomach, and kidneys. The *Atharvaveda*[6] refers to the heart as a "lotus with nine gates," a correct description of heart when held with its apex upwards. From this top view, one can observe nine openings in the heart: three in right atrium, four in the left atrium, and one in each of the right and left ventricles.[7] Similarly, perhaps there is a mention of blood circulation in the Atharvaveda:[8] "Who stored in him floods moving in all diverse directions and formed to flow in rivers pink, rosy red, and coppery dark running in all ways in a man, upward and downward."

## 8.1 Sacred Rivers and Mountains: Ecological Perspectives

Ecology is a branch of science that deals with the relations between organisms and their environments. In essence, it deals with the interaction between biological and physical processes. Ecology was an important part of the spiritual and materialistic worldview of the ancient Hindus. In their view, the mighty Earth is a living system that has several billion years of experience in developing processes that are sustainable and cyclic. The survival of plants and animals is dependent on these processes. Human beings are no exception either. The respect for animals prompted Yudhiṣṭhira to refuse to enter in heaven without his dog. It was their respect for the ecosystem that led the ancient Hindus to promote vegetarianism. Trees, mountains, rivers, and animals are worshipped by the ancient Hindus. Cutting trees is considered to be a sin, dumping waste products in a river is considered as a sin, and killing animals is also considered as a sin. It is a common phrase in India that the God lives in every particle of the universe. (*kaṇa kaṇa mei Bhagwān hai*). The *Bhagavad-Gītā*[9] suggests the omnipresence of God. Therefore, in the worldview of the Hindus, the ecology should be preserved and respected.

Ecology (*paryāvaraṇa*) is primarily considered as a branch of biology that deals with the relationships of organisms with their environment. Earlier, the main study under this discipline covered the environment or surroundings

---

[6] *Atharvaveda*, 10: 8: 43.
[7] Narayana, 1995; Rajgopal et al, 2002.
[8] *Atharvaveda*, 10: 2: 11.
[9] *Bhagavad-Gītā*, 7: 19; 13:13.

around the habitat. However, due to the far reaching effects of our actions, the discipline covers the whole Earth these days. The relationship of organisms to the environment, and with humans in particular, deals with issues that includes the domains of chemistry and physics too. Therefore, ecology is studied in physics and chemistry departments in many universities. In the context of this book, the relationships of flora, fauna, and water with the human actions is our concern.

The ancient Hindus considered nature (*prakṛti*) as a living sacred entity. Nature is not a property that is owned by the humans for their use; on the contrary, humans are a part of nature. Nature is represented by its five forces: *ākāśa* (space or sky), *vāyu* (air), teja (fire), ap (water) and *pṛthivī* (earth) – popularly known as *pañcā-bhūta* (five elements). All animate and inanimate objects in nature are made up of these five elements, including humans and animals. The ancient Hindus realized that they could neither control nature nor overpower its order; they just tried to live with it in a manner that was based on respect and appreciation for natural forces and order. Rivers, mountains, and oceans were paid reverence. It was not ethical to dump waste into a river or deplete a mountain of its trees. The ecosystem connects living beings with inanimate objects. For example, a little misuse of water, an inanimate object, has dire consequences to life forms who drink it. Ayurveda is the science of maintaining optimal balance of *pañcā-bhūta* in the human body for it to maintain vigor and good health.[10] The *Caraka-Saṁhita* tells us that the human (*puruṣa*) is equal to the universe. Whatever elements that form the universe also form a person and vice versa.[11]

The ancient Hindus considered the geographical region of the Indus Valley and the surrounding area as sacred. It was only natural for them to preserve the ecology of the region. All this allowed the ancient Hindus to develop ecological ethics that allowed interdependencies. "There is no fundamental difference between ourselves and others; both are under girded by the common substrate known as Brahman itself. This ethos signals deep concern for harmony among life forms and can be used to advocate a minimal consumption of natural resources," suggests Christopher Chappell.[12] It

[10]Roy, 2009

[11]*Caraka-Saṁhita, śarīrasthānam*, 4.13

[12]Christopher Key Chapple, Hindu Environmentalism: Traditional and Contemporary Resources, in the book by Tucker and Grim, 1994.

is this minimal consumption and health concerns, as written in the Hindu scriptures, that called for people to avoid overeating.

The interdependence between different life forms is a major area of activity for scientists. However, in the context of Hinduism, many topics are still not explored, as suggested by Gavin Van Horn.[13] He cites the relationship between depictions of human-animal divinities, treatment of animals, and moral consideration for animals as examples of these studies. David Gosling, a nuclear physicist who has studied South Asian ecology, suggests that the Hindu traditions serve as an important model in "raising social and environmental awareness, underscoring the continuities between past and present and their possible transformations within an environmental paradigm."[14]

Every day, hundreds of thousands of people visit Ganges [Gaṅgā] or Yamunā rivers and worship them. These rivers are considered to be purifiers of sins, and bathing in the sacred waters is a popular activity which is followed by a prayer. In the evenings, group *āratī* (prayer) is performed where people worship God by worshipping the river.

The Earth is worshipped as mother; it is considered to be a *devī* and has numerous names: *Bhūmi*, *Pṛthivī*, *Vasudha*, and *Vasundharā*. Even today, India is personified as Bhārat-mātā (mother India). The Earth is considered as mother and the inhabitants as her children, as suggested in *Atharvaveda*.[15] The *Atharvaveda* has a whole chapter in the praise of the Mother Earth. This chapter has a prayer for the Mother Earth to support people of all races and nations and remain fertile, arable and nourisher of all.[16] This verse does not discriminate between "us" and "them"; it includes everyone. Another verse in the same book suggests that the Earth caters to people of different religions and languages.[17]. A request is made for Mother Earth to provide medicines, medicinal plants and prosperity,[18] and prestige and prosperity.[19]

---

[13]Van Horn, 2006

[14]Gosling, 2001, taken from Van Horn, 2006.

[15]*Atharvaveda*, 12: 1: 12.

[16]*Atharvaveda*, 12: 1: 11

[17]*Atharvaveda*, 12: 1: 45

[18]*Atharvaveda*, 12: 1: 17 and 27

[19]*Atharvaveda*, 12: 1: 63

The practice of *bhūmi-pūjā* (prayer to the Earth) is fairly common among the Hindus at the time of a new construction, a new plantation, and marriage. O. P Dwivedi, in summarizing this chapter of *Atharvaveda*, wrote:[20] "These sentiments denote the deep bond between Earth and human beings, and exemplify the true relationship between the Earth and all living beings, as well as between humans and other forms of life. Such a comprehensive exposition depicting a theory of eco-spirituality as portrayed by this Hymn to the Mother Earth is not found in ay other religious tradition and beliefs. The Sukta [verse] provides us a guide to behave in an appropriate manner towards nature and our duty towards the environment."

## 8.2 Sacred *Tulsī* and Sacred Cow

Attitudes and behavior toward animals are interwoven with the intricate fabric of Hindu society which believes in the sanctity of all life forms. The *Vedas*, *Upaniṣads*, *Purāṇa*, *Mahābhārata*, and *Rāmāyaṇa* contain multitudes of stories related to plants and animals. In these scriptures, the turtle, mouse, snake, eagle, lion, elephant, and many other animals and birds live in harmony with humans.

Trees and plants are chemical factories. These "factories" are efficient and produce products of high quality for the human consumption. Human survival is dependent on the quality of the foods given by trees and plants. It is only natural to respect plant life and worship it.

The respect that the ancient Hindus bestowed on plants can be observed by noticing stone deities placed under many trees with flags and steamers adorning the branches. The mythological *kalpavṛkṣa*, a tree that fulfils all human desires, is well known in Hindu tradition. In their popular belief, each tree has a Vṛkṣa-devatā, or tree-deity, who is worshipped by the Hindus with prayers and offerings of water, flower, sweets, and encircled by sacred thread. The sacred thread symbolizes the wishes of the praying person. People are expected to untie the knot after their desire is fulfilled. This creates a unique problem as trees in Hindu temples have a multitude of such knots that can range into the thousands. It becomes impossible to recognize the exact knot

[20]Dwivedi, 1997

of an individual.

Plants or trees are used in most sacred ceremonies. Some are even worshipped and have God-like powers. For instance, in the *Mahābhārata*, Sakra says: "Rubbed with the astringent powder of the hanging roots of the Banyan tree (*Ficus bengalensis*) and anointed with the oil of *priyango* (*panicum italicum*), one should eat the *shashlika* paddy mixed with milk, By so doing one gets cleansed of all sins."[21] The *Padma-Purāṇa* advocated that planting trees was a simple way to reach *nirvāṇa*, to escape the cycle of birth and death.[22]

Since the Hindus realized the common connection between plants, animals and humans, and the importance of trees for human life, it became a religious code to preserve plant life. Aśoka (reigned 272 - 232 B.C.) issued a decree against the burning of forests.[23] It was a hunting practice during the period to set fire to grass and trees. As the animals got scared by the fire and ran away to protect themselves, they were trapped and killed by the hunters. Thus, Aśoka's decree simultaneously took care of plants as well as animals. This is in contrast to some cultures who treat the nature as a property that is waiting for exploitation and looting rather than as a sacred entity. As a result, plant and animal species are disappearing at an alarming rate.

*Tulsī* (*ocimum sanctum*) is perhaps the most sacred plant in Hindu households. It is like a small shrub from 30 - 60 cm in height. *Tulsī* is worshipped by the Hindus as a goddess, just like Lakṣmī, Sītā, or Rādhā. It is common for the Hindus to worship a *tulsī* plant in the morning and feed it water while facing the sun. When the Hindus die, people put *Tulsī* leaves, mixed with Ganges (Gaṅgā) water, in the mouth of the dead body for purification. In most *pujas* (Hindu prayer), a nectar made from milk, honey, curd, and *tulsī*, called *charṇāmrt* (meaning, nectar from the feet, the nectar having been used to wash idols before people drink it), is given to all participants.[24]

There is a general belief that *rudrākṣa* beads are helpful in cardiac ailments. People who suffer from high blood pressure wear jewelry made from

[21] *Mahābhārata*, *Anuśāsana Parva*, 10.

[22] *Padma-Purāṇa*, *Sṛṣṭi-Kāṇḍa*, Chapter 28: 19 - 22.

[23] Smith, 1964, p. 187.

[24] For the myths associated with *tulsī*, read Gupta, 1971, p. 74 - 82.

*rudrākṣa.*[25] According to famous folklore, Pārvatī, consort of Lord Śiva, wanted to wear jewelry to improve her feminine charms. As Lord Śiva was oblivious to such practices, he gave her the beads of *rudrākṣa* (*elaeocarpus ganitrus*) to make necklaces, bangles, and earrings.

The following is a list of some trees and their corresponding deities: Goddess Lakṣmī in *tulasī* (*Ocinlum sanctum*), Goddess Śītalā in *neem*, God Viṣṇu in *pīpal* (*Ficus religiosa*), and Lord Śiva in *Vata* (*Ficus indica*).[26] Similarly, in other interpretations, *tulasī* is beloved of Lord Kṛṣṇa; *pīpal* is inhabited by Brahmā, God Viṣṇu, and Lord Śiva; *aśoka* (*Saraca indica*) is dedicated to God Kāma; *palāś*, a tree, to the moon; *bakula* (*Mimusops elangi*) to Lord Kṛṣṇa; *rudrākṣa* (Eleaecarpus ganitrus) to Lord Śiva, and bananas to all married females. Worshipping banana plants ensures that the husband would survive his wife. This plant signifies prosperity and fertility. Most houses are decorated with banana plants during marriage ceremonies or other *pujas.*[27] "The Hindu spiritual heritage can provide new ways of valuing, thinking, and acting that are needed if respect for the nature is to be achieved and future ecological disasters are to be averted."[28]

In a Hindu wedding, the groom, his family members, and family friends generally go the bride's house in a procession. This is a joyous procession with people singing and dancing. However, in the midst of partying, they first go to a *pipal* or banyan tree and pray. Similarly, the bride and groom visit a tree as a couple and pray before they enter in their house after marriage. The sacred areas of most Hindu temples are surrounded by trees.

During the 1970s, Mr. Sunderlal Bahaguna started the *chipko* movement in India to protect trees. Any time developers came to a forest with their crews to cut trees, the local people hugged the trees to prevent the developers from cutting down these trees. The movement was created to protect the Himalayas from "deforestation and consequent erosion and landslides" and was pitted against the government and businesses, two powerful entities.[29] Although Bahaguna was the leader, it was mostly women who participated

[25]Gupta, Shakti, 1971, p. 39.

[26]Dwivedi, 1997.

[27]Bhatla, Mukherjee, and Singh, 1984, has provided a review on this topic.

[28]Dwivedi, 1997

[29]Robinson and Cush, 1997.

in the real act of *chipko* (meaning, to stick or hug). Females came out in large numbers from the nearby villages to protect mother earth and established this highly successful movement. The success of this movement gave it a worldwide recognition.

Feeding weak and old animals is considered a charity act. Like the household dogs and cats in the Western world, all animals were bestowed a proper treatment by the Hindus.[30] When Hindus eat food, they separate food for birds, dogs, and cows from their own plate. The first *chapāti* usually goes to a cow or dog in many Hindu households.[31] This is not to check the food for poison since it is given to the animal many times after the person has eaten his/her food.

The sanctity of the cow in Hinduism is well known and much publicized in the West. In legends and popular stories, the Earth is said to assume the figure of a cow, especially in times of distress, to implore the help of the gods. The sacredness of the cow represents an anomaly to social scientists who wonder why in a country with hungry population the cow is virtually untouched.[32] The *Mahābhārata* tells us that the killing of a cow is wrong and has its bad consequences and causes a person to go to hell.[33] The relationship between the cow and man is not competitive, but symbiotic. The cow augments the needs of human beings by providing them milk, bullocks, dung, and hides. Judging the value of Indian cattle by the western standards is inappropriate.[34] Due to inherent differences in the value systems, such an evaluation generally leads to confusion or misleading conclusions.

---

[30]Narayanan, 2001.

[31]Most Hindu families do not keep dogs as pets. Usually, each street has a few stray dogs that are patronized by the families living on that street. These dogs recognize people who live there and protect the street from strangers at night. The first *chapati* that was made by my mother in our household was always given to a cow. I was told as a child that this action was necessary to help my future. It did not make much sense to me as a child. However, I understand and appreciate it now. It was an effective way to ensure the safety and the protection of animals. Mahatama Gandhi used to say that the kindness of a culture can be judged by the way they treat animals. For more information, read Hammett, 1938 and 1950; Smith, 1991.

[32]Weizman, 1974.

[33]"All that kill, eat, and permit the slaughter of cows rot in hell for as many years as there are hairs on the body of the cow so slain." (*Mahābhārata*, 13: 74: 4)

[34]Harris, The Myth of the Sacred Cow, in Leeds and Vayda, 1965.

The ancient Hindus, in their analysis, observed a balance in *prakṛti* (primal nature); every species plays an important role in the balance. This balance of *prakṛti* was based on the knowledge gained through empiricism, intuition, and experimentation. Once a common connection between plants, animals, and humans was established, the coexistence of all living forms followed. Hanumāna (Monkey God), and Gaṇapati (Elephant God, or Gaṇeśa), as well as trees, are revered widely by the Hindus.[35]

## 8.3 Life in Plants

The story of human life on the Earth starts from the story of plants – our photosynthesizing kin. The story of plant evolution must remain central to any narrative on human life. It was, after all, a mass conversion of energy into biomass, using photosynthesis, that allowed life on Earth. Without the energy derived from photosynthesis by plants, animals and humans would not exist. The use of plants has been at the root of our dietary and medical practices; plants provide the nutrients essential to animals and humans.

In terms of their general dependence, humans need animals and plants, animals need plants, and plants do not need much help either from animals or humans.[36] Plants came before animals and humans on Earth and are likely to outlast both. In the event of global warming or air and water pollution, plants are likely to survive longer than the animals and humans. Plants may be the last survivors.

Earth has about 400,000 plant species that display a wide range of morphology (body form) and anatomy (internal cell structure). Like animals, plants have organs too; we call them stem, root, leaf, and flower. Because sunlight is the main energy resource for plants, they do not need a complex

[35]Mythological stories are associated with these gods. It may be demeaning to some that the Hindus gods have some animal attributes. However, most Hindus do not even think about that. Crowds in temples all over the India on Tuesday testify their reverence to Lord Hanumāna. Similarly, it is common for people to worship Lord Gaṇeśa) first on any auspicious occasion. For more information, see Nelson, Lance [Editor], 1998; and Chapple and Tucker [Editors], 2000; Roy, 2005.

[36]Abrahamson, 1989.

anatomy to hunt, chase, and digest food. A mere orientation or tallness can take care of their energy needs.

Animals and humans exhale carbon dioxide and inhale oxygen from the atmosphere during respiration. The importance of oxygen cannot be emphasized enough; respiration is a continual process and a mere three to four minutes of the absence of oxygen to the brain can put a person into a coma and eventual death. In reverse, plants release oxygen and consume carbon dioxide. A balance of oxygen and carbon dioxide is vital for the survival of all life forms.

The Human Genome Code brings us closer to animals and plants than ever before. Scientists have been searching for the "universal ancestor," the organism that created all life forms, including plants and animals. This search for the "universal ancestor" is inspired by the common connections between various life forms. Plants provide nutrition to their whole body like animals. Plants produce heat from their own metabolism and melt snow around them or entice insects to their flowers. Some plants hibernate like animals; they become dormant during cold periods. Humans do not have such power of hibernation although some *yogis* claim to have this power where they can apparently stop their breathing and slow down their vital organs.

The temperature regulation mechanism of animals provides the warmth that the cells need to carry out biochemical reactions efficiently. What about plants? As plants lack lungs, a circulatory system and hormones that step up respiration and circulation in animals, it is mystifying the ways they achieve this heat producing capability.[37]

Plants produce heat and even alter their heat production rate to maintain their body temperature at levels surprisingly constant in changing temperature conditions. Some plants produce extraordinary heat when they bloom, as much as for their weight as birds or insects. In scientific tests, a *Philodendrom selloum*'s temperature was found to be 38°C (100°F) when the environmental temperature was just 4°C (39°F). In this plant, a 125 gram spadix produces about nine watts of heat to maintain a temperature of 110°F. Skunk cabbage (*symplocarpus foetidus*) holds its temperature between 15 to

[37]For more information, Seymore and Schultze-Motel, 1996; Seymore et al, 1983.

20°C (59 - 72°F) for at least two weeks in cold weather during February and March, with snow melting around it. The sacred lotus (*nelumbo nucifera*) maintained its temperature near 32°C (90°F) for 2 - 4 days while the outside temperature dropped to 10°C (50°F).[38]

There are no neurons or pain conduits known in plants. However, plants "often act and react to various stimuli as effectively – and in some cases as quickly – as animals."[39] Plants even communicate with their neighbors by releasing a body odor that turns on the defensive mechanism of the nearby plants. As in animals, plant cells are hotbeds of electrical activity. As a duck bites a mouth full of *chara*, a relative of the green algae, it triggers an electrical mechanism in the plant, and ionic current rushes across the membrane of the nibbled cell. The protoplasm quickly jells at the periphery, preventing any leakage of cell fluid.[40] All plants respond to external stimuli: they respond to gravity and light. Most reorient their leaves and flowers for the maximum absorption of light. Similarly, most plants control their growth pattern and direction to have a stable configuration.

Aspirin, like in humans, somehow inhibits the production of jasmonic acid in plants which dampens down "pain."[41] In tomato seedlings, a mechanical wound induces electrical activity that causes the accumulation of proteins that limit further damage to the plant. Also, tomato leaves produce a hormone that, when eaten, blocks digestive enzymes in the insect's gut. The insect then has greater difficulty in digesting the leaves. Similarly, orchids produce needle-like crystals of calcium oxalate, the material that can cause kidney stones in humans. This is done to avoid insects.

Some plants produce padding so most pests do not bother them. Others prefer to poison their predators. Some of them are "intelligent" enough to play a political game of survival: the enemy of my enemy is my friend. These plants do not go one on one with their enemy; they find *someone else* to do this for them. For example, corn seedlings are vulnerable to attack by caterpillars. In response, as soon as the oral secretion of a caterpillar comes in contact with the damaged tissues of the plant, corn plants release airborne

[38]Seymore, 1997.
[39]Clarence Ryan in the article by Weiss, 1995.
[40]Wayne, 1993.
[41]Milius, 1998a.

N-(17-hydroxylinolenoy)-L-glutamine chemicals that attract parasitic wasps which are natural enemy of caterpillars.[42] These parasitic wasps lay their eggs in the caterpillars, and the wasp larvae eat the caterpillar.[43]

Plants release fragrant perfume as a sophisticated tool to entice pollinators, discourage microbes and fend off predators.[44] Some plants are carnivorous. For example, venus flytraps attract insects by secreting a type of nectar. Once insects are trapped, the plant digests them. Also, the sundew plant traps insects, secreting an enzyme to digest the insect.[45] Scientists even believe that plants care for their siblings as they provide obvious benefits to their seeds. Candace Galen and Anna Wied of the University of Missouri at Columbia have studied this relationship on high mountain slopes.[46] In some cases, the mother plant "packs a lunch" for their offspring in the nutrient-rich endosperm tissue inside a seed.

It was the similarities between animals and plants that prompted Jean Baptiste de Monet de Lamarck (1744 - 1829 A.D.) to suggest that plants and animals should be studied in one discipline. Perhaps he was the first person to unify botany and zoology under the single banner of biology.

The *Manu-Saṁhitā* suggests that plants are susceptible to pain and pleasure. "There are various kinds of shrubs and bushy plants, and various kinds of weeds and grass, creepers and trailing plants, some of which grow from seeds and others from graft. Variously enshrouded by the quality of *tamas* (ignorance), the effects of their own acts, they retain their consciousness inward, susceptible to pleasure and pain."[47]

The *Mahābhārata* mentions a discourse between two monks, Bhṛgu and Bhārdvāja, on the epistemology of plants. Bhṛgu asks: "Sir, all plant life and animals are guided by the same five primal elements (*pañcabhūta*). But I don't see it in the plants. Plants do not possess body heat, do not move their parts, remain at one place, and do not seem to have the five elements.

---

[42]Alborn, 1997.
[43]Hoffman, 1992.
[44]Pichersky, 2004.
[45]Wayne, 1993.
[46]Milius, 1998b.
[47]*Manu-Saṁhitā*, 1: 48 - 49.

Plants neither hear, nor see, nor smell, nor taste. They cannot feel the touch of others. Why then you call them a component of five elements? They do not seem to possess any liquid material in them, heat of body, any earth, any wind, and any empty space. How then can plants be considered a product of five elements?"[48]

To such questions, Bhārdvāja replied: "Though trees are fixed and solid, they possess space within them. The blooming of the fruits and flowers regularly take place in them. Body heat of plants is responsible for the dropping of flower, fruit, bark, leaves from them. They sicken and dry up. This proves the sense of touch in plants. It has been seen that sound of fast wind, fire, lightening affect plants. It indicates that plants must have a sense of hearing. Climbers twine the tree from all sides and grows to the top of it. How can one proceed ahead unless it has sight? Plants therefore must have the vision. Diseased plants may be cured by specific fumigation. This proves that plants possess sense of smell and breathe. Trees drink water from their roots. They also catch diseases from contaminated water. These diseases are cured by quality water. This shows that they have a perception of taste. As we suck water through a tube, so the plants take water through their roots under the action of air in the atmosphere. Trees when cut produce new shoots, they are favored or troubled by certain factors. So they are sensitive and living. Trees take in water through roots. Air and heat combine with water to form various material. Digesting of the food allows them to grow whereas some food is also stored."[49]

A close analysis of the above lengthy discourse between Bhrgu and Bhārdvāja depicts that the Hindus believed in life in plants, metabolism in plants, a need of food in plants, diseases in plants, and possible cures of diseases in plants using medicine. All these properties are also peculiar to animals and humans. The above quotation clearly indicates the Hindus' belief that plants are sensitive to sound, touch, and quality of irrigated-water just like humans.

The *Ṛgveda* mentions the hearing qualities in plants: "All plants that hear this speech, and those that have departed away, Come all assembled

[48] *Mahābhārata*, *Śānti-Parva*, 184: 6 - 18.
[49] *Mahābhārata*, *Śānti-Parva*, 184: 10 - 18.

and confer your healing power upon this herb."[50] Thus, though plants do not have organs like humans, they do have similar functions – they breathe, eat, sleep, listen, reproduce, and die similar to humans.

The *Bṛhadāraṇyaka-Upaniṣad* compares a man to a tree: "A man is indeed like a mighty tree; his hairs are like his leaves and his skin is its outer bark. The blood flows from the skin [of a man], so does the sap from the skin [of a tree]. Thus blood flows from a wounded man in the same manner as sap from the tree that is struck. His flesh [corresponds to what is] within the inner bark, his nerves are as though the inner fibers [of a tree]. His bones lie behind his flesh as the wood lies behind the soft tissues. Marrow [of a human bone] resembles with the pith [of a tree]."[51]

The *Chāndogya-Upaniṣad* recognized the presence of life in plants. "Of this great tree, my dear, if someone should strike at the root, it would bleed, but still live. If someone should strike its middle, it would bleed, but still lives. If someone should strike at its top, it would bleed, but still live. Being pervaded in *ātman* [soul or self], it continues to stand, eagerly drinking in moisture and rejoicing. If the life leaves the one branch of it, then it dries up. It leaves a second; it dries up . . . Verily, indeed, when life has left it, this body dies."[52]

Advocacy of life in plants in the Hindu literature persuaded Sir J. C. Bose (1858 - 1937 A.D.) to study the life essences in plants. Though Bose was trained in physics and had little training in botany, he ventured into the discipline and made valuable contributions. He demonstrated life essences in plants with his experiments during the late nineteenth century. His contributions are:

1. *Power of response* (reflex action)
   In a mimosa plant, if a single leaf is touched, all the leaves in the branch slowly close. Bose devised an instrument to study such a behavior in other plants. He concluded that all plants, like human beings, possess reflex actions.

---

[50] *Ṛgveda*, 10: 97: 21
[51] *Bṛhadāraṇyaka-Upaniṣad*, 3: 9: 28.
[52] *Chāndogya-Upaniṣad*, 6: 12: 1 - 3.

2. *Food Habits*
   Bose demonstrated that plants need food, like humans; nourishment of a plant comes from its roots as well as from its body. Bose experimentally demonstrated that, like humans, plants nourish themselves from a distribution of nutrients through a continuous cycle of contraction and expansion of cells. If a poison is administered to a plant, it may succumb to it. On the removal of poison, it may gradually revive and become normal. Plants also sway abnormally like a drunk person if treated with an alcoholic substance and become normal when the cause is removed.

3. *Mind Activity*
   A sun-flower plant adjusts itself throughout the day so that it always faces the sun. Many other plants close their leaves during the night and turn to face a brighter light source if the light is not uniformly distributed, indicating a mind activity in plants.

4. *Nerves*
   Plants also have nerves and feel sensation through leaves, branches, or trunks. If certain areas of a plant are made numb with ice, less pain is felt in the plant.

A popular science magazine, *Scientific American*, summarized the work of J. C. Bose as follows: "By a remarkable series of experiments conducted with instruments of unimaginable delicacy, the Indian scientist [J. C. Bose] has discovered that plants have a nervous system . . . With the possibilities of Dr. Bose's crescograph, in less than quarter of an hour the action of fertilizers, foods, electric currents and various stimulants can be fully determined."[53]

## 8.4 Biomimicry

Biomimicry is a relatively new term coined for the studies of nature's actions and its emulation/reproduction in solving human problems. Nature has fine-tuned herself for millions of years. Its processes are efficient and biomimicry

[53] *Scientific American*, April 1915.

allow us to learn these processes and turn them into human benefits or modern technologies. These uses range from agriculture, energy production, and from making medicine to making strong fiber and other materials, faster computers, new machine designs, and even business.[54]

Plants and animals deal with the environment every instant of their existence like human beings. As Darwin suggested, survival of the fittest is an important rule in nature. Thus, plants and animals demonstrate a variety of unique behaviors to survive. With careful observations, we can learn these behaviors and use them to our benefit. This is the science of biomimicry.[55]

The natural world synthesizes more materials with excellent qualities that can only make a designer, a manufacturer, a scientist, and an entrepreneur envious. Most pharmaceutical companies use plant products in their syntheses. The emphasis on biomimicry is not to control nature or exploit it; the emphasis is to imitate the nature. Thus, there is little harm to natural processes. Nature becomes a mentor, a model, or a teacher, and not the exploited one.

Velcro is an example of biomimicry and the term is derived from the French words: velvet (*Velour*) and hook (*crochet*). It was patented by Georges De Mestral in 1951 in Switzerland. Velcro mimics cockleburs which are round seedpods with a prickly surface that can attach to fabrics.[56] Similarly, the Japanese bullet train rail car was designed after the Kingfisher bird. This is one of the best fishing birds because of its aerodynamic beak that helps with their streamline transition from air to water.[57]

The concept of biomimicry is ancient although the term is new. It is so new that many dictionaries do not even list it yet. However, it is a new discipline that studies nature's best products and processes and imitates these to solve human problems. Thus, by observing an insect or animal, the ancient Hindus basically studied their natural world and themselves. The connections of various life forms was central to the philosophies of the ancient Hindus.

---

[54]Helms *et al*, 2009; Rosemond and Anderson, 2003; Wilson *et al*, 2010.

[55]Allen, 2010; Benyus, 1997; Kolb, 1998; Reed, 2003; and Tangley, 2000.

[56]DeYoung and Hobbs, 2009, p. 150.

[57]DeYoung and Hobbs, 2009. p. 64.

*Pīpal*, *neem*, and banyan trees along with peacocks, bears, lions, elephants, cats, dogs, and crocodiles became guides of the Hindus. Various herbal medicines, surgical instruments, and the science of yoga evolved from such studies of animals, plants and tribal people. The *Caraka-Saṁhitā* suggests the physicians that they contact tribal people to learn about the herbs.[58] The tribal people learn the therapeutic properties of herbs by observing animal life. Modern pharmaceutical industries invest resources to learn from tribal knowledge. Herbal medicines were selected by observing animals' diets and their physical conditions. Several hymns of *Atharvaveda* indicate that some herbal remedies were discovered by observing the actions of snakes, swans, cows, and sheep.[59]

A large number of *āsana* (postures) in yoga are named after the animals. (See Chapter 9) Even today, at a time when modern science has evolved so much, it is common to study medicinal plants by studying the insects they attract. Plants do not advertise their potential medicines; the insects that feed on these plants do. For example, aposematic insects (insects with bright colors) allow scientists to select medicinal plants. The aposematic butterfly and the diurnal moth species sequester unpalatable or toxic substances from their host plant as a defense mechanism against predators.[60]

Suśruta designed surgical instruments by observing the hunting techniques of birds and animals. Most animals have their own distinct way of grabbing and biting which provides a unique stability in specific situations. This allows them to use their abilities for hunting and eating. The examples of a crocodile's grab with its jaws and the grab of a swan with its beak are two useful applications that were used in designing surgical instruments. It is no coincidence that the surgical instruments described by Suśruta have animal names. Some had the beak of a bird, strong grip of a crocodile, or biting grab of a lion. (Chapter 10)

---

58 *Caraka-Saṁhitā*, *Sūtrasthānam*, 1: 121.
59 *Atharvaveda*, 8: 7: 22 - 28.
60 Helson, *et al*, 2008 and Nishida, 2002.

## 8.5 Metempsychosis; Kinship of All Life Forms

The concept of metempsychosis tells us that a soul is immortal and transmigrates or reincarnates from one life form to another – a fundamental tenet of Hinduism.[61] Our deeds of this life decide our future in the next life. Therefore, rebirth is associated with *karma* (action) that provides a thrust towards minimizing evil, inequality, and suffering.

There is nothing comparable to the soul in modern science. However, in most religions and cultures, the existence of the soul is central to their philosophies. Modern science deals with life as a sequence including inception, growth, decay, and death. It does not deal with life from one form to another. Although not in tune with the practices of modern science, I have decided to include this topic because it governs the biological views of the ancient Hindus. This worldview allowed them to treat plants and animals as sacred; it allowed them to regulate their eating habits and became vegetarians; and it allowed them to be tolerant and respectful of others and their religions.

The Hindu doctrine of transmigration of the soul considers life as a several-step-process and not just a single step. This essential aspect of the Hindu religion is well documented in the *Vedas* and *Upaniṣads*, and also curiously appears later in Greek thought. Pythagoras, Socrates, Empedocles, Orpheus, Plato, Plotinus, Apollonius, and other Pythagorean philosophers believed in the transmigration of the soul and the immortality of the soul.

In the Greek literature, to emphasize compassionate and merciful treatment of animals, Pythagoras suggested that the dog you beat may very well be a reincarnated friend or relative.[62] Pausianias (fl. ca 175 A.D.), a Greek traveler and geographer, wrote that the concept of metempsychosis was proposed by Indian and Chaledaean philosophers before it was introduced in Greek philosophy. He wrote: "I know that the Chaledaeans and Indian sages were the first to say that the soul of man is immortal, and have been followed by some to the Greeks, particularly by Plato, the son of Ariston."[63]

---

[61] *Atharvaveda*, 12: 2: 52.

[62] Kumar, Pythagoras, an article in *Cultural Encyclopedia of Vegetarianism*, by Margaret Puskar-Pasewicz, 2010.

[63] Pausianias, *Messenia*, 32: 4.

Plato, in his support of the theory of transmigration of soul, suggested that "there is a law of Destiny, that the soul which attains any vision of truth in company with a god is preserved from harm until the next period, and if attaining always is always unharmed. But when she is unable to follow, and fails to behold the truth, and through some ill-hap sinks beneath the double load of forgetfulness and vice, and her wings fall and she drops to the ground, then the law ordains that this soul shall at first birth pass, not into any other animal, but only into man . . . Then thousand years must elapse before the soul of each one can return to the place whence she came."[64]

In its support of the theory of reincarnation, Plato considered human knowledge as mere remembering or recollection.[65] Since we existed before, the humans mostly rediscover the prior knowledge we inherited.

"The idea that human souls enter animals after the death of the man is held in a simple form by a great many primitive peoples all over the world. But these peoples do not seem to think further than the existence which is to succeed the present one. They do not develop elaborate speculative systems which account for the soul's progress through many thousands of years. That is something found, outside Greece, only in India; and it is found there in a form strikingly like the Pythagorean," suggests West.[66]

The theory of the transmigration of souls helps to explain birth and death, it accounts for diversity among humans, and a hope of betterment of all. It is a moral philosophy of cause and effect that prompts people to become productive members of society. It unites various classes of a society as the lower class works to improve their status in their next life while the upper class is expected to treat the lower classes with kindness as it was the morally right action to ensure and maintain the same high status in their next life. People viewed difficulties in the present life as a result of their personal deeds and not the outcome of social exploitations. In Hindu society, people first look within to find solutions to their difficulties and do not focus on finding fault for their miseries in other people. People could seek the ultimate of their

---

[64] Plato, *Phaedrus*, 248 - 249.

[65] *Phaedo*, 72; *Phaedrus*, 247 - 249; *Theatetus*, 149 - 151.

[66] West, 1971, p. 61.

abilities for growth through the generations, to attain salvation. All human beings were treated with equal respect by the ancient Hindus. Of course, they defined specific roles to different people based on their traits.

Metempsychosis is explained in the *Bhagavad-Gītā*, in the sermon of Lord Kṛṣṇa to persuade Arjuna to fight against his cousins in the battle of *Mahābhārata*. "O Arjuna, both you and I have had many births before this; only I know them all, while you do not. Birth is inevitably followed by death, and death by rebirth. As a man casting off worn-out garments and takes on new ones, so does the dweller in the body, casting off worn out bodies, enter into others that are new."[67]

The doctrine of the transmigration of the soul also played an important role in the growth of science among the ancient Hindus. The idea of rebirth occupied a central role in Hindu philosophy. Our future lives are led by the actions in this life, as defined in the Law of *Karma*. According to the doctrine of *Karma*, every action, whether good or bad, has its own consequences. In this doctrine, even fate is the result of actions in prior births. Our actions lead us to innumerable life forms, from a god to a cow, insect, or even a plant. This made the ancient Hindus cognizant of every act, big or small; it also made the ancient Hindus cognizant of all life forms, big or small. The *Bṛhadāraṇyaka-Upaniṣad* suggests that "According as one acts, according as one conducts himself, so does he become. The doer of good becomes good. The doer of evil becomes evil. One becomes virtuous by virtuous action, bad by bad action."[68] The *Śvetāsvetara-Upaniṣad* tells us that our present life is an outcome of our actions in the previous lives.[69]

The following are a few excerpts from the *Upaniṣads*, defining the philosophy of metempsychosis:

"The wise one [i.e. the soul, the atman, the self] is not born, nor dies. This one has not come from anywhere, has not become anyone. Unborn, constant, eternal, primeval, this one is not slain when the body is slain. If the slayer think to slay, if the slain think of himself slain, Both these under-

[67] *Bhagavad-Gītā*, 2: 11 - 13.
[68] *Bṛhadāraṇyaka-Upaniṣad*, 4: 4: 5.
[69] *Śvetāsvetara-Upaniṣad*, 5: 11-12.

stand not. This one slays not, nor is slain."[70]

"According unto his deeds the embodied one successively assumes various forms. The embodied one chooses forms according to his own qualities. . . . Those who know Him [God], have left the body behind."[71]

"Either as a worm, or as a moth, or as a fish, or as a bird, or as snake, or as a tiger, or as a person, or as some other in this or that condition, he is born again here according to his deeds, according to his knowledge."[72]

"People are born with bodies of various kinds, or they become various fixed things – according to *karma*, according to knowledge."[73]

"Accordingly, those who are of pleasant conduct here – the prospect is, indeed, that they will enter a pleasant womb, either the womb of a *Brāhmaṇa*, or the womb of a *Kṣatriya*, or the womb of a *Vaiśa*. But those who are of stinking conduct here – the prospect is, indeed, that they will enter a stinking womb, either the womb of a dog, or the womb of a swine, or the womb of an outcast [*caṇḍāla*]."[74]

The *Upaniṣads* have prescribed methods to avoid the cycle of transmigration. As births and deaths continue until a person attains the knowledge of the Imperishable [God], knowing the God is the state that takes a person to a state of contentment or desirelessness. Such a state is called *mokṣa*–a person goes beyond the cycle of birth and death: "But they who seek the *ātman* [self] by austerity, chastity, faith and knowledge . . . they do not return [no rebirth]."[75]

Al-Bīrūnī suggests that the Hindus believed in a common life force between human, animals, and plants. "They [Hindus] consider the wandering about in plants and animals as a lower stage, where a man dwells for punishment for a certain length of time, which is thought to correspond to the

[70] *Kathā-Upaniṣad*, 2: 18, 19.
[71] *Śvetāsvetara-Upaniṣad*, 5: 11 - 14
[72] *Kaushītaki-Upaniṣad*, 1: 2.
[73] *Kathā-Upaniṣad*, 5: 7.
[74] *Chāndogya-Upaniṣad*, 5: 10: 7.
[75] *Praśna-Upaniṣad*, 1: 10.

wretched deeds he has done."[76] In the above quotation "the wandering" references the Hindu belief in metempsychosis (transmigration of soul). According to Hindu beliefs, all living forms have a soul that transcends from one living form to another. Based on our actions in one life, after death, we are born in another form of life. A human being can transcend into a dog, a cow, or a tree. Similar views were also proposed by Pythagoras and other Pythagorean philosophers in Greece.

As society is a collection of a large number of connected souls, the mass uplift of souls became equally important as the uplift of an individual. Thinking this way, the whole world becomes a single family (*Vasudhaiva kuṭumbakam*). This family included all life forms, not just humans. It is for this reason, it is a common prayer for the Hindus to chant: "May everyone be happy, may everyone be free of diseases, may everyone see what is noble, may no one suffer from misery." (*"Sarve bhavantu sukhinaḥ, sarve santu nirāmayā sarve bhadrāṇi paśyantu, ma kaschita dukha bhāg bhaveet."* )

Many Gods in the Hindu religion were closely associated with other life forms: Lord Śiva with snakes and bulls, Lord Rāma with monkeys, Lord Gaṇeṣa with mice, Goddess Durgā with lions, Lord Kṛṣṇa with cows. The *Srimad-Bhāgvatam* tells us that "one should look upon deer, monkeys, donkeys, rats, reptiles, birds, and flies as though they were their own children."[77] Similarly, The *Bhagavad-Gītā* tells us that the person of knowledge "sees no difference between a learned Brahmin, a cow, an elephant, a dog, or an outcast."[78]

The transmigration of soul theory of the Hindus unites one generation to another, one individual to another, one culture to another, and one life form to another. This theory implies that all human beings are created equal, as they can go from one level to another with their deeds. Various life forms are just the various evolutionary stages and are connected among themselves. A monkey, cow, dog, or human finds a common connection between themselves. They are just one great family of life-forms. On one level, this is similar to Darwin's theory of evolution where various forms evolved from each other

[76] Sachau, vol. 1, 1964, p. 61.
[77] *Srimad-Bhāgvatam*, 7: 14: 9
[78] *Bhagavad-Gītā*, 5: 18

and are connected. The Darwin theory of evolution leads us to believe that all life forms evolved from a single source. Even plants and animals have common ancestors. In this sense, all lives are connect with each other, as also advocated by the ancient Hindus.

## 8.6 *Ahiṁsā* and Vegetarianism

The following diets are generally considered under the vegetarian categories: Lacto-Ovo (milk-egg) vegetarians don't eat any fish or meat, but do eat eggs, milk, and other dairy products such as cheese, butter, and yogurt. Lacto vegetarians eat dairy products, but not eggs, fish, or meat. Vegans do not consume any animal products whatsoever.[79] Most of the ancient Hindus were lacto-vegetarian where they used milk, curd, and cheese, along with cereals, lentils, beans, and vegetables.

People become vegetarians for many reasons including religious or cultural traditions and customs, health, societal trends, economic, or moral (against the killing of animals and animal cruelty) reasons. As nutritional science expanded from the mid-20th century, vegetarianism acquired general recognition as a healthful dietary alternative. Pamela Anderson, Paul McCartney, George Harrison, Ellen DeGeneres, Steve Jobs, George Bernard Shaw, Leo Tolstoy, Mark Twain, Henry David Thoreau, Leonardo da Vinci, Apollonius of Tyana, Porphyry, Confucius, Socrates, and Buddha are some of the well-known vegetarians.

Animals have associated with humans from the earliest period. Dogs serve humans by watching out for their welfare, cows give milk to humans, sheep give wool to help us survive in the winter season, horses allow us to move far distances, elephants fight for humans, bees polinate plants that provide fruits, etc.

The doctrine of *ahiṁsa* supports the viewpoint that all life forms, includ-

[79]The word Vegan was coined in 1944 by Elsie Shrigley and Watson, the founders of the UK Vegan Society. Both of them were against the consumption of eggs and dairy products by the vegetarians. They took the first three letters of the word vegetarian and combined with the last two letters (Veg + an = Vegan)

ing plants and animals, are equal, and possess souls. The term *ahiṁsa* means nonviolence. *Ahiṁsa* (non-violence) is one of the *yamas* (moral rules or codes of conduct) of Patañjali.[80] "Non-harming, truthfulness, non-stealing, chastity and greedlessness are the restraints."[81] Vegetarianism is a way to practice *ahiṁsa*. The purchaser of flesh performs *hiṁsa* (violence) by his wealth; he who eats flesh does so by enjoying its taste; the killer does *hiṁsa* by actually tying and killing the animal. Thus there are three forms of killing. He who brings flesh or sends for it, he who cuts of the limbs of an animal and he who purchases, sells, or cooks flesh and eats it – all of these persons are to be considered meat-eaters. The *Manu-Saṁhitā* tells us that people who give consent to killing, who dismember a living body, who actually kill, who purchase or sell meat, who purify it, who serve it, and who eat the meat are all sinners.[82]

The Hindus enforced sanctity for the animals by associating them with gods and goddesses. Swans are associated with Sarasvatī, mice with Lord Ganesha, snakes and bulls with Lord Śiva, lions with the Goddess Durgā, Lord Kṛṣṇa with snakes and cows, and monkeys with Lord Rāma. Cows are worshipped, so are snakes, and many other life-forms.

In India today, cereals and lentils are perhaps the most common food. Cereals are turned into flour and made into bread, common known as *roti* or *chapati*. Wheat and rice are the most common cereals. A large variety of lentils, peas, or beans are cooked almost daily in a Hindu household. The combination of roti, rice, and dal provide most of the amino acids. Milk is the main source of protein. Ghee (refined butter) is the preferred cooking medium and yogurt is commonly eaten during lunch time.

The Hindu religion prescribed a vegetarian diet for humans; killing animal for food was prohibited. It is a curious connection that almost all Pythagorean philosophers were also vegetarian; Pythagoras, Empedocles, Socrates, Porphyry, Plato, Plotinus, Apollonius, Plutarch, Clement of Alexandria, and several others advocated vegetarianism.[83] Diogenes Laertius mentioned an incidence in which Socrates told his wife, Xanthippe, not to be

---

[80] *Patañjali's Yoga-Sūtras*, 2: 30.
[81] Patañjali's *Yoga-Sūtra*, 2: 30; Feuerstein, p. 80.
[82] *Manu-Saṁhitā*, 5: 51 - 52.
[83] Dombrowski, 1984a and 1984b.

ashamed of serving vegetarian meals to a rich man visiting their house. According to Socrates, serving vegetables was not due to financial reasons, meat being somewhat more expensive than vegetables, but due to a philosophical conviction.[84]

In Plato's *Republic*, Socrates suggested a model city where all food needs were fulfilled by vegetarian foods.[85] Plato suggested several food items to eat: barley and wheat loafs, roots and herbs, salt, olives, cheese, onion, greens, desserts of figs, chick-peas, beans, berries, and acorns. Plato considers vegetarian food as the foods of "health and peace." Here "peace" indicates a peace with animals.[86]

Plutarch argued that by nature the human body is not designed to digest meat. He points out that human beings do not have the bodily traits of predatory animals. According to him, if it were natural for us to eat meat, our physical body would be designed to catch and kill animals without the help of a knife or other such weapon.[87]

Apollonius of Tyana (fl. 1st century A.D., a well known Pythagorean philosopher) promoted vegetarianism with the following arguments: "Earth grows everything for mankind, emperor, and those who are willing to live at peace with the animals need nothing: they can gather with their hands or with the plough depending on the seasons, what the earth produces to feed her sons. But some . . . disobey the earth and sharpen knives against animals to gain clothing and food."[88]

In the Hindi language, the word *ācar-vicār* (*ācar* = conduct, *vicār* = thought) signifies the connection of mind and physical body. Here conduct is connected to diet also. Diet and the state of mind are closely related. In India, *ācar-vicār* is the combination of words that is used to describe the personality of a person. The *Vaiśeṣika-Sūtra*, a holy book of the Hindus suggests that "improper diet instigate violence."[89] Caraka suggested that

---

[84]Diogenes Laertius, 2: 34.

[85]*The Republic*, 369d- 373e.

[86]For the eating habits of Plato, see Dombrowski, 1984a.

[87]Regan and Singer, 1976, p. 113.

[88]Philostratus, 1970, p. 212.

[89]*Vaiśeṣika-Sūtra*, 6: 1: 7.

obesity as well as bullemia are wrong and can cause sicknesses. A person must be vigilent of his/her weight and other organs.[90]

The ancient Hindus, in their astute observation of *prakṛti* (nature) noticed a distinct role of each life-form and the importance of the "balance" in *prakṛti* was realized: "A man who does no violence to anything obtains, effortlessly, what he thinks about, what he does, and what he takes delight in. You can never get meat without violence to creatures with the breath of life, and the killing of creatures with the breath of life does not get you to heaven; therefore, you should not eat meat. Anyone who looks carefully at the source of meat, and at the tying up and slaughter of embodied creatures, should turn back from eating any meat, " suggests the *Manu-Saṁhitā*.[91]

Bhīṣma explains to Yudhiṣṭhira that the meat of animals is like the flesh of one's own son, and a person who eats meat is considered a violent human being.[92] The *Mahābhārata*[93] also suggests that "dharma exists for the general welfare of all living beings; hence by which the welfare of all living creature is sustained, that is real dharma."

Most religions advocate the so-called Golden Rule of Morality. In the *Bible*, this rule is provided in *Matthew* (22: 37 - 40) where people are instructed to treat other people like they want themselves to be treated. For the Hindus, they extended the Golden Rule of Morality even to animals and plants – implying no injury to any living being was allowed.

Aśoka (reigned 272 - 232 B.C.), grandson of Chandragupta (Candragupta, reigned 324 - 300 B.C.), tried to sanctify animal life as one of his cardinal doctrines. The Greek and Aramaic edicts in Kandhar in Afghanistan show Aśoka's resolute resolve against animal killing. "The King abstains from the slaughter of living beings, and other people including the king's hunters and fishermen have given up hunting. And those who could not control themselves have now ceased not to control themselves as far as they could. . ."[94]

---

[90] *Caraka-Saṁhitā*, *Sūtrasthānam*, 21: 15 - 17.

[91] *Manu-Saṁhitā*, 5: 47 - 49.

[92] *Mahābhārata*, *Anuśāsana Parva*, 114: 11.

[93] Śānti Parva, 109: 10

[94] Sircar, 1957, p. 45.

Aśoka issued the following decree as known in Rampurva text: "Those she-goats, ewes (adult female sheep) and sows (adult female pig), which are either pregnant or milch, are not to be slaughtered, nor their young ones which are less than six months old. Cocks are not to be caponed. Husks containing living beings should not be burnt. Forests must not be burnt either uselessly or in order to destroy living beings."[95] These were the moral codes, not only the administrative one for ruling. Kinship with nature became a basis of their kinship with God.

Hindu nonviolence made such an impact on the Muslim King Jahāngīr (r. 1605 - 1627 A.D.) that, in 1618 A.D., he took a vow of nonviolence. In 1622 A.D., this vow was broken when he had to pick up his gun to save his own life and his reign against his own son Khurram. This was the period when rulers used cruelty against animals and humans to demonstrate their imperial authority. Jahāngīr took this vow after killing about 17,167 life forms in 37 years. He ordered Thursday and Sunday to be holidays against the slaughter of animals and animal eating – Sunday being the birthday of his father Akbar and Thursday as the day of his accession to the throne.[96]

Some people become vegetarians because they strongly believe in the principle of utility. These people live a lifestyle which can be sustained and inflict minimal harm to ecology and nature. Peter Singer, author of the book *Animal Liberation*, is one such proponent.[97] He believes that the ethical rights of all life-forms suggest that we become vegetarians. Also, inflicting minimum pain or harm to nature, or animals specifically, should also lead us to vegetarianism. Kauṭilaya, the teacher of Chandragupta (Candragupta, r. 321 - 297 B.C.), long before Peter Singer, assigned "*ahiṁsā*, truthfulness, cleanliness, compassion and tolerance" as basic values (*dharma*) for all people.[98] Here *ahiṁsā* was extended to animals also. Currently, about a billion people are vegetarians. The highest concentration of vegetarians is in India.

A vegetarian diet among the Seventh-day Adventists in America is asso-

[95]Sircar, 1957, p. 73 - 74. For more information, please also see, Smith, 1964; Barua, 1946.

[96]Findley, 1987.

[97]Singer, 1990 and 1980.

[98]*Arthaśāstra*, 1: 3: 13.

ciated with greater longevity.[99] This reason alone is good enough for many to adopt a vegetarian lifestyle. Similarly, the unhealthy impact of dietary animal fat intake on cardiovascular diseases and certain types of cancer is now well established.[100] Fraser[101] found a reduction by a factor of two among vegetarians in comparison to meat eaters, for colorectal and prostate cancers. On top of this, the use of animal fat in diet amounts to over 6% of the total U.S. greenhouse gas emissions.[102] Also, meat production requires 6 to 17 times as much land as soy for the same amount of protein produced.[103] Water consumption is about 4 to 26 times higher with meat diet in comparison to a vegetarian diet.[104] All this in the situation when the human population is at an all-time high.

Obviously, the ancient Hindus were not concerned with energy or water consumption issues when they propagated vegetarianism. However, they were concerned with long life, good health for themselves, morality, animal rights, and sustaining ecology. They studied nature and came up with relationships between various life forms: plants, animals, and humans. These scientific studies led them to define a life style that led them to vegetarianism. Their concerns have a newly found meaning in today's world. Their life style can easily take care of the energy crisis that we have today, improve the quality of water, completely erase the problem of hunger for about half a century with the current population growth, and conserve ecology and nature. This is definitely a worthwhile cause.

A research project was carried out at Loma Linda University, California, whose goal was to compare the environmental effects of vegetarian and non-vegetarian diets. The State of California was chosen as a model because of its large amount of agricultural and food products. California state agricultural data was collected and used to calculate different dietary consumption patterns. The results led to the conclusion that a non-vegetarian diet required 2.9 times more water, 2.5 times more primary energy, 13 times more

---

[99] Singh, Sabate, and Fraser, 2003

[100] Eschel and Martin, 2006, and references therein. This article has an excellent survey on this issue.

[101] Fraser, 1999.

[102] Eschel and Martin, 2006.

[103] Reijnders and Soret, 2003.

[104] Reijnders and Soret, 2003.

fertilizer, and 1.4 times more pesticides than a vegetarian diet.[105]

Even today, several vegetarian dishes are popularly mentioned as "Buddha delight" in many Chinese restaurants. This is due to the reason that the Indian Buddhists were largely vegetarians and they propagated the idea of vegetarianism to China and Japan. Vegetarianism is a virtue even for many Christians as they do not eat meat, except for fish, during the period of Lent.

### 8.6.1 Vegetarianism and Global Warming

Although the emphasis on vegetarianism by the ancient Hindus was not done due to global warming, it does have indirect impact on the environment. The life style recommended and lived by the ancient Hindus is relevant today in view of the global warming that the planet Earth is facing these days. Most activities that use a lot of energy are obvious examples, like the emissions that comes from a vehicle. When we see cooked steak on a dinner plate, we do not think the energy used into the raising of that animal for human consumption. It may surprise some people to learn that raising cattle is one of the leading causes of global warming, even more so than the use of automobiles. Vegetarianism is a more humane, environmentally friendly and efficient way to cut back on the spending of our nation's resources.

Global warming is a phenomenon caused by certain gases in the atmosphere that allows the incoming shorter wavelength solar radiation to enter the atmosphere and heat the earth, causing longer wavelength radiation from the warmed earth to be radiated outward, which causes an increase in the temperature of the atmosphere. More importantly, the greenhouse gases block these longer wavelengths and reflect them back to the earth, thereby causing melting of the polar ice caps and other wide-ranging climatic and environmental changes. Carbon dioxide ($CO_2$), methane ($CH_4$), nitrogen oxides, and water vapor are typical greenhouse gases. The livestock sector plays a significant role in the anthropogenic release of these greenhouse gases. Thus, our dietary choices play a significant role in the global warming of the earth.

[105]Marlow *et al*, 2009.

The basic physics of global warming has been known and well-established by research dating back over a century and a half. Another example of the greenhouse effect is the common glass greenhouse, so prevalent in colder climates. In this case, glass plays the role of the atmosphere, since glass, like the greenhouse gases, reflect infrared radiation. On a hot summer day, the interiors of closed cars warm up significantly due to the greenhouse effect.

Livestock production accounts for 70% of all agricultural land use and 30% of the land surface of the planet, excluding the polar ice-caps. The livestock business contributes between 4.6 and 7.1 billion tons of greenhouse gases each year to the atmosphere, which accounts for 15% to 24% of the total current greenhouse gas production, according to a report from the United Nations.[106] The greenhouse gases released from livestock production is higher than releases of greenhouse gases from all forms of transportation combined. Yes, the climate change that the world is facing is directly connected to our dinner plate.

To produce one pound of beef, a steer must eat sixteen pounds of grain and soy, the remaining fifteen pounds are used to produce energy for the steer to live.[107] S. Subak calculated the environmental impact of beef production by focusing on methane and $CO_2$ production. It was found the 1 kg of beef in a US feedlot causes the emission of about 14.8 kg of $CO_2$. In comparison, 1 gallon of gasoline emits about 2.4 kg of $CO_2$.[108] An average American's meat intake is about 124 kg per year, the highest in the world. In comparison, the average for a global citizen is 31 kg a year.[109] If the world can change it dietary habits and turn to vegetarianism, world hunger would soon disappear and there would be few $CO_2$ issues related to the dietary component of global warming.

### 8.6.2 Vegetarianism and World Hunger

A shortage of food is a looming crisis in view of the exponential population growth. At present, an estimated 850 million people are undernourished.

[106] Steinfeld, *et al*, 2006.
[107] Kaza, 2005.
[108] Subak, 1999 and Fiala, 2008.
[109] Fiala, 2008.

With the growth of the human population, this number is likely to grow in view of the limited availability of fertile land and fresh water resources. Food security will become even a bigger issue in the future than it is today, and we need to find solutions for this crisis. Promoting vegetarianism as a possible solution is definitely one of the quick solutions to the problem.

When Julius Caesar (ca. 50 B.C.) ruled Rome, the world population was about 250 million people. It grew to about 500 million when Christopher Columbus arrived in America in 1492. The world population became about 750 million when Thomas Jefferson signed the Declaration of Independence in 1776. It became 2 billion at the time of World War II. Presently, the world population is approximately 6 billion and is growing exponentially at the rate of about 1.5% a year with a doubling period of 47 years, adding about 80 - 90 million people (roughly the population of Germany today) to the earth annually. According to most estimates, however, the growth rate is expected to become lower and our population will be about 9 billion by 2050. This increase of population by 50 percent, from 6 billion to 9 billion, will require a corresponding increase of food production to feed people.

Because the earth is a closed system, we need processes that will sustain us in the long run. We need to find effective methods to deal with the population growth. We also must develop new technologies for food production, new family planning practices, and we must change our dietary choices.

It has been suggested that we can save more than half of commercially grown corn by changing our dietary practices toward being more vegetarian. This is because more than half of the grains currently being produced are fed to livestock. Similarly, we currently consume about 56% of all freshwater for irrigation and raising livestock. Farm animals consume about 1 trillion gallons of water per year; much of it can be saved, however, by adopting a meatless diet.

Past methods of increasing food production have had limited success. We are already destroying fertile cropland at an alarming rate, by about one-third in the last 50 years. The problems of erosion are difficult to resolve; it can take several centuries to convert eroded land into fertile soil suitable for agriculture production. We can no longer afford to continue on this destructive path. Eating a vegetarian diet is one way to significantly sustain our environment.[110]

[110]Kumar, World Hunger, an article in *Cultural Encyclopedia of Vegetarianism*, by Margaret Puskar-Pasewicz, 2010.

# Chapter 9

# Yoga

Our physical body is the basic instrument for all our actions, perceptions, and thoughts. It is the body and mind (popularly defined as two different entities) that are the basic tools of our knowledge, virtues, and our happiness. In the modern culture, more emphasis is provided on body. Gyms are popularly becoming a place where people go for exercise and sculpt their bodies for self improvement. Our icons are the supermodels and superathletes. What about the mind? Or, more importantly, a unison of body and mind?

The practice in yoga perhaps provides the best answer for an optimum unison of body and mind. Yoga provides a cost-effective and non-invasive treatment in many medical conditions that has minimal adverse impact. It also provides an overall fitness to the body and allows the body to heal itself. It is the most integrated science of self improvement where body and mind are both nourished that allow people to reach their fullest potential. The Hindus saw the relationship between mind and body and created practices that combined both.[1] The earth's gravitational field has an influence on every movement the human body makes. The first organizing principle underlying human movement and posture is our existence in a gravitational field.[2] The combination of the nervous system and skeletal muscles functioning together in gravity forms the basis of various yoga postures.[3]

[1]Atkinson, 2003.
[2]Coulter, 2001, 21.
[3]Coulter, 2001, 23.

The term yoga means "union", "join" or "balance" in the Sanskrit language. The term "union" defines the union of our physical self and mind for some, while implies a union of self with the divine for others. There is not much of a difference between the two. In both unions, a person transcends from everyday mundane existence to his/her fullest potential that leads to salvation – an ultimate goal for all Hindus. Body and mind are the vehicles for the Hindus that allowed them to liberate their soul from the cycle of reincarnation. Body and mind are the Siamese twins that create synergy for this ultimate goal of salvation.

Yoga is also a philosophy (*darśana*) that was originated by the ancient Hindus; it is a tool to overall stabilize an individual's physical body and mind. Not only does practicing yoga improve overall health, it is also used as therapy for a variety of medical conditions. Researches prove that practicing yoga can immensely affect the body in many different ways.

Yoga is all about balance; it is a balance of mind and body, a balance of strength and flexibility, a balance in a particular posture. It is a philosophy; it is a way of life. It helps to avoid sickness. Its medical benefits are recognized by the modern medicine.

Yoga is one of the six schools in the Hindu traditions that accept the *Vedas* as authority. These schools (*darśana*) are: *Nyāya*, the school of logic; *Vaiśeṣika*, the school of atomism; *Sāṁkya*, the school of Vedic Exegesis; and *Vedānta*, the school based upon the end of the *Vedas* (i.e. the Upaniṣads).[4] The term *darśana* is used for philosophy. The literal meaning of this word is "seeing," the ability to see what is, as it is. Although yoga is one of the six philosophical systems of India, it is not a religion. Rather, yoga is a philosophy of living.

These *darśana* (philosophies) cover even natural philosophy and deal with questions that are a part of physics, chemistry, and mathematics. This is the reason *Vaiśeṣika* came up with the concept of the atom while yoga eventually moved into the domain of medicine and the science of personal development. It allowed people to design breathing techniques and *āsana*, along with meditation to reach the ultimate goal of salvation. Yoga, thus, is a ladder to

[4]King, 1999, p. 45.

divinity – merging of the self with the divine.

Yoga is perhaps the oldest system of personal development that is effective.[5] Although the *Vedas* were the first to mention the importance of yoga, the *Upaniṣads* were the first to provide a systematic form of yoga. The first major treatise of yoga known to us is Patañjali's *Yoga-Sūtra*. The Indus seals of Harappa and Mohenjo-daro provide archaeological support for the existence of yoga where the human figures are shown in lotus postures in a meditative state.[6]

In Patañjali's *yoga-sūtras*, yoga is defined as a tool for the "control or cessation" of thought processes.[7] Failure of such control can cause restlessness, suffering, frustration, irregular breathing, and diseases. *Prāṇāyāma* or controlled breath is helpful in achieving tranquility.[8] Only a person with tranquil thoughts can achieve mastery of the micro to macro-universe.[9]

Yoga is an outcome of biomimicry practiced by the ancient Hindus. As mentioned in Section 8.3, the ancient Hindus observed various life forms – small and big, their life styles, the ways they exercised, the ways they cured themselves, the ways they relaxed, and the ways they avoided sickness. These studies evolved into a system of medicine called ayurveda. It also led to a system of personal development as defined in yoga. It is no wonder that most *āsana* are named after animals.

In yoga, people sometimes arch their spine (back) like cats, stretch their chest upward from the ground like cobra, or balance the body on two hands only. However, yoga is far more than just a few sitting postures. It is a philosophy of self realization whereby a person can reach to his/her fullest potential. The self [*ātman*] cannot be realized by the weak. In the Hindu religion, any kind of weakness has no place as it hampers progress in achieving salvation. As *Muṇḍaka-Upaniṣad* states: "The Self is not to be gained

---

[5]Coulter, 2001; Cowen, 2010; Cowen and Adams, 2005; Eliade, 1969; Feuerstein, 1989; Iyengar, 1966; and Kulkarni, 1972. For a beginner who is interested in the medical aspects of yoga, Coulter's book is good. For the philosophical aspects, consult Eliade's book.

[6]Worthington, 1982, p. 9.

[7]Patañjali's *yoga-sūtras*, 1: 2.

[8]Patañjali's *yoga-sūtras*, 1: 30 - 34.

[9]Patañjali's *yoga-sūtras*, 1: 40.

by the weak."[10]

Yoga is also a science that is cognizant of the bones, muscles, joints, organs, glands and nerves of the human body (biology). It uses the physics of balance in designing postures, the physics of motion and balance to allow changes in postures and the strength of various muscles to make the body stronger. It uses the power of mind (psychology) in controlling the thought process (meditation) for optimum results.

The word yoga has appeared in the *Ṛgveda* to define yoking, connection, achieving the impossible.[11] By yoga, one gains contentment, endurance of the pairs of opposites, and tranquility, tells the *Maitreyī-Upaniṣad.*[12] "When cease the five senses, together with the mind, and the thoughts do not stir. That, they say, is the highest course. This they consider as yoga, firm holding back of the senses. Then one becomes undistracted. Yoga, truly, is the origin and end," suggests the *Kathā-Upaniṣad*.[13]

Here "origin and end" means that yoga is essentially involved in knowledge and experience; it is a process at all stages. Yoga is advocated for the knowledge or realization of the self (*ātman*).[14] Patañjali equated yoga with *samādhi* (tranquil state) in the very first verse of his book. Yoga is not a mere abstract speculation of human minds; it is real with a concrete referent: "[This supreme ecstasy] is near to [him who is] extremely vehement [in his practice of Yoga]."[15]

According to the *Vaiśeṣika-Sūtra*, yoga is a state where the mind becomes steady and our senses are beyond the pain and pleasure of the phenomenal world.[16] A person in such a state is described as a "yogi".

Yoga is a science of health, while modern medicine is largely a science

---

[10] *Muṇḍaka-Upaniṣad*, 3: 2.

[11] *Ṛgveda*, 1: 34: 9; 3: 27: 11; 7: 67: 8; 10: 114: 9; Dasgupta, 1963, vol. 1, p. 226.

[12] *Maitreyī-Upaniṣad*, 6: 29.

[13] *Kathā-Upaniṣad*, 6: 10 - 11.

[14] *Kathā-Upaniṣad*, 2: 12; *Muṇḍaka-Upaniṣad*, 3: 2: 6; *Śvetāśvatara-Upaniṣad*, 1: 3: 6 - 13; *Maitreyī-Upaniṣad*, 6: 18, 19, 27.

[15] Patañjali's *Yoga-Sūtra*, 1: 21.

[16] *Vaiśeṣika-Sūtra*, 5: 2: 16.

of disease and symptoms. While modern medicine suppresses the symptoms of disease in general, yoga is a Hindu practice of disease prevention/control. The mind learns how to rest and refresh itself from the practice of yoga. Its quickness of apprehension and its retentive power are considerably increased. Its ordinary activities thus becomes efficient with significant positive results.

Our body and mind are intimately linked. If the muscles of our body are toned and relaxed, it is easier for our mind to relax. Similarly, if our mind is anxious, it directs stimuli to our physical body and changes the chemical composition and physical state of our muscles. The outcome is stress to our mind and physical body. It drains our energy physically and emotionally. Stress in modern living is inevitable. It is related to the demands of time. We need to learn to manage stress in a meaningful way. The practice of yoga helps in such management.

The ancient Hindus realized the intimate link of mind and body, and formulated exercises for both. The physical body received its vital energies through yogic postures (*āsana*), while mind gained its vital energies through meditation. "Restraint (*yama*), observance (*niyama*), posture (*āsana*), breath-control (*prāṇāyāma*), sense-withdrawal (*pratyāhāra*), concentration (*dhāraṇā*), meditative-absorption (*dhyāna*) and enlightenment (*samādhi*) are the eight members [of Yoga]."[17] These eight members (limbs) of yoga are collectively called *aṣṭāṅga-yoga*.

"Sickness, languor, doubt, heedlessness, sloth, dissipation, false vision, non-attaining of the stages [of yoga] and instability [in these stages] are the distractions of consciousness; these are the obstacles. Pain, depression, tremor of limbs, [wrong] inhalation and exhalation are accompanying [symptoms] of the distractions. In order to counteract these [distractions] [the yogi should resort to] the practice [of concentration] on a single principle," suggests Patañjali.[18]

Many styles of yoga follow this eight-fold path, but the term that is most often referred to is *hatha* yoga, which is a generic term for a wide range of styles: stretching, breathing, movement, balance, meditation, and strength

[17]Patañjali's *Yoga-Sūtra*, 2: 29.
[18]Patañjali's *Yoga-Sūtra*, 1: 30 - 32.

practices.

## 9.1 Some Basic Processes of Hatha Yoga

Hatha yoga can be broken down into five steps, with each step preparing the body for the next. The first two steps create moral awareness in the individual: *yama* (restraint) and *niyamas* (observance). The third step, *āsanas*, are the postures. *Āsanas* can be broken up into six categories: standing, balancing, sitting, backward bending, forward bending and twisting and inverting. These postures lead to increased blood circulation, internal massing of the organs, increased digestion, flexible muscles and joints, improved oxygen rates, increased lymphatic circulation and a better general mood.[19] The fourth step involves *prāṇāyā*, the breathing control aspect that brings greater awareness to oneself by integrating the mind, body and spirit. Lastly, *Pratyāhāra*, or sensory withdrawal allows the mind to become calm while ignoring sensory information. The final three steps of concentration, meditation and enlightenment make a complete total of eight steps.

### 9.1.1 Āsana (Posture)

The postures in yoga are called *āsana*. Postures prescribed in yoga are not movements; they need to be attained and held. Multitudes of postures provide enough selections for people of all ages and athleticism. Balance must be good; it is rhythmic alternation rather than a static balance.[20] The practice of yoga *āsanas* helps in the development of a healthy body. These *āsanas* allow exercises for various parts of the physical body, to stretch and tone the muscles, to make joints flexible, to improve the cardiovascular system, and to firm the entire skeletal system.

The ancient Hindus designed these *āsanas* by observing various life forms. The names of various postures are either based on their geometry or their similarity to an object, bird or animal. The following are some popular

[19] Mackenzie and Rakel, 2006, p. 200.
[20] Gilbert, 1999a and 1999b.

*āsana*: *dhanura-āsana* (bow posture), *garuḍa-āsana* (eagle posture), *krounc-āsana* (heron posture), *makar-āsana* (crocodile posture), *maṇḍūka-āsana* (frog posture), *mayur-āsana* (peacock posture), *padma-āsana* (lotus posture), *trikoṇa-āsana* (triangle posture), *vakra-āsana* (curved posture), and *hala-āsanas* (plough posture).

To guide a beginner, Patañjali suggests that "the posture [should be] steady and comfortable. It is [accompanied] by the relaxation of tension and coinciding with the infinite [consciousness-space]."[21] The practitioner must be able to practice breathing control and meditation in these postures.

An active organ receives a larger flow of blood than an inactive organ. Blood is an essential nutrient for the proper functioning of the various organs of body, and these organs get enriched due to a higher and more efficient transfusion of oxygen through our lungs. *Āsanas* work on the body frame as well as on the internal organs, glands, and nerves. Most joints and organs are put into isometric or other ranges of motions. They help to maintain a stable mind and a healthy body, and our bodies become a place for the ecstatic experiences.

### 9.1.2 *Prāṇāyāma* (Breathing)

Most people take short and shallow breaths throughout the day using their chest. This is not the optimal way to breathe. This kind of breathing does not allow our "lungs to expand and soak up oxygen."[22] Although hyperventilation or heavy breathing can be useful in the short term by boosting sympathetic nervous system activity, a better way to breathe is using the abdomen and diaphragm, called the belly breath and take slow and steady breaths. You need to use your diaphragm, which is the muscle underneath your lungs. When the diaphragm flexes, it pulls down and opens the lower lobes of your lungs, allowing more air inside.[23] You breathe through your nose and you can place your hand on your belly when you are learning. You should feel your belly push your hand out as you breathe in. Breathe slowly,

[21] Patañjali's *Yoga-Sūtra*, 2: 46 - 47.

[22] Dollemore, Giuliucci, Haigh, Kirchheimer and Callahan, 1995, p. 152; Gilbert, 1999a and 1999b.

[23] Dollemore, Giuliucci, Haigh, Kirchheimer and Callahan, 1995, p. 152

in and out, about 3 seconds each. Chest breathing comes from stress; its a reaction to stress. But breathing from the belly is a natural way to relax and spread more oxygen to your entire body.[24]

Patañjali suggest that the precise regulation of the deep rhythms of inhalation, retention, and exhalation of breath is called *prāṇāyāma*.[25] Citing benefits of *prāṇāyāma*, Patañjali suggests that with ample practice one can destroy the obstacles to wisdom and the mind is fit for meditation.[26]

*Prāṇa* can be vaguely defined as life breath; it is the power within breath. The term is mostly used for "the inhaled air." *Apāna* is the Sanskrit term used for "the exhaled air." Together, it forms the respiration process. The conscious practice of breathing is called *Prāṇāyāma*.

The term *prāṇa* is used in the *Ṛgveda*.[27] In the *Atharvaveda*, *prāṇa* is associated with life and the promotion of longevity.[28] In the *Chāndogya-Upaniṣad*, *prāṇa* is equated to "the chiefest and best."[29] and a Brahman:[30] "*prāṇa* gives life; it gives life to a living creature. . vital breath is one's father. . . mother . . . brother . . . sister . . . teacher . . . and Brahman." Breathing is perhaps the most essential external form of energy that is essential to our body.

A human being can last for weeks without food, several days without water. However, only 4 - 5 minutes deprivation of air can put a person into a coma. Yet, the role of lungs in the well-being of a person is not well known to the general public. Otherwise, we would not pollute the environment the way we do. We would not design chairs and tables in a way that constrain our breathing process. Our lungs are like any other muscle that needs to be exercised. This is crucial for increased oxygen consumption that enriches our blood – a chief source of energy to our brain. Improvements in the mood and stress are the common after-effects of using proper breathing exercises.

[24]Dollemore, Giuliucci, Haigh, Kirchheimer and Callahan, 1995, p. 152

[25]Patañjali's *Yoga-Sūtra*, 2: 49 - 52.

[26]Patañjali's *Yoga-Sūtra*, 2: 52 - 53.

[27]*Ṛgveda* 1: 48: 10; 3: 53: 21; 10: 121: 3.

[28]*Atharvaveda*, 2: 15: 1-6; 3:11: 5-6; 8: 2: 4; for a longer list, consult Zysk, 1993.

[29]*Chāndogya-Upaniṣad*, 5: 1: 1.

[30]*Chāndogya-Upaniṣad*, 7: 15: 1.

All yoga exercises start with basic deep breathing techniques along with proper *āsana*. It serves two purposes: first, to bring an optimum amount of oxygen into the lungs; second, to control the mind by controlling the breath. Deep breathing is not the fast pumping action of our lungs where we take a fast deep breath and puff it out; it involves a controlled rhythmic action to fill the lungs with air, to retain it for a brief period, and slowly exhale. The whole process has four parts: inhalation (*pūraka*), retention (*kumbhaka*), exhalation (*recaka*), and suspension (*kumbhaka*).

Nasal breathing is preferred in yoga, unless specifically instructed since it allows warming, filtering, and humidifying the inhaled air and recovering heat and moisture from the exhaled air. It is true that nasal breathing requires more efforts than breathing through the mouth, but this effort is what keeps the diaphragm strong. Also, it is easier to control the breathing pattern to establish slow rhythmic breathing using nasal breathing than mouth breathing. Oxygen absorption is better through nasal breathing, and the lungs are able to keep air in for a longer period.[31]

The inhalation of air should be done by extending the lower as well as the upper parts of the lungs. The upper abdomen should be used during the inhalation process along with the chest. It should be a slow but steady process to inhale air. The second stage is retention, in which enough time is provided for the lungs to transfer fresh air into the fine capillaries of the lungs. Through fine capillary tubes, oxygen is absorbed in the blood by our lungs during the inhalation and retention process of breathing.

In the third stage, it is important to get rid of all the waste carbon dioxide trapped in the capillaries of the lungs so that it can be replaced with fresh oxygen. It is done with a proper deep exhalation where the diaphragm or the upper abdomen is lifted up to squeeze the lungs to a smallest possible volume. The trapped carbon dioxide is released and the person is ready for a new breath. The last process of "suspension" allows the empty lungs to squeeze well and to be ready for another inhalation. The ancient Hindus developed various kinds of breathing – to increase the body temperature, to decrease the body temperature, to calm down the mind, to slow down the

[31]Gilbert, Christopher, 1999a.

physical body, to energize the physical body, etc. The controlled breathing helps to focus the mind and achieve relaxation while meditation aims to calm the mind.[32]

As mentioned above, it is better to breathe slowly in a normal routine. It promotes the absorption of oxygen. For normal air quality, the inhaled air has usually 21% oxygen while the exhaled air has about 16%. Thus, we absorb about 5% of oxygen and the rest is unused.[33] Similarly, an average person produce around 200 milliliters of $CO_2$ which need to be excreted in totality. Any imbalance that can keep $CO_2$ in the system can create "strain in the muscles, tendons and joints."[34] Hyperventilation creates a shortage of $CO_2$ in the system and produces alkalosis which raises the pH of blood and makes it more alkaline. The presence of $CO_2$ in right amount maintains the acidity of blood, which should be around 7.4. Any increase in alkalinity increases calcium that constrict coronary blood vessels and arteries. This can create all sorts of physical ailments: migraine, cerebral vasospasm, bronchial asthma, laryngeal spasm, spastic colon and constriction of coronary arteries.[35] People who suffer from chest pain due to minor physical activities may be suffering from this restricted blood flow to cardiac muscles. This pain is an indication that the nutritional needs of the heart are not being met. The absorption of oxygen per breath can be improved by slowing down the stages of respiration. A prescribed ideal rate is about 6 breaths per minute which can only be achieved with conscious training for an extended period of years.[36]

Another rule is that the exhalation period should be twice of the inhalation period. Such a pattern can bring calmness and vitality to the person. Anxiety, anger, and exercise can easily change this ratio. The perfect pattern of breathing is one period for inhalation, four periods for retention, and 2 periods for exhalation. Alternate breathing from each nostril where you inhale the air from one nostril and exhale it from the other and vice versa helps to balance the nasal air flow and has a balancing impact on both hemispheres of our brain. According to Christopher Gilbert, just breathing properly can

---

[32]Riley, 2004.
[33]Gilbert, 1999a
[34]Gilbert, 1999b
[35]Gilbert, 1999b
[36]Gilbert, 1999a.

take care of the back pain since the lower back muscles are guided by the abdomen muscles and can prevent muscular spasms and promote relaxation and alertness.[37]

### 9.1.3 *Dhyāna* (Meditation)

The word meditation derives from the Latin, *mederi*, to heal. It is the healing of a mental affliction caused by psychological stress. To deal with stressful situation, it is imperative to first achieve an inner calm that will allow you to have peaceful state of mind. In this state of mind, one should be contemplating the problem, its cause and the way to resolve it. Meditation allows you to achieve this. It is a self-directed approach for relaxation and self-control.

Managing our thought processes is a key to managing the stress that affects our lives. Ancient seers realized that we become slaves to our thoughts if we don't control them. Our thoughts are a result of past experiences that our memory traces continually. These memories come to our consciousness and sometime create egotistical or selfish behaviors.

Meditation is a process of *knowing* for the Hindus. The Sanskrit words that reflect meditation are: *cintan*, *dhyāna*, or *manana*. Nowhere in the world was the art of meditation as perfected as was done by the ancient Hindus. The *Śvetāsvetara-Upaniṣad* tells us that meditation and yoga are the way to know the self-power (*ātma-śakti*) of God within us.[38] The *Chāndogya-Upaniṣad* tells us that all people who achieved greatness in the past did so with the help of meditation.[39]

Meditation can help a person find new ideas and practical answers to problems. It gives ample stillness to the mind to think properly in order to make proper judgments. It helps the mind to control emotion without suppression but with an outlet where the emotional waste could be discarded in order to become more at peace with the world. For this reason, the *Maitreyī-Upaniṣad* considers meditation as essential to achieve God, along with knowledge (*vidyā*) and austerity (*tapas*).[40]

---

[37] Gilbert, 1999a.

[38] *Śvetāsvetara-Upaniṣad*, 1: 3.

[39] *Chāndogya-Upaniṣad*, 7: 6: 1.

[40] *Maitreyī-Upaniṣad*, 4: 4.

The practice of concentration (*ekāgratā* or *dhārṇa*) is the endeavor to control the two generative sources of mental fluidity: sensory activity (*indriya*) and subconscious activity (*saṃskāra*).[41] It is difficult to achieve pursuit of one object (*ekāgratā*, single-mindedness) with a tired body or restless mind, and with unregulated breathing.

Meditation can decrease our reaction time, increase alertness and improve the efficiency of a person. Insomnia, headache, lack of appetite, shaking of the hands and other symptoms can be either reduced or nearly gone. It also helps in asthma, anxiety, high blood pressure, back pain, heart disease, etc. Concentration of the mind acts like a focusing lens (convex lens) of sun's rays. It concentrates thought and invokes miraculous powers to the mind's activities. The only way to understand the impact of yoga is to go through the experience.

There are different techniques that you may practice, preferably in a quiet and comfortable room with little distractions. You can sit on a floor, in a chair or lay on the floor. You can either relax all of your limbs at your side or place your arms on your lap, palms up, keep you hands open, and place the end of your thumb and index finger together in both hands. You should concentrate on your breathing. Some say you should think about each body part starting with your face all the way down to your feet, and let them relax individually till your whole body is relaxed.

Many present day exercises, such as cardiopulminary exercises, for example, require rapid movements, maximum effort, and are tedious. Yoga involves being stable, with slow dynamic movements, with endless opportunities to strengthen and calm your body and mind. Often times after a physical workout you may feel exhausted and fatigued because of the lack of balance between the different parts of the body and the mind. Since the goal of yoga is to achieve that balance, relaxation and breathing techniques are incorporated, which improve energy levels.

[41] Eliade, 1975, p. 62.

## 9.2 Benefits of Yoga

Researches have shown that yoga can help people suffering from asthma, heart disease, high blood pressure, type 2 diabetes, and obsessive-compulsive disorder, lower their dosage of medications, and sometimes eliminate the use of medication.[42] Practicing yoga not only relieves temporary symptoms such as headaches, sinus pressure and hot flashes, but can also improve more extreme medical conditions such as cancer, diabetes, anxiety, and heart disease, to name a few.

The medical benefits of yoga are endless; yoga is practiced not only to improve a person's general health, but also to assist in aiding serious medical conditions. One of the most terrifying disease, with no known cure, is cancer. Cancer patients face many symptoms; pain and stress are the most prevalent. Since patients are often weak after chemotherapy, restorative postures and simple *prāṇāyāma* exercises can be practiced to increase energy and relaxation.

Stress is the body's response to a situation or thought, that, if not managed properly, can lead to other serious medical conditions. It has been proven that yoga relaxes the nervous system and is a stress reducer. Although some stress can be a good thing, too much stress can have harmful effects on the body. When experiencing high amounts of stress, the adrenal glands get stimulated by the brain resulting in increased heart rate and blood pressure, faster and shallower breathing, dilated pupils, tense muscles, sweating, heightened senses, and increased blood flow to the brain. Problems such as depression, heart attacks, headaches, strokes, obesity, and multiple sclerosis can occur due to high stress levels. Since stress is more than often induced by your thoughts, yoga teaches you the ways to avoid those thoughts that brings you down. Guided imagery is a yogic tool that uses thoughts to change the body and mind.[43] This is done by placing yourself in a relaxed position and guiding your imagination through thoughts. Yoga also uses breathing techniques to calm the mind. Erratic breathing is a cause of stress, so by slowing and regulating your breathing you can reduce stress. By taking slow, deep breaths you relax the nervous system which in return calms the mind.

[42] McCall, 2007, p.43.
[43] McCall, 2007, p. 51

"Yoga is not a panacea, but it is powerful medicine indeed for body, mind, and spirit," suggests Dr. McCall, a medical doctor who studied and practiced the effects of yoga.[44]

According to McCall, "this single comprehensive system [yoga] can reduce stress, increase flexibility, improve balance, promote strength, heighten cardiovascular conditioning, lower blood pressure, reduce overweight, strengthen bones, prevent injuries, lift mood, improve immune function, increase the oxygen supply to the tissues, heighten sexual functioning and fulfillment, foster psychological equanimity, and promote spiritual well-being."[45]

In a study carried by Cara Rosaen and Rita Benn from the University of Michigan at Ann Arbor, ten seventh-grade African-American students (ages 12 - 14 years), five males and five females, from a Detroit charter school, were recruited to study the effect of meditation. It was observed that meditation provided a greater capacity for self-reflection, self-control, and flexibility as well as improved academic performance among these children. The state of restful alertness induced by meditation allowed growth in social/emotional capacities, academic performance, and in emotional responses.[46]

People suffering from obesity not only suffer from weight problems but emotional and mental issues as well. Physical results such as weight loss, flexibility, strength, and endurance have a positive correlation with practicing yoga. Most importantly yoga can change one's body, and also one's attitude about his or her body. Since yoga incorporates breathing techniques into its stretching exercises, mental stability can help a person to gain control over bad eating habits and eventual weight loss. Other benefits associated with weight loss include the large amount of calories that are burned on a yoga mat, a raised heart rate that can fall into the aerobic range, and the decreased stress level which often causes overeating if not properly reduced. Meditation can also help you respond to hunger pangs effectively.[47] In a study conducted by Jon Kabat-Zinn and others at the University of Massachusetts Medical School at Worcester, a practice of meditation significantly reduced anxiety

---

[44] McCall, 2007, p. XIX.
[45] McCall, 2007, p. 1.
[46] Rosaen and Benn, 2006.
[47] McCall, 2007, p.479.

and depression in 20 out of 22 patients that suffered from panic symptoms.[48]

Flexibility is another physical benefit of yoga besides weight-loss. Improving flexibility is a challenge but can be done through lengthening nerves and muscles. Nerves and muscles are the two kinds of extendable anatomical structures that run lengthwise through limbs and across joints.[49] By studying muscles that have been held in casts in stretched positions, muscle fibers can grow in length by the addition of little contractile units called sarcomeres. Sarcomeres are composed of fibrous proteins that slide past each other during contraction or relaxation. Mind plays a pivotal role in causing muscles to either relax or tighten up.[50] Hatha yoga stretches are a safe way to for the connective tissues to follow the lead of muscle fibers to develop a flexible body.

A group of scientists from Harvard Medical Center and Boston Medical Center, both in Boston, studied the clinical application of yoga for the pediatric population that ranged from 0 to 21 years of age.[51] The group collected a total of thirty-four controlled studies that were published from 1979 to 2008. It was found that the practice of yoga is helpful in physical fitness, cardiorespiratory effects, motor skills/strength, mental health and psychological disorders, behavior and development, irritable bowel syndrome, and birth outcomes following prenatal yoga. Also, no adverse effects were noticed in the reported trials.

About 108 firefighters from the New York City fire department were recruited to practice *prāṇayāma*, *āsana*, and *śavāsana* on a regular basis. These firefighters had no prior exposure to yoga. After a period of six weeks, a significant improvement in the "Functional Movement Screen" was observed. Out of the participants who filled out post-yoga assessment, 56% reported that the total experience was positive in terms of their job performance, 62% claimed to be more flexible, 17% reported less muscular pain, 16% reported better breathing control, and 15% reported improved balance/core strength.[52]

---

[48]Kabat-Zinn *et al*, 1992.
[49]Coulter, 2001, p. 60
[50]Coulter, 2001, p. 61.
[51]Birdee *et al*, 2009.
[52]Cowen, 2010.

### 9.2.1 Cardiovascular

Moderate forms of yoga practice burn energy that is equivalent to moderate forms of exercise. Cowan and Adams studied the response of the heart rate with practices of postures (*āsanas*), breathing exercises (*prāṇayāma*), and relaxation (*śavāsana*). A practice of yoga is found to be good for cardiovascular fitness and muscular strength.[53] By age 55, about 30 percent of White American men and 20 percent of White American women have blood pressure readings above the recommended maximum of 140/90.[54] Again, it is suggested that yoga in conjunction with medical treatment and supervision can help reduce high blood pressure.[55] The following *āsanas* are suggested: the corpse pose, the knee squeeze, the child pose, the moon pose, meditation and breathing exercises.[56] These can help improve blood circulation, relieve tension and stress.

The diaphragmatic movements provide a "messaging action to the heart" and the *vena cava* passes through the diaphragm and is squeezed in a periodic way with breathing and pushes the "blood flow toward the heart" and can be labeled as the "second heart."[57] In the Universidade Federal de São Paulo, Brazil, Danucalov *et al* investigated the changes in cardiorespiratory and metabolic intensity resulting from *prāṇāyā* and meditation during the same *hatha* yoga session. Nine yoga instructors were subjected to analysis of the air exhaled during the three periods, each of 30 minute duration: rest, respiratory exercises, and meditation. The oxygen uptake and carbon dioxide output were proven to be statistically different during the active sessions (*prāṇāyāma* and meditation) compared to the rest phase. In addition, the heart rate showed decreased levels during rest as compared to meditation. Therefore, the results from this study suggest that meditation reduces the metabolic rate while the *prāṇāyāma* techniques increases it.[58]

Herur, Kolagi, and Chinagudi studied the impact of yoga and mind at S. Nijalingappa Medical College in Karnataka, India, with 50 subjects, 28

[53] Cowen and Adams, 2007.
[54] Dollemore, Giuliucci, Haigh, Kirchheimer and Callahan, 1995, p. 360.
[55] Dollemore, Giuliucci, Haigh, Kirchheimer and Callahan, 1995, p. 360.
[56] Dollemore, Giuliucci, Haigh, Kirchheimer and Callahan, 1995, p. 363.
[57] Thomas, 1993; taken from Gilbert, 1999a
[58] Danucalov *et al*, 2008.

males and 22 females, over the age of 30. This study concluded a significantly high improvement in the mental status of the subjects. It was observed that the poses, minimizing movement, and breathing exercises can help a person achieve optimal physical and mental relaxation. This leads to the improvement of psychological and psychomotor components. They also tested heart rate, mean arterial pressure, and systolic and diastolic pressure. The researchers concluded that yoga practice can be used as an intervention in aging people to reduce the morbidity and mortality from cardiovascular diseases, irrespective of age and gender.[59]

The prevalence of coronary heart disease is about 50% higher and mortality rates for stroke are 4 to 5 times higher in African Americans than in Anglo Saxons. Hypertension is also disproportionately high in African Americans and is a major contributor to their risk for atherosclerotic CVD mortality.[60] Carotid Intima-Media (IMT) thickness is a valid surrogate measure for coronary atherosclerosis (hardening of arteries), a predictor of coronary outcomes and stroke, and is associated with psychosocial stress factors. A team of doctors from the University of California, Los Angeles and the Center for Natural Medicine and Prevention in Maharishi University of Management, Fairfield, Iowa, tried to reverse this trend among the hypertensive African Americans by introducing them to the practices of transcendental meditation (TM). A reduction in the thickening of artery walls was noticed. This reduction demonstrates the physiological effect of meditation in reducing cardiovascular and stroke risk.[61] The team of doctors divided their control group into two sections: one was trained in transcendental meditation while the second group was provided health education related to these diseases. This preliminary randomized control trial suggests that stress reduction with the TM program is associated with reduced carotid arteriosclerosis in African Americans with hypertension compared with a health education comparison group.

### 9.2.2 Cancer

In Hacettepe University in Turkey, Ülger, Özlem, and Naciye Yağl studied the effect of yoga on twenty female breast cancer patients. The ages of these

[59]Herur *et al*, 2010.
[60]Castillo-Richmond *et al*, 2000.
[61]Castillo-Richmond *et al*, 2000.

twenty women ranged from 35 to 48 years. All patients were examined before and after to determine their quality of life. The participants of the study did eight sessions of a classical yoga program, twice a week. In particular, the results had significantly affected patients' pain and stress levels when comparing before and after scores. The results of this study were encouraging as it "improved patients cancer-related anxiety levels as well as their quality of life."[62] The physical characteristics of the patients were recorded and general physiotherapy assessments performed. Eight sessions of a yoga program including warming and breathing exercises, *āsana*, relaxation in supine position, and meditation were applied to participants. It was concluded that yoga is valuable in helping to achieve relaxation and diminish stress, helps cancer patients perform daily and routine activities, and increases the quality of life in cancer patients. This result was positively reflected in patients satisfaction with the yoga program.

A celebrity lifestyle guru and cardiologist, Dr. Dean Ornish, studied patients with prostate cancer. This program included gentle *āsana*, breathing techniques, and meditation. After the program the blood serum of patients was added to a line of prostate cancer cells. The results were encouraging, as "the serum inhibited their growth by 67 percent, compared to only 12 percent for controls"[63] Also, there was a 3 percent decrease in these patients of prostate-specific-antigen levels as contrasted to a 7 percent increase in the subjects in the control group. The study concluded that the overall health of cancer patients who incorporate yoga into their life, as a form of medicine, improves significantly.

Fibromyalgia (FM) is a debilitating condition affecting 11 - 15 million people in the US. This disease carries an annual direct cost of more than $20 billion.[64] FDA-indicated drug therapies are generally only 30% effective in relieving symptoms and 20% effective in improving function. Multiple position statements recommend that medications be accompanied by exercise and coping skills approaches. A group of doctors from the Oregon Health and Science University in Portland, Oregon, studied the effect of yoga on patients suffering from FM. They formed a group of 53 patients and in-

---

[62] Ülger and Yağl. 2010.
[63] McCall, 2007, 210.
[64] Carson *et al*, 2010.

troduced yoga poses to them, complemented by meditation and breathing exercises, yoga-based coping presentations, and group discussions. All participants were women with age above 21 years and were subjected to yoga practice for 8 weeks. The results suggests that the yoga intervention led to a beneficial shift in how patients cope with pain, including greater use of adaptive pain coping strategies (i.e., problem solving, positive reappraisal, use of religion, activity engagement despite pain, acceptance, relaxation) and less use of maladaptive strategies (i.e., catastrophizing, self-isolation, disengagement, confrontation). It was concluded from this study that yoga promotes both balance and strength – two functional deficits targeted by FM exercise interventions. At post-treatment, women assigned to the yoga program showed significantly greater improvements on standardized measures of fibromyalgia symptoms and functioning, including pain, fatigue, mood, stiffness, poor sleep, depression, poor memory, anxiety, tenderness, poor balance, environment sensitivity, vigor, and limited strength. Similar results were also obtained in another study conducted at Kaiser-Permanente Northwest in Portland, Oregon.[65] Forty-five patients, who were recruited for the study, practiced two-hour sessions three times a week for six months. Again, patients experienced a reduction of pain and of pain severity with the practice of yoga.

### 9.2.3 Mental Illnesses and Addictions

A review study on yoga-based interventions for the treatment of depression was made by a group of researchers from the Research Council for Complementary Medicine in London.[66] The aim of this study was to systematically review the research evidence on the effectiveness of yoga on controlling depression. Five randomized controlled trials were chosen, each of which utilized different forms of yoga interventions and in which the severity of the condition ranged from mild to severe. All trials reported positive findings and no adverse effects were reported with the exception of fatigue and breathlessness among participants in one study. Overall, the initial indications are of potentially beneficial effects of yoga interventions on depressive disorders.

Katya Rubia, of the Institute of Psychiatry, King's College University,

[65]Hsu *et al*, 2010.
[66]Pilkington *et al*, 2005.

London, reviewed the evidence for the effect of meditation on body and brain physiology and for clinical effectiveness in psychiatric disorders.[67] At present there is no long-term cure for mental disorders. In her studies, Katya Rubia observed evidences for long-term improvement in cognitive skills with meditation, mainly in the domains of attention, inhibitory control and perceptual sensitivity. The combined evidence from neurobiological and clinical studies for an improved regulation of attentional and motivational networks and for a reduction in symptoms of anxiety, depression and attention, respectively, bears promise.

Tricia L. da Silva, Lakshmi N. Ravindran from Toronto, and Arun V. Ravindran from San Diego reviewed the existing clinical studies on the effect of yoga in the treatment of mood and anxiety disorders.[68] The group found that the practice of yoga is comparable to medication in mood and anxiety disorders. Once combined with medication, the overall effect was even superior to just medication alone. The practice of yoga found to increase GABA levels in the brain by 27% in a single yoga session among practitioners using magnetic resonance spectroscopic (MRS) imaging. Low GABA levels are known to be associated with anxiety and depression. The practice of yoga provides a considerable improvement.

The practice of meditation in the context of yoga can help patients reduce addictive tendencies both in the short term and over the long term, as suggested by Morton Kissen and Debra A Kissen-Kohn.[69] In their views, "most addicts do not regularly exercise their ability to create high levels of energy and the positive affects associated with self-soothing. They instead rely on various kinds of external substances such as alcohol, drugs and food or highly externalized experiences compulsively enacted (e.g., gambling, shopping, sexuality, overeating) to produce sought-after positive affects." The expansion of neuroscientific tools such as PET and fMRI for studying various cognitive, emotional and even more complex ego states has allowed for a closer look at how the neurological correlates of different states of consciousness may impact addictive disorders. Dr. Morton Kissen and Dr. Debra A. Kissen-Kohn concluded in their studies that the emptying of the mind of ordinary

---

[67] Rubia, 2009.
[68] da Silva *et al*, 2009.
[69] Kissen, and Kissen-Kohn, 2009.

thoughts and feelings for the sake of a focus upon the body is both ego enhancing and stress reducing. The multiple prompts toward a continuous and active scanning of different aspects of bodily experience is linked to enhanced capacities for self-soothing and positive body narcissism. These prompts may insulate against the tendency to damage the body and the failures to protect the body against the deleterious effects of various addictive substances.

### 9.2.4 Pain

Chronic lower back pain is a typical problem among adults around the world. This can result in missed work, substance abuse due to the pain medications, and in disability. About 1% of the U.S. population (both men and women) are chronically disabled due to pain, functional disability, sleep disturbances, fatigue, and medication abuse.[70] Most American's either try to get prescription medication or go through painful surgery with a long recovery. However, there are studies that demonstrate that yoga can help such patients. There was a significant reduction in pain for those in the yoga group compared to the control group. Subjects in the yoga group also showed a significant improvement in their flexibility and rotation compared to those in the control group. Spinal flexion, right lateral flexion, and left lateral flexion improved in the yoga group more that in the control group.

Yoga is found to be helpful in managing carpal-tunnel syndrome. This is a syndrome that is common among females and is commonly associated with rheumatoid arthritis, diabetes, hypothyroidism, haemodialysis, infection, tumor, or trauma. People who suffer from carpal-tunnel syndrome have difficulty in gripping objects, doing simple typing on a computer, or doing simple household chores. In a clinical study, Garfinkel and colleagues compared the use of splints with yoga in the treatment of carpal-tunnel syndrome.[71] In the group with yoga, 11 yoga postures for the arm were practiced for 30 seconds with a relaxation practices in between. For example, the popular greeting, *namaste*, was practiced by gently extending and stretching the wrists and fingers and some isometric resistance exercise of the extensors and flexors. After 8 weeks, grip strength of the group was significantly better and pain

[70] Tekur *et al*, 2008.

[71] Garfinkel, M. S., A. Singhal, W. A. Katz, A. A. David, and R. Reshetar, Yoga-based intervention for carpal tunnel syndrome: a randomized trial, *JAMA*, 280, 1601 - 03, 1998.

was reduced. It was found that yoga dramatically reduced pain and increased grip strength in the patients. In contrast, people who used wrist braces and splints did not experience similar benefits.[72]

Three Thai scientists, Songporn Chuntharapata, Wongchan Petpichetchian, and Urai Hatthakit, studied the impact of yoga on pain and comfort issues during pregnancy. In this study, 74 pregnant women were divided into two groups: one group followed routine practices while the other group practiced 30 minutes of yoga at least three times per week for 10 weeks. It was found that practicing yoga is an effective complementary means for facilitating maternal comfort, decreasing pain during labor, and shortening the length of labor. With the help of diaphragmatic breathing, directive methods of emotional control using *āsana* and meditation, the threshold of pain was considerably improved among practitioners.[73]

---

[72]Sequeira, 1999.

[73]Chuntharapata, 2008.

# Chapter 10

# Medicine

Diseases are as old as life itself. Manu (progenitor of humanity), or Adam and Eve, must have needed the science of medicine. The problems of colds, fever, headache, and exhaustion has been experienced by all peoples at one time or another. The ancient peoples must have tried to cure their problems by changing their diet, by taking rest, by performing minor surgeries, or by performing rituals. They must have used whatever worked to alleviate the suffering. Humans were not the only ones to have tried this; animals and plants have also suffered from diseases and tried to cure themselves. Medicine, therefore, was bound to, and ultimately did evolve in all cultures.

In the words of Suśruta, Brahmā (creator of the universe) was the first to inculcate the principles of ayurveda and taught it to Prajāpati. The Aśvins learned from Prajāpati and taught it to Indra. Indra taught it to Dhanvantari who later taught it to Suśruta.[1] Dhanavantari's status in India is similar to Aesculapius in the Western world.

Care of a sick individual became a fundamental need of community life in all societies. For this reason, Caraka, the Hippocrates of Hindu medicine, considered ayurveda eternal, and without a specific beginning, for the purpose of preserving good health of a person and curing a sick person.[2] "The science of life has always been in existence, and there have always been people who understood it in their own way: it is only with reference to its first systematized comprehension or instruction that it may be said to have a be-

[1] *Suśruta-Saṁhitā*, *Sūtrasthānam*, 1: 16.
[2] *Caraka-Saṁhitā*, *Sūtrasthānam*, 30, 26 - 28.

ginning."[3]

Most animals instinctively try to cure themselves by changing their diet. For example, dogs do not eat grass when healthy. However, in certain sicknesses, dogs instinctively eat grass. This knowledge is somehow ingrained in them. Similarly, dogs, cats, and other animals do proper stretching exercises after they wake up from sleep or rest. Careful observations of such animals led the ancient Hindus to invent various *āsana* (postures) in yoga. Their keen observations of animals also led them to the discovery of various medicinal herbs.

The *materia medica* of the Indus-Sarasvatī region is particularly rich due to a large number of rain forests around the country. The Indian climate is unique, as all four seasons are experienced. Due to substantial rain and large rain forests, herbs are in abundance. The abundance of *materia medica* has been recorded by several foreigners who visited India during the ancient and medieval periods. Although India's total land area is only 2.4% of the total geographical area of the world, it accounts for 8% of the total global biodiversity with an estimated 49,000 species of plants of which 4,900 are endemic.[4] According to the World Health Organization (WHO), 80% of the population in developing countries relies on traditional medicine, mostly in the form of plant drugs, for their health care needs. Additionally, modern medicines contain plant derivatives to the extent of about 25%. On account of the fact that the derivatives of medicinal plants are non-narcotic without having much side effects, the demand for these plants is on the increase in both developing and developed countries. There are estimated to be around 25,000 effective plant-based formulations available in Indian medicine. It is estimated that there are over 7,800 medicinal drug manufacturing units in India, which consume about 2,000 tonnes of herbs annually.[5] The *Caraka-Saṁhitā* listed 341 plants and plant products for use in medicine. Suśruta described 395 medicinal plants, 57 drugs of animal origin and 64 minerals and metals as therapeutic agents. [6]

[3] *Caraka-Saṁhitā, Sūtrasthānam*, 30: 26 - 27.

[4] Ramakrishnappa, K., 2002. This report can be accessed at http://www.fao.org/DOCREP/005/AA021E/AA021e00.htm.

[5] Mukherjee and Wahile, 2006. Also, for some statistical information, consult the World Health Organization website: http://www.who.int/mediacentre/factsheets/fs134/en/

[6] Mukherjee and Wahile, 2006. Krishnamurthy has given quite a different number for

The *Ṛgveda* suggests the human life expectancy to be 100 years.[7] The *Yajurveda-White* tells everyone to aspire to have an active life till they reach 100.[8] The *Atharvaveda* mentions prayers that can make a person as "strong and hard as stone" and lists a large number of diseases and drugs.[9] The *Upaniṣads* also suggest us that we should expect to live an active life for hundred years. "Even while doing deeds here, one may desire to live a hundred years," suggests the *Iśā-Upaniṣad.*[10]

The ancient Hindus understood that disease is a process that evolves over time. Our genes, our environment, and our lifestyle play an important role in this process. The ancient Hindus tried to prevent diseases by using proper hygiene, a healthy lifestyle, a balanced diet, yoga, internal and external medicine, surgery, and prayer. Good health was considered as essential to achieve salvation.[11] Detailed experiments were performed to evolve the existing knowledge. For example, Suśruta selected a body that was not damaged. The body was cleansed and wastes were removed. The body was then wrapped in grass and placed in a cage to protect it from animals. The cage was then lowered into a river or a rushing stream of water. This was done to ensure constant cleaning. Once the decay process took a toll on the body, it was taken out of water. Skin, fat, and muscles were brushed away from the corpse with grass brushes, and the anatomy was observed.[12]

---

the medicinal plants. According to Krishnamurty, Suśruta defined 793 plants for medical usages. Krishnamurthy (1991) has compiled the list of these plants in the ninth chapter of his book. Krishnamurthy, K. H., *The Wealth of Suśruta*, International Institute of Ayurveda, Coimbatore, 1991.

[7] "That Indra for a hundred years may lead him safe to the farther shore of all misfortune." *Ṛgveda*, 10: 161: 3. "With the most saving medicines which thou givest, Rūdra, may I attain a hundred winters." The *Ṛgveda*, 2: 33: 2.

[8] *Yajurveda-White*, 40: 2.

[9] *Atharvaveda*, 1: 2: 2; read Narayana, 1995 for diseases and cures.

[10] *Iśā-Upaniṣad*, 2.

[11] *Caraka-Saṁhitā*, *Sūtrasthānam*, 1: 15 - 16.

[12] Rajgopal *et al*, 2002.

## 10.1 The Biological Nature of Diseases

The ancient Hindus knew the biological origin of diseases and used complex drug formulations, comprehensive diagnostic tools, and properly cared for patients with proper medicine. Suśruta wrote on the causes of various diseases: "Man is receptive to any particular disease, and that which proves a source of torment or pain to him, is denominated as a disease. There are four kinds of diseases such as traumatic or of extraneous origin (*āgantuka*), bodily (*śarīra*), mental (*mānasa*) and natural (*svābhāvika*). A disease due to an extraneous blow is called *āgantuka*. Diseases due to irregularities in food, drink, or incidental to a deranged state of blood, or the bodily humors, acting either singly or in-concert are called *śarīra*. Excessive anger, grief, fear, joy, despondency, envy, misery, pride, greed, lust, desire, and malice are included within this category of mental distemper; whereas hunger, thirst, decrepitude, imbecility, death, and sleep are called the natural derangements of the body. The mind and body are the seat of the above said distemper according as they are restricted to either of them, or affect both of them in unison."[13]

For the ancient Hindus, medicine was more than biology or biochemistry, in the modern terms. It included the roles of the mind and spirit. The role of the mind in the well-being of our physical health is discussed often in Hindu scriptures. This role of the mind and physical body in the well-being allowed the ancient Hindus to develop yoga. Pregnant females were advised to keep their emotions checked, avoid anger, fright or other agitating emotions. According to Suśruta, a pregnant female should be a positive thinker, avoid anger, avoid carrying a heavy load, and is advised to even avoid criticisms of others or talk loudly.[14] Caesarian section was an option for child-birth where the mother is deceased or if the child is in danger.[15]

Caraka defined good health as "a state in which *dhātu* are in balance." Good health is defined as "happiness" and an imbalance of *dhātu* is "melancholy."[16] A similar definition of good health was also provided by Suśruta. According to him, a person is healthy when "the three functions of the body

---

[13] *Suśruta-Saṁhitā, Sūtrastānam*, 1: 20.
[14] *Suśruta-Saṁhitā, Sārīrstānam*, 10: 2.
[15] *Suśruta-Saṁhitā, Nidānstānam*, 8: 8 - 19.
[16] *Caraka-Saṁhitā, Sūtrasthānam*, 9: 4.

(*doṣa*) are in a state of equilibrium" and "sense organs and the mind are clear and bright."[17] The purpose of medical treatment is to bring about a balance of seven *dhātu* in the body.[18] *Rasa*, blood, flesh, fat, bone marrow, and semen are the seven *dhātus*.[19] "A person with uniformly healthy digestion, and whose bodily humors are in a state of equilibrium, and in whom the fundamental vital fluids course in their normal state and quantity, accompanied by the normal processes of secretion, organic function, and intellection, is said to be a healthy person."[20]

The hereditary or genetic effects of diseases were also noticed. It was advised in *Manu-Saṁhitā* to avoid marriages with people where family members suffer from hemorrhoids, weak digestion, epilepsy, leucoderma, and leprosy.[21]

### 10.1.1 Caraka

Caraka belongs to the pre-Buddhist period around 600 B.C..[22] This assignment is based on the literary style of writing which resembles that of of the *Brāmaṇas* and *Upaniṣads*, and the fact that the book does not mention Buddha (566 - 480 B.C.). This sets the writing period of this book prior to the period of Buddha. Some other scholars place Caraka around the beginning of the Christian era, particularly to the period of the Kuśāṇ dynasty. However, several texts that were written before Kuśāṇ's period refer to the work of Caraka.

The *Caraka-Saṁhitā* consists of 120 chapters in eight sections (branches)– one for each branch of ayurveda. This book is still taught in all medical schools in India for ayurvedic medicine even after more than 2,000 years of its first existence – an impressive feat.

---

[17] *Suśruta-Saṁhitā*, *Sūtrasthānam*, 15: 41.
[18] *Caraka-Saṁhitā*, *Sūtrasthānam*, 9: 5.
[19] For more information, Section 10.4.
[20] *Suśruta-Saṁhitā*, *Sūtrasthānam*, 15: 43 - 44.
[21] *Manu-Saṁhitā*, 3: 7.
[22] Dash, in Selin, Kluwer Academic Publishers, Boston, 1997, p. 183. I have a suspicion that this date assignment may be revised in the coming years.

The eight branches of ayurveda are:[23]

1. *Kāyā-cikitsā* (internal medicine)
2. *Śalya-tantra* (surgery)
3. *Śālākya-tantra* (diseases of eyes, ears, nose, and throat)
4. *Agada-tantra* (toxicology)
5. *Bhūta-vidyā* (treatment using occult and similar practices)
6. *Bāla-tantra* (pediatrics)
7. *Rasāyana-tantra* (geriatrics, including the preparation of elixirs and alchemy)
8. *Vājikaraṇa-tantra* (sexology and aphrodisiacs)

The *Caraka-Saṁhitā* documents a symposium in which some 52 noted sages of the Hindus, along with many forest-dwelling hermits, gathered to find remedies of various diseases: "Good health is the root of virtue, wealth, enjoyment, and salvation. Diseases are the destroyer of health, of the good life, and even of life itself. Diseases are the great impediment to the progress of humanity. How can we find a remedy of this problem? With this thought in mind, they meditated."[24] *Caraka-Saṁhitā* was compiled based on the collective knowledge of these sages, rather than the observation or revelation of just one person. This topical symposia tells us about the organized system that was in place to investigate knowledge.

The *Caraka-Saṁhitā* deals with fetal generation and its development, anatomy, function, and malfunction in human body; it describes pathology, diagnosis, prognosis, cure and rejuvenation of the human body. Caraka's descriptions of various diseases and their cures include even minute instructions. In his instructions of the child's room in a hospital, it is suggested to place a variety of toys to entertain a child. The shape, color, and weight of the toys are also prescribed for the children.

---

[23] *Caraka-Saṁhitā*, *Sūtrasthānam*, 30: 28; *Suśruta-Saṁhitā*, *Sūtrasthānam*, 1: 7.
[24] *Caraka-Saṁhitā*, *Sūtrasthānam*, 1, 15 - 17.

Hundreds of diseases are mentioned in the *Caraka-Saṁhitā*. Kanjiv Lochan has compiled a list of various diseases with proper citations. The list includes dyspepsia, squint eye, opthalmitis, acute constipation, stiff neck, hernia, epilepsy, anorexia, hemorrhoids, urinary infections, dysentery, dilatation of blood vessels, tumors, swelling inside throat, tonsillitis, proctitis of the anus, fever, herpes, deafness, pus discharge from ear, frontal headache, chronic abortion, hypersomnia, jaundice, pulmonary consumption, vascular thrombosis, asthma, abscess of the palate, gingivitis, angina or quinsy, and diseases of the reproductive organs.[25] For all diseases, proper drugs and other precautions are described in the *Caraka-Saṁhitā*.

### 10.1.2 Suśruta

Suśruta was another celebrated physician/surgeon in the ayurvedic tradition. He was the son of Viśvāmitra and a student of Divodāsa Dhanvantarī, the King of Kāśī, and flourished prior to 1000 B.C.[26] His work, *Suśruta-Saṁhitā*, is still a pioneer work in surgery (*śalya-tantra*) after 3000 years of its publication.

The *Suśruta-Saṁhitā* has five books with 120 chapters. The five books are:

1. Fundamental Postulates (*Sūtrasthāna*) has 46 chapters dealing with the origin of medicine, the initiation of a medical student, the different branches of medical knowledge, and preliminary preparations for surgery.

2. Pathology (*Nidānasthānam*) has 16 chapters dealing with the diseases of the nervous system, hemorrhoids, urinary track problems, skin diseases, scrotal and genital diseases.

3. Embryology and anatomy (*Śarīrsthānam*) has ten chapters covering the cosmic origin of life, eugenics, pregnancy, fetal development, human anatomy, care of a pregnant women, and postnatal care of children.

[25] Lochan, 2003.

[26] Dash, in Selin, 1997, p. 927.

4. Medical treatment (*Cikitsāsthānam*) has 40 chapters dealing with fractures, dislocations, diabetes, tumors, ulcers, tonics, enemas, douches, and therapeutic applications of fumes, snuffs, and gargles.

5. Toxicology (*Kalpasthānam*) has eight chapters dealing with poisoning of food and drinks, vegetable and mineral poisons, animal poisons, snake venoms, poison treatments, and rabies treatment.

## 10.2 Food as Medicine: Curry for Cure

Plants produce roots, flowers, seeds, and bark to survive and propagate. These constituents of plants are basically some specific "chemical cocktail" that play specific roles for that plant. Humans used them as curative medicines in some cases and preventive medicines in other. It was only natural to eat these plant products and get benefit of their antimicrobial, antifungal, and antiviral properties. Some benefits include better digestion, lower blood pressure, higher energy levels, and a better mind.[27] The healing power of herbs, trees, and vegetables and their products were known to most ancient civilizations. The ancient Hindus were no exception. Their reverence for plant life is symbolized in their worship of specific plants; coconut, banana, *pīpal*, fig, *tulsī*, banyan, and mango trees are used in religious ceremonies.[28]

Spices are used in cuisines all over the world to provide colors, aromatic tastes and odors. They are also used as cosmetics, fragrances, and medicines. They are derived from various parts of plants: flower, bark, fruits, seeds, roots, saps, and leaves. Some of the common medicinal plants that are used as spices in India are: cumin, coriander, fenugreek, cardamom, ginger, fennel seeds, cinnamon, mustard, and nutmeg. These spices have been used as a household therapy as well as in Ayurvedic medicine. Mradu Gupta of the Institute of Post Graduate Ayurvedic Education & Research in Kolkata has compiled a summary of the medical benefits of the Indian spices that are supported by clinical researches.[29]

---

[27]Dalby, 2000; Hirasa, and Takemasa, 1998; Johns, 1990; Parry, 1953; Sherman, 2002; Sherman and Billings, 1999; and Sherman and Flaxman, 2001; Sherman and Hash, 2001.

[28]Simoons, 1998.

[29]Gupta, 2010.

The search for new drugs from the ancient pharmacopoeia is an intense activity in the pharmaceutical industry today. The use of turmeric (*haldī*), the yellow-orange powder from *Curcuma longa* plant, as an antibiotic is a prime example. Turmeric is used in skin diseases, blood impurities, worm manifestation, cough, asthma, deficient lactation, obesity, and diabetes. It is used for antibacterial, anti-inflammatory, hypocholesteremic, antihistaminic, antihepatotoxic, antifungal, and antiarthritic actions. It also has potential activity against cancer, diabetes, arthritis, and Alzheimer's disease.[30] In just 2005, nearly 300 scientific and technical papers referenced curcumin, an ingredient of turmeric, in the National Library of Medicine's PubMed database.[31] In India, there is hardly a day in a Hindu household, when this herb is not eaten in food. It is practically an essential ingredient in most food recipes in India.

The *Caraka-Saṁhitā* mentions *haldī* in a variety of cures that are similar to what modern researchers are finding. Turmeric is the poster child of biopiracy issues that the West is accused of by the Indian government. The University of Mississippi received a patent for *haldī* as a compound for wound healing. This is nothing new. In India, my mother used turmeric often on my cuts as a child. The indigenous knowledge that has been there for millions to use, how can one patent it? Neem is another example of such an issue between indigenous knowledge and modern pharmaceutical industries. *Mucuna pruriens* (Sanskrit name, *atmagupta*) is prescribed in ayurveda for Parkinson's disease. *Brāhmī* (*Cacopa monnieri*) significantly reduces beta amyloid deposits that are the underlying cause of human Alzheimer's disease. Similarly, reserpine, an alkaloid of the plant *rauwolfia*, is used in the treatment of hypertension and stress.

*Elettaria cardamomum Maton* (*Elaichi*) has been traditionally used to cure bad breath, vomiting, thirst, and pyrexia. Cineol, terpenene, and terpeneol are some of the primary active ingredients in it and have anti inflammatory, antipyretic, carminative, and digestive pharmacological properties. Similarly, clove (*yzygium aromaticum*, *lavang* in Hindi) is useful in dental and oral diseases, throat infections, cough, asthma, and colic pain. It is also useful as a local anesthetic, analgesic, antimicrobial, and for its anticarcino-

[30]Gupta, 2010; Stix, 2007.
[31]Stix, 2007.

genic pharmacological properties.[32]

Most people pop over-the-counter pills from a pharmacy, or call their family doctor, schedule an appointment, get a prescription for medicine, and consume the medicine. Very few people have the knowledge of right kind of food and the right portion sizes when they make choices of their food. Unfortunately, our educational system does not provide us with this valuable information. When people eat food that is high in calories but low in nutrients, they are depleting their body of nutrients that help sustain wellness. Once their body becomes sick, they are usually assigned drugs to help. These drugs don't heal the body like maintaining a healthy diet would. It is estimated that obesity alone "is associated with a twofold increase in mortality, costing society more than $100 billion per year."[33]

A main reason for the difference between medicine and food is smell and taste. Herbal medicines may have a very distinctive smell or taste, sometimes proving to be unpleasant. Biomedical pills are prepared to de-emphasize smell and to make them more appealing to people. Therefore, some people would rather take a laxative pill than to eat celery and broccoli.

The best way to replenish your body is to have a high diet rich in nutrients, but low in calories. Vegetables and fruits are the best. These foods contain phytonutrients which are "chemical compounds in plants that act on human cells and genes to bolster your body's innate defense against illness. Put simply, phytonutrients can save your life."[34] Phytonutrients are something the body craves for, be it a healthy body or a body in need of healing. They have been proven in studies to prevent cancer, which is a remarkable feat. In his studies, Khalsa concludes that the "future of healing is not in drugs, vaccines, or stem cells . . . it is in food."[35]

A proper diet offers a variety of different nutrients that the cells of the body need to stay strong. This prevents damage from occurring within the body. With all the cells working correctly in your body, the chances of living a longer life are increased. Broccoli is a vegetable that is very well known

[32] Gupta, 2010.
[33] Bender *et al*, 1998; Wolf and Colditz, 1998.
[34] Khalsa, 2003, p. 26.
[35] Khalsa, 2003, p. 26.

for its ability to heal the body and prevent sickness. Broccoli contains the phytonutrient called sulforaphane. "Sulforaphane triggers natural cancer-blocking agents in the body" and it "protects genes from damage by helping rid your body of harmful stress and alarm signal molecules". It is a natural cancer-blocking agent[36] and may significantly lower the risk of breast, bladder, lung, skin, colon, prostate, and liver cancer.[37]

Blueberries are a fruit that is quite beneficial to the human body. Among all fruits and vegetables, blueberries contain the highest antioxidant capability, according to the United States Department of Agriculture and this helps to reduce inflammation in the brain, a condition thought by many experts to play a major role in Alzheimers, Parkinsons, and other diseases of aging.[38] Blueberries not only have been proven to prevent these diseases of aging, but they have been researched and proven to "actually reverse some of the age-related neuronal and behavioral dysfunction."[39] The phytonutrients in berries are effective in maintaining (or improving) the health of eyes, brain, heart, and immune system.[40] Berries contain ellagic acid and quercetin. Ellagic acid is known to suppress cancer while quercetin is a powerful anti-cancer, anti-inflammatory, and cardioprotective antioxidant.[41]

Watermelon has lycopene and glutatione that makes it an anticancer, antioxident, and anti-aging fruit.[42] Similarly, an orange has vitamin C, terpene and limonene. These protect the cardiovascular and immune systems.[43] Similarly, cantaloupes have a vast amount of zinc, potassium, and vitamins A and C.[44]

The ancient Hindus understood the importance of plants in driving away diseases from the human body. The *Ṛgveda* emphasizes the importance of trees in preserving human life: "May Plants, the Water and the Sky pre-

---

[36]Khalsa, 2003, p. 27
[37]Khalsa, 2003, p. 29.
[38]Khalsa, 2003, p. 42.
[39]Khalsa, 2003, p. 43.
[40]Khalsa, 2003., p. 40.
[41]Khalsa 2003, p. 40.
[42]Khalsa, 2003, p. 46
[43]Khalsa, 2003, p. 45.
[44]Khalsa, 2003, p. 46.

serve us . . ."[45] A whole chapter in the *Ṛgveda* (10: 97) is written on the therapeutic properties of plants, describing their usefulness in achieving good health. "Over all fences have they passed, as steals a thief into the fold. The plants have driven from the frame whatever malady was there."[46] "Let fruitful Plants, and fruitless, those that blossom, and the blossomless, urged onward by Bṛhaspati, release us from our pain and grief."[47]

The use of herbal medicine (*oṣadhi*) was restricted not only to human beings but was extended to cure diseases in animals. For example, herbs were used to increase milk production in cows, as indicated in the *Atharvaveda*. "First O *Arundhati*, protect our oxen and milky Kline: Protect each one that is infirm, each quadruped that yield no milk. Let the plants give us sheltering aid, Arundhati allied with Gods; avert consumption from our men and make our cow-pen rich in milk."[48] The *Atharvaveda* has multiple hymns about the herbal medicines.[49] The *Atharvaveda* tells us that mustard is eaten to get rid of disease and mustard oil makes a person strong. Suśruta mentioned some 760 kinds of medicinal plants.

*Āmlā* (*Emblica officinalis*) was long known as a cure for lost vitality and vigor. In the ancient texts of ayurveda, it is labeled as "the best among rejuvenative herbs" or "useful in relieving cough and skin diseases."[50] In a recent study by Mani Vasudevan and Milind Parle, *āmlā* is found to improve the memory of young and aged mice. It also reduced cholesterol levels among mice by simply taking the oral dose of *Āmlā-chūraṇ* (powder of *Āmlā*) for 15 days.[51] It is indeed a magnificent achievement that must be tried on humans.

---

[45] *Ṛgveda*, 5: 41: 11.

[46] *Ṛgveda*, 10: 97: 10.

[47] *Ṛgveda*, 10: 97: 15.

[48] *Atharvaveda*, 6: 59: 1 - 2. *Arundhati* is a climbing plant that is not presently known. It was used to increase milk production in cows, and to heal broken bones. "Thou art the healer, making whole, the healer of the broken bone: maker thou this whole, *Arundhati*! Whatever bone of thine within thy body hath been wrench or cracked, may *Dhatar* set it properly and join limb by limb with marrow be the marrow joined, thy limb united with the limb. Let what hath fallen of thy flesh, and the bone also grow again." *Atharvaveda*, 6: 59: 1, 2.

[49] *Atharvaveda*, 8: 7: 1 - 10; 6: 16: 1; 6: 24: 2.

[50] Vasudevan and Parle, 2007.

[51] Vasudevan and Parle, 2007.

Figure 10.1: Lord Hanumāna carrying "a mountain of herbs" for Jambavān.

During the battle between Lord Rāma and Rāvaṇa that is compiled in two famous epics, *Rāmāyaṇa* of Vālmīki and the other of Tulsī Dās, Rāvaṇa's son, Indrajit, hits Rāma's brother, Lakṣmaṇa, with a weapon and puts him in a coma. Hanumāna, in his search of specific medicinal herbs, *viśalyakarṇī, suvarṇkarnī, saṁdhānī,* and *mrt-Sanjīvanī-būtī,* went to the Himalayan mountains and fetched an entire mountain of herbs so that Jāmbavān, the medicine man, could select the required herbs.[52] This episode is commonly painted on the walls of Hindu temples in India. (Fig. 10.1) This may or may not be an entire mountain that Hanumāna carried. However, it certainly indicates a broad selection of herbs that he brought with him.

According to an ancient legend, Jivaka Kumarabhacca was a revered physician who provided medical care to Lord Buddha and his disciples. Jivaka studied medicine under the legendary and revered physician Ātreya, in

[52] *Rāmāyaṇa* of Vālmīki, *Yudh-Kāṇḍa*, 33.

the city of Taxila. After many years of training, one day Jivaka shared his desire to start his own practice with his teacher. It was a common practice in India to test the knowledge of the disciples, similar to the modern evaluation system, before they were allowed to graduate. Ātreya asked Jivaka to go around the *āshram* (monastery) and collect all plants with no medicinal value. Jivaka walked around and toiled to look for a plant with no medicinal value. After considerable effort, he went back to Ātreya with no plant in his hand and told Ātreya that he could not find even a single plant with no medicinal value. Ātreya became pleased with the answer and gave his blessing to Jivaka to start his own practice for the welfare of human beings.

Food provides us with essential nutrients and is a necessity of life. It is integral to the *materia medica* of ayurvedic medicine.[53] It is food that allows a person to get energy, avoid diseases, and through metabolic functions provides blood, bone, and flesh to the body. Aloe vera (*kumāri*), *śatāvarī* (*asparagus racemosus*), *neem* (*azadirachta indica*), *brahmī* (*bacopa monnieri*), *amalāki* (*phyllanthus emblica*), *aśvagandhā* (*withania somnifera*), *harītakī* (*terminalia chebula*) are some of the popular herbal plants that are used in ayurveda.[54]

Suśruta recommended that water should be boiled if contaminated. Contaminated water can give skin diseases, cough and breathing problems, indigestion, jaundice, colic pain, and other diseases. He described numerous water purification techniques that can filter particle contamination, mineral contaminations, and bacterial contaminations. For example, boiling was suggested as an important tool to take care of contaminations.[55] Similar recommendation of boiling was prescribed for milk that was stored for a period.[56] In the summer season to avoid the inconvenience of hot temperature, it was advised to first boil the milk and then cool it for drinking.[57]

Indians use more spices per recipe than most other countries.[58] In a re-

[53] Alter, 1999

[54] Arivazhagan et al, 2004; Chopra, Nayar, and Chopra, 1956; Govindarajan, Vijayakumar, and Pushpangadan, 2005; Jain, 1991, and Khalsa, 2003.

[55] *Suśruta-Saṁhitā, Sūtrastānam*, 45: 11 - 17.

[56] *Suśruta-Saṁhitā,Sūtrastānam*, 45: 62 - 63.

[57] *Suśruta-Saṁhitā,Cikitsāstānam*, 24: 102 - 106.

[58] Sherman and Hash, 2001.

cent study, Indians were found to use 9.3 spices per recipe for meat-based dishes and 6.5 spices per recipe for vegetarian dishes.[59] Foods of India are well known around the world for their spicy taste. There is hardly any city in the Western world without an Indian restaurant. The smell of curry fills the air of cities like New York, Los Angeles, London, Paris, and Frankfurt.

Some foods contain microorganisms that can be deadly. "Cooking, salting, and use of spices are cultural responses to this dilemma."[60] Spices are the plant products that are used in various foods and beverages. Typical examples are chilies, garlic, coriander, ginger, cumin, and onion. Spices also kill parasites and pathogens that are in plants and, thus, reduce the health risks of food poisoning and other food borne illnesses.[61] As cells of dead plants are better protected against bacteria and fungi than the cells of dead animals, it is imperative to use more spices for meat-based recipes to kill parasites and pathogens. Plant cells are better protected from microorganisms as the cell walls contain cellulose and lignin which are difficult to decompose.[62] Sure enough, the meat-based recipes use more spices than vegetarian recipes to counter the ill-effects of meat.[63]

Spices are prominent in traditional dishes from the tropical and subtropical regions while people from the cold climatic regions use less spices. Many common spices inhibit bacteria that can contaminate food. Garlic, onion, and black pepper are established bacteria inhibitors, as they cleanse food of pathogens and thereby contribute to health. This allowed people to preserve food for an extended period.[64] In India, ginger, onion, chilies, and coriander are the most common spices used these days, as indicated in a study by Shermon and Billing.[65] Almost all spices are used in various drugs. In their experiments, Billing and Shermon found that onion and garlic inhibited the growth of all 30 microorganisms considered in the study.[66] Thus, recipes for Indian food are a record of the cultural co-evolutionary race between us and

---

[59] Sherman and Hash, 2001.
[60] Shermon, 2002.
[61] Nakatani, 1994; Sherman and Hash, 2001.
[62] Barnwart, 1989; Sherman and Hash, 2001.
[63] Sherman, 2002.
[64] Billing and Shermon, 1998.
[65] Shermon and Billing, 1999.
[66] Billing and Shermon, 1998.

microbes.

Garlic, onion, and allspice (*garm-masālā*) are found to be the most potent spices in their antibacterial and antifungal powers. These spices inhibited or killed every bacterium on which they were tested. Several spices such as garlic, ginger, cinnamon, and chilies are used against a wide spectrum of diseases, including dysentery, kidney stones, arthritis and high blood pressure.[67]

In a recent computer simulation, the cost of saving a diabetic patient with drugs or with lifestyle changes were compared. It was found that while a lifestyle program costs about $8,800 per year, the cost with drugs was $29,000. And while diet and exercise delayed the onset of diabetes by 11 years, the drug only held it off for three years.[68]

Diabetes is a deadly nutritionally related disease that is both preventable and, in many cases, reversible. A diet rich in refined carbohydrates has a quick glucose release rate and can cause Type II diabetes. Eating a big bowl of refined cereal or a large portion of refined pasta, for instance, will cause sudden peaks of blood glucose, triggering the release of extra insulin to deal with it.[69] By maintaining a healthy diet and weight, your body is able to work at a normal rate at its utmost abilities. Diabetes can also trigger other diseases such as kidney failure or heart attack.

"Foods work in way drugs can't because the body's ecosystem has been designed to work with it. That's why they are called nutrients – because they feed us and keep us healthy and functional," suggest Holford and Burne.[70] Vegetarians in the U.S. have a lower incidence of heart disease than the general U.S. population. This may be because most vegetarian diets are low in total fat, saturated fat and cholesterol. Vegetarian or non-vegetarian diets that are low in fat, saturated fat and cholesterol may decrease blood cholesterol levels. Decreased fat and/or increased fiber, along with a decreased incidence of smoking, increased physical activity, and/or less obesity (often associated with vegans) could also aid in a lower rate of high blood pressure and a lower rate of Type 2 diabetes in vegetarians as opposed to

[67] Johns, 1990.
[68] Holford and Burne, 2007, p. 99.
[69] Holford and Burne, 2007, p. 135.
[70] Holford and Burne, 2007, p. 88.

non-vegetarians.

As aromatic plants and herbs of the East became a part of the luxurious life in Rome, the medicinal uses of these plants also became known to them. The spice trade became crucial to economics and politics.[71] During the period of Augustus (63 B.C. - 14 A.D.), aromatic spices and juices from the Indus-Sarasvatī region, such as pepper, cinnamon, and cardamom, were in common use in Rome. Other imported products such as ginger, bdellium-myrrh, sugar, indigo, cotton, rice, peaches, and apricots were also used by households in Rome.[72] The use of pepper as a table spice and medicine made it the most important of all the foreign plant products used by the Romans.[73] The spice trade from India became crucial for European countries especially during the early colonial period. Expeditions were made from the European countries to raid spice-growing countries and several wars were precipitated, as an outcome.[74] According to Warmington, "[t]he excessively high price of pepper during the Middle Ages was one of the reasons why the Portuguese were led to seek a sea-route to India, and when such a route was made available by the discovery of the passage round the Cape of Good Hope in 1498 [A.D.] a considerable fall in the price of pepper took place as a result."[75]

## 10.3 Doctors, Nurses, Pharmacies, and Hospitals

In the tradition of Ayurveda, sickness was considered as a physiological imbalance and was treated with proper drugs. During the Vedic times, doctors were divided into surgeons (*śalya-vaidya*) and physicians (*bhiṣak*).[76] These doctors were well respected in society. "He who stores herbs [medicine man] is like a King amid a crowd of men, Physician is that sage's name, fiend-slayer, chaser of disease," suggests the *Ṛgveda*.[77]

[71]Parry, 1953.
[72]Warmington, 1974 , p. 40.
[73]Warmington, 1974 , p. 222.
[74]Dalby, 2000.
[75]Warmington, 1974 , p. 233.
[76]Mukhopādhyāya, 1994.
[77]*Ṛgveda*, 10: 97: 6.

The inscriptions of King Aśoka (third century B.C.) refer to the cultivation of medicinal plants and the construction of hospitals as an organized activity. He built hospitals throughout the country and made a law to forbid the cutting down of trees. The rock edict of Girinar in India, erected by Aśoka (r. 269 - 232 B.C.) indicates that separate hospitals for humans and animals were built. Aśoka suggested *ahiṁsā* in the treatment of all animals: "Meritorious is abstention from the slaughter of living beings."[78] Also in the Girnar text, Aśoka wished to achieve heaven by helping all living forms. "There is verily no duty which is more important to me than promoting the welfare of all men. And whatever effort I make is made in order that I may discharge the debt which I owe to all living beings, that I may make them happy in this world, and that they may attain heaven in the next world."[79]

Caraka defines following qualities of a doctor: good knowledge of the medical texts; experienced in curing various diseases, skillful in the preparation of drugs, and their uses; good in quick thinking in difficult situations; and good in hygiene.[80] The doctor should also have control over his hands when in surgery or difficult situations. A person with shaking hands is not a good surgeon.[81] Caraka and Suśruta emphasized that a good doctor must have all needed equipment with him.[82]

In his list of items needed in practice, Suśruta suggested multitude of surgical tools, leeches, fire for sanitizing tools, cotton pads, suture threads, honey, butter, oil, milk, ointments, hot and cold water, etc. For the tools, roundhead-knife, scalpel, nail-cutter, finger-knife, lancet that is as sharp as the leaf of the lotus, single-edge knife, needle, bistoury, scissors of various shapes, curved bistoury, axe-shaped hammer, trocar, hooks, narrow-blade knife, toothpick, etc. Surgeon must know to hold these instruments properly for surgery. Suśruta provides ways to hold these instruments and the way they should be used.[83] Suśruta suggests that surgery should be avoid if one could effectively use leeches to do the same job as the procedure is simpler

[78] Sircar, 1957, p. 47.
[79] Sircar, 1957, p. 51.
[80] *Caraka-Saṁhitā*, *Sūtrasthānam*, 9: 6.
[81] *Caraka-Saṁhitā*, *Sūtrasthānam*, 29: 5 - 6.
[82] *Caraka-Saṁhitā*, *Sūtrasthānam*, 16: 1; *Suśruta-Saṁhitā*, *Sūtrasthānam*, 5: 6.
[83] *Suśruta-Saṁhitā*, *Sūtrasthānam*, 8: 3 - 4.

than the surgical process.[84]

"One who knows the text but is poor in practice gets confounded on seeing a patient; he is like a coward on the battle field. One who is bold and dexterous but lacking in textual knowledge fails to win the approval of peers and risks capital punishment from the king. Having half-knowledge, implying either textual or practical knowledge, is a physician who is unfit for the job and is like a one-winged bird."[85] Suśruta suggested that to be limited to theoretical knowledge in ayurveda only, with no clear exposition of the same, is "like an ass that carries a load of sandal wood [without ever being able to enjoy its pleasing scent]."[86] "Drugs are like nectar; administered by the ignorant, however, they become weapons, thunderbolt or poison. One should therefore shun the ignorant physician."[87] Emphasizing constant practice in diagnosis, Suśruta suggests that "the physician who studies the science of medicine from the lips of his preceptor, and practices medicine after acquired experience in his art by constant practice is the true physician."[88]

Caraka advised people to avoid ignorant doctors,[89] as it can even aggravate their condition.[90] Caraka has provided a detailed code of conduct for doctors which is similar to the Hippocratic Oath in the Western world.[91] This code of conduct forbids doctors to indulge in money-making and lust: "He, who practices not for money nor for caprice but out of compassion for living beings is the best among all doctors."[92] And, most importantly, a doctor must continue to help his patients, despite adversity. This is the only way to save the life of a person.[93] Physicians were instructed to be friendly, kind, eager to help underprivileged people, and remain calm in difficult diseases.[94] "The underlining objective of treatment is kindness. Kindness to another human being is the highest *dharma*. A doctor becomes successful

---

[84] *Suśruta-Saṁhitā*, *Sūtrasthānam*, 13: 3 - 4.
[85] *Suśruta-Saṁhitā*, *Sūtrasthānam*, 3: 48 - 50
[86] *Suśruta-Saṁhitā*, *Sūtrasthānam*, 4: 2
[87] *Suśruta-Saṁhitā*, *Sūtrasthānam*, 3: 51.
[88] *Suśruta-Saṁhitā*, *Sūtrasthānam*, 4: 7
[89] *Caraka-Saṁhitā* *Sūtrasthānam*, 9: 15 - 17.
[90] *Caraka-Saṁhitā*, *Sūtrasthānam*, 16: 4.
[91] *Caraka-Saṁhitā*, *Vimānasthānam*, 8: 13.
[92] *Caraka-Saṁhitā*, *Vimānasthānam*, 8: 13.
[93] *Caraka-Saṁhitā*, *Sūtrasthānam*, 1: 134.
[94] *Caraka-Saṁhitā*, *Sūtrasthānam*, 9: 26.

and finds happiness with this objective."[95] However, doctors were advised to concentrate only on the treatment of the patient and not on materialistic rewards, the patient was advised to reward the doctor with money that he can afford and to be respectful toward the doctor.[96]

Suśruta advised that doctors must involve all five senses in the physical examination of the patient: inspection, palpation, auscultation (listening to the sound of the heart, lungs, etc.), taste, and smell.[97] For incisions, surgeons were advised to use the tool rapidly only once to desirable depth.[98] Obviously, this was done to avoid or minimize pain to patients. It was also advised that the surgeon must be courageous, must be quick in action and in making decisions, self confident, should not sweat in tense situations, his hands should not shake, and he should have sharp instruments.[99]

Caraka advised doctors to differentiate the diseases that can be cured or not.[100] He advised doctors to not treat patients with incurable diseases in advanced stages.[101] Since disease is a process with various stages, he suggested that even serious diseases can be cured if caught in earliest stages. Proper medication at the early stage and proper lifestyle is essential to control these serious ailments.[102] It is essential for a doctor to recognize the disease by knowing the symptoms of the ailment. However, since the number of diseases is very large, a doctor should not feel ashamed if he cannot identify the disease from the symptoms.[103] This is an example of honesty that is expected of all physicians.

---

[95] *Caraka-Saṁhitā*, *Cikitsāsthānam*, 4: 63.

[96] *Caraka-Saṁhitā*, *Cikitsāsthānam*, 4, 55.

[97] *Suśruta-Saṁhitā*, *Sūtrasthānam*, 10: 4.

[98] *Suśruta-Saṁhitā*, *Sūtrasthānam*, 5: 7.

[99] *Suśruta-Saṁhitā*, *Sūtrasthānam*, 5: 5 - 6.

[100] *Caraka-Saṁhitā*, *Sūtrasthānam*, 18: 39.

[101] *Caraka-Saṁhitā*, *Sūtrasthānam*, 18: 40 - 44. This is a sensitive issue. Most of our medical expenses are incurred in the last five years of our lives. Should we treat a patient with no or little chance of recovery? This has become a issue in America where patients with terminal illnesses are kept alive with life-supporting systems. The *Edwin Smith Papyrus* has dealt with the same issue more than 3000 years ago in Egypt. Both ancient documents are in agreement, and are in contrast to the current medical practices.

[102] *Caraka-Saṁhitā*, *Sūtrasthānam*, 18: 38.

[103] *Caraka-Saṁhitā*, *Sūtrasthānam*, 18: 45 - 46.

Doctors were responsible for treatment and were punished for wrongdoings. The penalty for such wrongdoings depended on the extent of damage. Humans received more attention than animals in their medical treatment.[104] Physicians were not above the law; their greed and carelessness were checked with laws during the period of Chandragupta Maurya (Candragupta, reigned 322 - 298 B.C.) in northern India.[105]

According to Caraka, the following are the four constituents that are crucial for a quick recovery of the patient: a qualified doctor, a quality drug, a qualified nurse, and a willing patient.[106] This clearly signifies the importance of nursing in medical care. The roles of a nurse are: the complete knowledge of nursing, skillful, affection for the patient, and cleanliness.[107] A good patient is defined as someone with a good memory, follows the instructions of the doctor, free of fear, and can explain the symptoms of the disease well. This combination of doctor, nurse, drug, and patient is essential for a successful cure. Otherwise, the cure is incomplete.[108]

Patients were advised to ignore doctors who practice just for money, were not qualified, or were too proud of their status.[109] Patients should go to a doctor who has studied *śāstra* (texts), is astute, is pure, understands the job well, has a good reputation, and has good control of his mind and body, suggested Caraka.[110] A good doctor should be respected like a guru.[111] "The patient, who may mistrust his own parents, sons, and relatives, should repose an implicit faith in his own physician, and put his own life into his hands without the least apprehension of danger; hence a physician must protect his patient as his own begotten child," suggest Suśruta.[112]

Pharmacies associated with the hospitals prepared medicines as powders, pastes, infusions, pills, confections, or liquids using chemical techniques of

[104] *Manu-Saṁhitā*, 9: 284.
[105] Kauṭilaya's *Arthaśāstra*, 144.
[106] *Caraka-Saṁhitā*, *Sūtrasthānam*, 9: 3.
[107] *Caraka-Saṁhitā*, *Sūtrasthānam*, 9; 8.
[108] *Caraka-Saṁhitā*, *Sūtrasthānam*, 9: 9 - 13.
[109] *Caraka-Saṁhitā*, *Sūtrasthānam*, 29: 12.
[110] *Caraka-Saṁhitā*, *Sūtrasthānam*, 29: 13.
[111] *Caraka-Saṁhitā*, *Sūtrasthānam*, 4: 51.
[112] *Suśruta-Saṁhitā*, *Sūtrasthānam*, 25: 42 - 45.

oxidation, fermentation, and distillation.

In providing the design of a hospital, Caraka suggests that it should have good ventilation and the rooms should be big enough for patients to walk comfortably. The rooms should not experience the intense glare of sunlight, water, smoke, and or dust. The patient should not experience any unpleasant sound, touch, view, taste, or smell. Expert chefs, masseuses, servants, nurses, pharmacists, and doctors should be hired to take care of the patients. Birds, deer, cows and other animals, along with singers and musicians should be kept in the hospital compound.[113]

Hospitals should have enough quality food in the storage room, stored drugs, and a compound with drug plants. The doctors must be intelligent in knowing the books as well as equipment. Only then, they were allowed to help a patient.[114]

## 10.4 Ayurveda

Ayurveda is the Hindu science of healing and rejuvenation. It is a holistic system of medicine which is curative as well as preventive. A person needs to practice the doctrines of ayurveda when healthy to avoid sickness. It is a holistic approach which focuses on mind and body. Due to ayurveda's success, it is becoming increasingly popular in the western world. Dietary measures and lifestyle changes are recommended to delay the aging process at the cellular level and to improve the functional efficiency of body and mind. The basic idea is to create a balance in the body that will allow the body to cure itself.

Ayurveda is a combination of two words–*āyur* and *veda*. *Āyur* means the span of life while *veda* means unimpeachable knowledge or wisdom. Ayurveda signifies a life science for the well being of body and mind. This stemmed out of Hindus' belief that health and disease co-inhere in body and mind. Ayurveda has its origin in the *Vedas* for the "principal elements of its general doctrines."[115]

[113] *Caraka-Saṁhitā*, *Sūtrasthānam*, 15: 7.
[114] *Caraka-Saṁhitā*, *Sūtrasthānam*, 16: 1.
[115] Filliozat, 1964, p. 188.

In the Hindu tradition, Ayurveda is considered as an *Upa-Veda* (subordinate *Veda*) that deals with medicine and medical treatment for good health, longevity, and elimination of disease. It is thought to have evolved into a well-developed form at the time of *Atharvaveda*. Suśruta wrote: "The Ayurveda originally formed one of the subsections of the *Atharvaveda* and originally consisted of 100,000 verses."[116]

Ayurveda uses herbal medicine, dietetics, surgery, psychology, and spirituality to cure diseases. *Pañca-karma* (detoxification of body), laxatives, herbal oil messages, and nasal therapies are some of the treatments. Though ancient in origin, it is still the most popular system of medicine in India. Several Indian universities provide doctoral degree programs in ayurveda. These universities train doctors who practice throughout India and provide low cost medical care to poor people. These days, Banaras, Chennei, Hardwar, Lucknow, Patna, and Trivandrum are some of the prominent cities that house universities to train ayurvedic doctors.

Caraka suggested that physical ailments are cured by proper diet, changing daily routines that included moderate physical exercise,[117] and medicine, while the mental diseases are cured by self control, meditation, and even by charms (*bhuta-vidyā*).[118]

The purpose of Ayurveda is to restore harmony of the mind and physical body. Ayurveda deals primarily with questions about the way diseases originate, their symptoms, their cure, and the way to prevent them from happening. Caraka defined that the purpose of ayurveda was to preserve the well-being of a healthy person and to get rid of the diseases of a sick person.[119]

In the ayurvedic tradition, the human body is considered as a conglomeration of five elements, known as *mahābhūtas* – earth (*pṛthivī*), air (*vāyu*), water (*apaḥ*), fire (*tejas*), and space (*ākāśa*). The five *mahābhūtas* produce seven *dhātus*: *rasa* (juice-plasma), *rakta* (blood), *māṃsa* (flesh), *medas* (fat), *asthi* (bone), *majjā* (marrow), and *śukra* (reproductive fluids in male and fe-

---

[116] *Suśruta-Saṁhitā*, *Sūtrasthānam*, 1: 3.
[117] *Caraka-Saṁhitā*, *Sūtrasthānam*, 7: 30.
[118] *Caraka-Saṁhitā*, *Sūtrasthānam*, 1 - 8.
[119] *Caraka-Saṁhitā*, *Sūtrasthānam*, 30: 26.

males).[120] The five *mahābhūta*, or the seven *dhātu*, constitute the *prakṛti* (attributes) of a person. These *dhātu* are a product of the five elements. When in equilibrium, these *dhātus* make a person happy and healthy. The various combinations of these *dhātu* constitute the tri-*guṇas* (three-attributes) in a person: *sattva*, *rājas*, *tāmas*. A person with dominant *sattva* values truth, and honesty; *rājas* value power, prestige, and authority. A *tāmas* dominant person lives in ignorance, fear, and servility.

When a patient visits a doctor, the diagnosis includes examination of body temperature by feeling the wrist, counting the pulse rate by feeling pulsations at the wrist, pulse observation at different places of the body, observing skin color and eye color, examination of the tongue, examination of feces and the color of the urine. If needed, doctors ask the patients to lie down and press their kidneys, livers, and other areas to observe softness and the reactions of the patients. In therapeutics, a proper diet and regimen are carefully detailed, and baths, emetics, inhalations, gargles, blood-letting, urethral and vaginal injections are employed. Most ayurvedic medicines contain one or more substances that are "active ingredients." These ingredients are mostly taken with water, milk, or honey.

The basic theory of Ayurveda rests on the three humors (*tridoṣa*; *tri* = three, *doṣa* = the literal meaning is defect, popularly known as elements)–*Kapha* (phlegm), a heat regulating system, and mucous and glandular secretion; *vata* (wind), a nerve force; and *pitta* (bile), metabolism and heat production. These three *doṣa* control the normal functions of a human body and a balance is necessary. The term *doṣa* is generally translated as "corrupting agents," "defect," or "imperfection." However, in ayurveda, these *doṣa* are the keys to good health. The theory of *doṣa* is pretty complex as it "is affected by an almost infinite number of exogenous factors and combinations of factors that make up human ecology: diet, rate, time, and context of food consumption, climate, direction of wind, age, behavioral patterns, rest, accidents, exercise, state of mind, and so forth."[121] According to ayurvedic principles, life and the biological processes are dependent on the production of heat inside the organism. The body-heat comes from the food that maintains the metabolism and nourishes the body. The digestive process is carried

[120]For the roles of these *dhātu*, read *Suśruta-Saṁhitā*, *Sūtrasthānam*, Chapter 15.
[121]Alter, 1999; Lad, 1984; and Zysk, 1991.

out by *agni* (fire or heat) that converts food into its different constituents.[122]

An imbalance of these three *doṣas* causes most diseases. It is the purpose of all treatments to keep the balance of the *doṣa* by balancing the seven *dhātus*.[123] The best way to treat a patient is to recognize the imbalance and provide a cure.[124] The *Suśruta-Saṁhitā* gives various symptoms of the deficiency of these *tridoṣa* and their effect on body.[125]

The aim of ayurveda is to restore and maintain the equilibrium of the constituents for all body types. This equilibrium is maintained or restored by formulations that oppose the disturbed *doṣa*. For example, *Vata* properties relate to cold, non-lubricant, light, mobile, and rough. *Vata* manifests itself as the motor systems of the human body, including movement and transmission within the body. It includes the motor activities of respiration, speaking, digestion, and various coordination: *Prāṇa* (breathing for respiration), *udāna* (breath of stomach for digestion), and *apāna* (breath of the excretory system that also includes sexual activity).[126] *Pitta* is mildly lubricating, hot, sharp, sour, mobile, and stinging. *Pitta* is responsible for digestion, heat, visual perception (keenness), lustrous skin (oiliness), intelligence, courage, and softness of the body.[127] In modern terms, *pitta* refers to the nutritional and thermogenetic system of the body. Similarly, *Kapha* is heavy, cold, soft, lubricating, sweet, greasy and immobile. *Kapha* is for the firmness, sweetness, dullness, heaviness, softness, peace, love, forgiveness, and lubrication of the joints. *Kapha* shows itself in mucus and plasma, and holds together the limbs and act as an integrator system.[128] Thus, *kapha* includes mental as well as bodily functions.

We need drugs to oppose these specific dominant properties to bring about equilibrium. The deficient *dhatu* are supplied using drugs and the excesses of *dhātu* are removed using changes in diet or with drugs. This equilibrium

[122]Larson, 1987.
[123]*Caraka-Saṁhitā*, *Sūtrasthānam*, 16: 35 - 36.
[124]*Caraka-Saṁhitā*, *Sūtrasthānam*, 18: 45 - 46.
[125]*Suśruta-Saṁhitā*, *Sūtrasthānam*, Chapter 15.
[126]*Caraka-Saṁhitā*, *Sūtrasthānam*, 18: 49.
[127]*Caraka-Saṁhitā*, *Sūtrasthānam*, 18: 50.
[128]*Caraka-Saṁhitā*, *Sūtrasthānam*, 18: 51.

is synonymous with health and happiness. [129] In *tridoṣa* theory, earth and water constitute *kapha*, water and fire constitute *pitta*, and air and ether constitute *vata*. *Kapha* is cold, wet, heavy, and static while *pitta* is hot, sharp, clear, and light. A *vata* person is dry, light, subtle, rough, and changeable.

The *tridoṣa* theory is not just a hypothetical construct; these *doṣa* are actual substances (*dravya*) in the human body. The excess or deficiency of these substances can only be diagnosed through bodily functions. These elements nourish the body and are in some way the *chi* or *qi* in the Chinese medicine. However, both are not generally recognized in the modern medicine as yet.

Balance is essential in the natural world and is a key concept in Hindu philosophy. It is the balance of *āsana* that is crucial in yoga. This involves a balance of body and mind. It is the balance of *tridoṣas* in ayurveda that is crucial in the well being of a person, just like the *yin* and *yang* in Chinese philosophy. Therefore, before ayurvedic doctors look for the cure of a disease, they inquire about the patient, his/her constitution, eating habits, and nature to know the imbalance in *dhātu*. For example, in describing the cause of diabetes and its cure, Suśruta and Caraka attribute lack of exercise, laziness, sweet foods, alcoholic foods and beverages, elevation of *kapha*, and excess of newly harvested food grains as the cause.[130]

According to Caraka, human beings have five *indriya* (senses): hearing, touch, vision, taste, and smell. These senses are created by *dravya* (materials): space, air, water, earth, and *tejas* (aura). The senses are located in five places: ear for hearing, skin for touch, eye for vision, tongue for taste, and nose for smell.[131]

A continual balance of *tridoṣa* results in contentment, longevity, and enlightenment. We eat food to nourish the seven *dhātus* in our body. By changing diet, people can change the equilibrium of *dhātus* and thus their attributes. Therefore, among the Hindus, it is a customary to describe the attributes of a person from a combination of two words, *ācar-vicār*, describ-

---

[129] *Caraka-Saṁhitā*, *Sūtrasthānam*, 1: 53.
[130] Tewari, 2005.
[131] *Caraka-Saṁhitā*, *Sūtrasthānam*, 8: 3.

ing the thinking and eating habits of a person. Suśruta suggested eating in moderation for a healthy life.[132] Caraka clearly regulates the quantity of food that a person should eat: "The food must be digested in time and should not cause any inconvenience in activity of the person."[133]

Patients should seek effective drugs immediately after recognizing a sickness.[134] Thus, quick diagnosis is the key in fighting against a disease. Following are the typical drugs used in ayurveda:

1. Animal products (honey, milk, blood, urine, fat, horn, etc.)
2. Minerals and metals (salt, gypsum, alum, gold, lead, copper, iron, phosphorous, sulfur, potassium, calcium, iodine, lime, mud, precious stones, etc.)
3. Herbs and plants (for example, aloe vera, barley, beans, cinnamon, clove, coriander, cumin, black pepper, fenugreek, neem, etc.)

In Ayurveda, metals are oxidized, reduced, and transformed into a chemical compound that is non-toxic, to be used as a drug.[135] Several elements like gold, lead, copper, iron, phosphorous, sulfur, potassium, calcium, and iodine are processed with physio-chemical techniques. In ayurvedic technical terms, these metals were killed (*bhasm*) before they could be used for drugs. Killing defines oxidation, reduction, or transformation into a complex compound to be used as effective drugs. Killing sometimes means that a metal is deprived of its well-characterized physical properties such as color, lustrousness, and hardness. The techniques that are defined in Suśruta- and *Caraka-Saṁhitā* for various diseases are still commonly used in ayurvedic medicines in India, and are effective in their use.

Water, sunlight, milk, honey, and fermented extracts of fruits are also used as medicine in Ayurveda. For example, *Drākṣaśava*, that is an extract of grape and apple-vinegar along with several minerals, with a specific fermentation technique, is perhaps the best natural medicine for indigestion. It cleans the blood, improves the functioning of kidneys and liver, improves the

---

[132] *Suśruta-Saṁhitā, Sūtrasthānam*, 46: 145.
[133] *Caraka-Saṁhitā, Sūtrasthānam*, 5: 4.
[134] *Caraka-Saṁhitā, Sūtrasthānam*, 11: 63.
[135] For more information, read Zimmer, 1948.

pH value of urine, and provides minerals. It is not only a curative medicine but could also be used on a regular basis as a preventive medicine for good health.

Minerals are also used for curative purposes; *śilājīta*, perhaps the most important one, is a gelatine substance secreted from mountain stones during the scorching heat of June and July in northern India, and contains traces of tin, lead, copper, silver, gold, and iron. *Śilājīta* has heat-producing and body purifying properties and is used in the treatment of diabetes, leprosy, internal tumor, and jaundice.[136]

Caraka catalogs all drugs into 50 groups. The list is detailed and provides ample evidence of the various cures of the ancient Hindus:[137]

1. *Jīvanīya*, to prolong life
2. *Brinhanīya*, to promote nutrition
3. *Lekhanīya*, to thin the tissues
4. *Bhedanīya*, to promote excretion
5. *Sandhānīya*, for the union of fractured bones
6. *Dīpanīya*, to increase appetite and digestive power
7. *Balya*, to increase strength
8. *Varnya*, to improve complexion
9. *Kanthya*, to cure voice or hoarseness
10. *Hṛdya*, to promote cheerfulness
11. *Triptighna*, remove phelgm
12. *Arsoghna*, to cure piles
13. *Kushtāghna*, cure skin disease

[136] *Suśruta-Saṁhitā*, *Cikitsāstānam*, Chapter 13.
[137] *Caraka-Saṁhitā*, *Sūtrasthānam*, 4: 9 - 19; taken from Kutumbiah, 1962, p. 122 - 123.

14. *Kandūghna*, cure pruritis
15. *Krimighna* cure worms
16. *Viṣagna*, antidotes to poison
17. *Stanyajanana*, to promote secretion of milk
18. *Stanyasodhana*, to improve the quality of milk
19. *Sukrajanana*, increase secretion of semen
20. *Sukraśodhana*, purify the semen
21. *Snehopaga*, emollients
22. *Swedopaga*, diaphoretics
23. *Vamanopaga*, emetics
24. *Virechanopaga*, purgatives
25. *Āsthapanopaga*, medicines for enemas
26. *Anuvāsanopaga*, medicines for oily enemas
27. *Sirovirechanopaga*, to promote discharge from the nose
28. *Chhardhinigrahana*, relieve vomiting
29. *Trishnānigrahana*, relieve thirst
30. *Hikka nigrahana*, relieve hiccups
31. *Purīshasangrahanīya*, render the feces consistent
32. *Purishavirechanīya*, alter the color of the feces
33. *Mutrasangrahanīya*, reduce the secretion of urine
34. *Mutravirajanīya*, alter the color of urine
35. *Mutravirechanīya*, increase the secretion of urine
36. *Kāsahara*, cure coughs

37. *Svāsahara*, asthma
38. *Sothahara*, cure anasarca or swellings
39. *Jvarahara*, febrifuges
40. *Sramahara*, remove fatigue
41. *Dahaprasamana*, relieve burning or heat of the body
42. *Sītaprasamana*, relieve a sense of coldness
43. *Udardhaprasamana*, cure urticaria
44. *Angamarddaprasamana*, relieve pain in the limbs
45. *Sūlaprasamana*, cure pain in the bowels
46. *Sonitāsthāpana*, styptics
47. *Vedanāsthāpana*, anodynes
48. *Samjnāsthāpana*, restore consciousness
49. *Prajāsthāpana*, cure sterility
50. *Vayasthāpana*, prevent the effects of old age

The daily routine (*dinacarayā*) is crucial for good health. However, for the Hindus, it is not only a form of preventive medicine, it is also for self-development.[138] Caraka deals with the issue of life-style in great detail in the first chapter of *Caraka-Saṁhitā*.[139]. Suśruta also devoted a long chapter to prescribe a good lifestyle for a healthy person.[140] Almost as a rule, it was considered good to get up early in the morning and practice yoga and meditation after toiletry functions and bath. To optimize health and as preventive measures for various body organs and for general health, dental cleaning with herbs, mouthwash, tongue scraping, daily bath, massage, clean clothes, good shoes, daily exercise, and good thought processes were suggested. The role of the mind was known to the ancient Hindus. Thus, they suggested good

[138] Alter, 1999. This article includes several excellent references on this topic.
[139] *Caraka-Saṁhitā*, *Sūtrasthānam*, sections 5, 7, and 8.
[140] *Suśruta-Saṁhitā*, *Cikitsāstānam*, 24: 3 - 132.

thought process as essential to good health.

Mouthwash received much attention in ancient writings. Caraka suggested to brush the teeth twice a day[141] and clean the tongue using a metallic scraper.[142] He also suggests many herbs such as clove, cardamom, beetle nut, and nutmeg to keep in the mouth for a pleasant breath.[143] Keeping fresh breath is a multi-million dollar industry in the West which can easily be taken care of using these ancient techniques which are quite inexpensive, effective, and good for health. Just tongue scraping can do a major job in keeping the breath fresh. Suśruta provides the following instructions for the hygiene of mouth. "A man should leave his bed early in the morning and brush his teeth. The tooth-brush should be made of a fresh twig of a tree or a plant grown on a commendable tract and it should be straight, not worm-eaten, devoid of any knot or at most with one knot only (at the handle side), and should be twelve fingers in length and like the small finger in girth."[144]

"The teeth should be daily cleansed with (a compound consisting of) honey, powdered *trikaṭu*, *trivarga*, *tejovatī*, *saindhave* and oil. Each tooth should be separately cleansed with the preceding cleansing paste applied on (the top of the twig) bitten into the form of a soft brush, and care should be taken not to hurt the gum any way during the rubbing process. This tends to cleanse and remove the bad smell (from the mouth) and uncleanliness (of the teeth) as well as to subdue the *kapha* (of the body). It cleanses the mouth and also produces a good relish for food and cheerfulness of mind," suggests Suśruta.[145] Afterward, the person should cleanse the tongue by scraping it with gold, silver, wood, foil.[146] The eyes and mouth should be washed with cold water, hair must be combed, and the body should be anointed with oil to guard against the aggravation of dryness and air. People must exercise and experience massage afterward. Bathing must be practiced every day.

---

[141] *Caraka-Saṁhitā*, *Sūtrasthānam*, 5: 71.
[142] *Caraka-Saṁhitā*, *Sūtrasthānam*, 5: 74.
[143] *Caraka-Saṁhitā*, *Sūtrasthānam*, 5: 76.
[144] *Suśruta-Saṁhitā*, *Cikitsāstānam*, 24: 3.
[145] *Suśruta-Saṁhitā*, *Cikitsāstānam*, 24: 6 - 7.
[146] *Suśruta-Saṁhitā*, *cikitsāsthānam*, 24: 11 - 12.

### 10.4.1 *Pañca-karma*

*Pañca-karma* is an ayurvedic treatment to cleanse the body system and remove toxins, that can get rid of diseases using five actions. This treatment has three phases: *purva-karma* (preparation), *pradhā-karma* (main treatment), and *paśca-karma* or *Uttara-karma* (post-treatment).[147] Before the actual process, a dietary and herbal treatment is prescribed to cleanse the intestinal system, reduce the excess or vitiated *doṣas*, and increasing the gastric fire. In the oleation process, oily substances like sesame, flaxseed, or mustard oil or ghee are taken internally or externally in massage and sudation (sweating) techniques are used. Ghee is clarified butter by removing its water contents and milk solids. It becomes lactose-free and can be used by people with lactose-allergies. Ghee is antimicrobial and antifungal and, therefore, can be preserved without refrigeration for an extended period.[148] The oleation process lubricates bodily tissues and should be taken in moderation especially people with *kapha*. The exact prescription in *purva-karma* depends on the individual and the diet includes simple foods such as *khicaṛī*, a combination of basmati rice, split mung lentils and some mild spices, such as turmeric, cumin and coriander.

The five actions of *Pañca-karma* are as follows: *vamana*, *virecana*, *vasti*, *nasya*, and *rakta-moksha*.[149] These actions are provided by Suśruta. First, the body is detoxified by inducing vomiting using herbal mixture (*vamana*), then an herbal mixture is taken for numerous bowel movements (*virecana*), oil or herbs are administered anally for some time and expelled (*vasti*). *Nasya* is practiced by using *neti-kriyā* (luke warm water passed through the sinus) and oil is applied inside to clean the sinus system, and lastly blood is purified or even taken out using needles, incisions, or by leeches.[150]. In the post-treatment, dietary and lifestyle lessons are provided to have balanced *dinacaryā*. These suggestions take into account the needs of the patient. *Pañca-karma* is a preventive measure and commonly practiced by healthy people also.

Caraka as well as Suśruta suggested *pañca-karma*. However, there is a

---

[147]Ninivaggi, 2008, p. 207
[148]Ninivaggi, 2008, p. 209.
[149]Ninivaggi, 2008, p. 212.
[150]Ninivaggi, 2008, p. 219

small difference between the two. Caraka suggested *vamana* (emesis or vomiting), *virechana* (purging), two kinds of *vasti* (therapeutic enema), *nasya* (nasal therapy).[151] Suśruta defined a single category for *vasti* and added *rakta-moksha* (blood letting).

*Pañca-karma* is advised for a variety of medical conditions: arthritis, rheumatism, respiratory disorders, gastrointestinal disorders, menstrual problems, obesity, etc. A team of Harvard Medical School in Massachusetts conducted a clinical study of this practice to observe the role of psychosocial factors in the process of behavior change and salutogenic process. The 20 female participants underwent in a 5-day ayurvedic cleansing retreat program. Measurements were taken before the program, immediately after the program, and after three months for the quality of life, psychosocial, and behavioral changes. This study indicates that *pañca-karma* "may be effective in assisting one's expected and reported adherence to new and healthier behavior patterns."[152]

## 10.5 Modern Research on Ayurvedic Prescriptions

Prescriptions used in traditional ayurveda are about two thousand years old. Since medicine has evolved during the period, are these prescription still valid in today's world? Scientists in India as well as in the West are trying to investigate the properties of various herbs mentioned in ayurvedic texts. In most cases, there is a concordance between the ancient practices and modern uses of these herbs. Two detailed reviews of the ancient ayurvedic practice of various herbal drugs and the current scholarship on some of these drugs are provided by Govindarajan *et al*[153] and Mukherjee *et al.*[154]

The juice of leaves of *andrographis paniculata*, known as *kalmegh* in ayurveda, is a household remedy for flatulence, loss of appetite, bowel complaints of children, diarrhea, dysentery, dyspepsia, and general debility. Decoction or infusion of the leaves gives good results in sluggish liver, neuralgia, in general

[151] *Caraka-Saṁhitā*, *Sūtrasthānam*, 2:1.
[152] Conboy, Edshteyn, and Garivsaltis, 2009.
[153] Govindarajan *et al* 2005.
[154] Mukherjee, Maiti, Mukherjee, and Houghton, 2006a.

debility, in convalescence after fevers, and in advanced stages of dysentery. In clinical trials, administration of *andrographis paniculata* showed that it is antioxidant and has hepatoprotective activity.[155]

*Nimba* (*Azadirachta indica*), commonly known as *neem* in India is very useful in blood disorders, eye diseases, intermittent fever, as well as persistent low fever. Oil is very useful in leprosy, skin diseases, ulcers, and wounds. The bark of the neem tree contains a resinous bitter compound and is usually prescribed in the form of a tincture or an infusion. It is also regarded as beneficial in malarial fever. Ethanolic neem leaf extract exerts protective effects against genotoxicity and oxidative stress by augmenting host antioxidant defense mechanisms. This explains the spread of *neem* trees and its products around the world. In America, particularly in the hot and humid climates of Florida and Hawaii, this tree is planted in large numbers. Several *neem*-based products such as toothpastes, soaps, shampoos, facial creams, and other products are commonly sold in health food stores.

The liver is often the first organ to be infected by metastasizing cancer. Hepatocarcinogenesis is one of the most prevalent and deadly cancers worldwide, and ranks seventh among cancers in order of frequency of occurrence. The effect of *neem* was studied in male rats by Hanachi et al. While the cells of cancer group without treatment were severely necrotic at week 12, cells of cancerous rats treated with *neem* appeared nearly normal. There was no evidence of side effects of neem towards normal cells. This predicts that *neem* is a potential preventive agent for cancer.[156]

*Brahmī* (*Bacopa monnieri*) is a popular herb that is used by young people to improve their memory. It is also used for epilepsy, insomnia, and serves a mild sedative. A poultice made of the boiled plant is placed on the chest in acute bronchitis and other coughs of children for relief. Leaves are also used in asthenia (loss of strength), nervous breakdown, and other similar conditions. In clinical trials, treatment with *Brahmī* extract significantly increased the antioxidant enzymes and inhibited lipid peroxidation and reduced the tumor markers.[157] Roodenrys *et al* studied the effects of *Brahmī*

---

[155] Govindarajan *et al*, 2005.
[156] Hanachi *et al*, 2004.
[157] Rohini, Sabitha, and Devi, 2004.

on human memory in seventy-six adults aged between 40 and 65 years in a double-blind randomized clinical trial. There were three testing sessions: one prior to the trial, one after three months on the trial, and one six weeks after the completion of the trial. The results show a significant effect of *Brahmī* on the retention of new information. *Brahmī* decreases the rate of forgetting of newly acquired information.[158]

The starch obtained from the roots and stems of *Tinospora cordifolia*, known as *guduchi* in ayurveda, is useful in diarrhea and dysentery. *Guduchi* is reported to possess anti-spasmodic, anti-inflammatory, anti-allergic, anti-diabetic, and anti-oxidant properties. Its watery extract, known as Indian quinine, is very effective in fevers due to cold or indigestion; the plant is commonly used in rheumatism, urinary diseases, dyspepsia, general debility, syphilis, skin diseases, bronchitis, spermatorrhea, and impotence. In reviewing the medicinal properties of *guduchi*, Singh *et al* noticed it to be anti-diabetic, anti-periodic, anti-spasmodic, anti-inflammatory, anti-arthritic, anti-oxidant, anti-allergic, anti-stress, anti-leprotic, anti-malarial, hepatoprotective, immunomodulatory and anti-neoplastic. *Guduchi* improves the immune system and significantly reduces the mortality from E. coli induced peritonitis in mice. In a clinical study, it has afforded protection in cholestatic patients against E. coli infection. The hepatoprotective action of *tinospora cordifolia* was reported in one of the experiments in which goats treated with *Guduchi* have shown significant clinical and hemato-biochemical improvement in hepatopathy. Extract of *guduchi* has also exhibited properties against Hepatitis B and E surface antigen.[159]

*Ashwagandha's* roots (*Withania somnifera Dunal*) are known to: be anti-aging, augment resistance of the body against disease and diverse adverse environmental factors, revitalize the body in debilitated conditions and increase longevity. In Ayurveda, it is used as nerve tonic, aphrodisiac, sedative, antirheumatism, and for the treatment of constipation. It relieves inflammation, pain, backache and it stimulates sexual impulses and increases sperm count. It is considered to be good for strength, vigor and rejuvenation. Its roots are used as an application in ulcers and rheumatic swelling. It infuses fresh energy and vigor in the patient who is worn out owing to any

[158]Roodenrys *et al*, 2002.
[159]Singh, Pandey, Srivastava, Gupta, Patro, and Ghosh, 2003.

constitutional disease like syphilis, and in rheumatic fever. The importance of *ashwagandha's* root extract in the regulation of lead toxicity with special reference to lipid peroxidative process has been investigated in liver and kidney tissues.[160] Gupta *et al* investigated the effects of *ashwagandha* on copper-induced lipid peroxidation and antioxidant enzymes in aging spinal cord of rats. It was found to successfully attenuate GPx activity and inhibited lipid peroxidation in a dose-dependent manner. The results indicate the therapeutic potential of *withania somnifera* in aging and copper-induced pathophysiological conditions.[161]

*Triphala* is a well-known polyherbal formulation in Ayurveda. It modulates the neuro-endocrine-immune systems and has been found to be a rich source of anti-oxidants. It prevents aging, reestablishes youth, brain power, and prevent diseases.[162] *Triphala* is the mixture of three (*tri*) fruits (*phala*) which is composed of the dried fruits of *Emblica officinalis Gaertn* (known as Indian gooseberry), *Terminalia belerica Linn* (*Behara*), and *Terminalia chebula Retz* (*Haritaki*) in equal proportions. Traditionally this formulation has been prescribed as a laxative in chronic constipation, a detoxifying agent of the colon, for food digestive problems and as a rejuvenator of the body.[163] Its antidiabetic,[164] cardiovascular activity including blood pressure control, cholesterol controlling activity, and liver protective activities are well established.[165] It is used traditionally for its purgative activity and to cure bleeding and piles. Further, it has been reported to possess anticancer, antimutagenic potential and inhibits local anaphylaxis. Gallic acid is a common phytoconstituent present in all the three fruits used in the *triphala* and is reported to possess hepatoprotective and antioxidant activity.[166]

## 10.6 Surgery

Surgery is as ancient as any other sciences known to humans. The extraction of arrows, cleaning the wound, and the support of broken limbs by splints

[160] Chaurasia, Panda, and Kar, 2000.
[161] Gupta, Dua, and Vohra, 2003.
[162] Ponnusankar *et al*, 2011.
[163] Mukherjee *et al*, 2008.
[164] Sabu and Kuttan, 2002
[165] Mukherjee *et al*, 2006b.
[166] Govindarajan, Vijayakumar, and Pushpangadan, 2005.

were tried since the ancient period. The *Suśruta-Saṁhitā* describes some one hundred medical surgical instruments of metals in various shapes and sizes. These include probes, loops, hooks, scalpels, bone-nippers, scissors, needles, lancets, saws, forceps, syringes, rectal speculums, and probes. Many surgical operations including those for hernia, amputation, tumor, restoration of nose, skin grafting, extraction of cataracts, and caesarean section are described in the *Caraka-Saṁhitā*. In both books, blunt as well as sharp tools depending on the needs are described. Some of these instruments are sharp enough to dissect a hair longitudinally.[167] Suśruta performed lithotomy, cesarean section, excision of tumors, and the removal of hernia through the scrotum.[168]

Suśruta classified surgery into eight parts: extraction of solid bodies, excising, incising, probing, scarifying, suturing, puncturing, and evacuating fluid.[169] He designed his surgical instruments based on the shapes of the beak of various birds and the jaws of animals. He stressed the importance of observation and practical experience to become a better surgeon.[170] Suśruta suggested that the bookish knowledge was not sufficient for surgery; internship with an expert was advised. "A pupil, otherwise well-read, but uninitiated into the practice (of medicine or surgery) is not competent to take in hand the medical or surgical treatment of a disease."[171]

In his attempt to explain surgery, Suśruta wrote: "The scope of surgery, a branch of medical science, is to remove an ulcer and extraneous substance such as fragments of hay, particles of stone, dust iron, ore, bone; splinters, nails, hair, clotted blood or condensed pus, or to draw out of the uterus a dead fetus, or to bring about safe parturitions in cases of false presentation, and to deal with the principle and mode of using and handling surgical instruments in general, and with the application of cautery and caustics, together with the diagnosis and treatment of ulcers."[172]

All surgical procedures were defined in terms of three stages of the process: Pre-operative measures, operation, post-operation measures. This included

---

[167] For a detailed account of the tools, see Mukhopādhāya, 1979.

[168] Singh, Thakral, and Deshpande, 1970. It is an excellent article with basic information.

[169] *Suśruta-Saṁhitā, Sūtrasthānam*, 5: 5

[170] *Suśruta-Saṁhitā, Sūtrasthānam*, 9; 2; *Suśruta-Saṁhitā, Śarīrsthānam*, 5: 47.

[171] *Suśruta-Saṁhitā, Sūtrasthānam*, 9: 2.

[172] *Suśruta-Saṁhitā, Sūtrasthānam*, I: 4.

the preparation of the surgical room, the patient, and the tools. Starvation or mild dieting was advised on the day of surgery. Intoxicating beverages were advised to avoid the pain of surgery. In the post-operative measures bandaging, redressing, healing and cosmetic restoration were also involved.[173] Anesthesia, antiseptic, and sterile techniques are the landmarks of surgery in India.

Suśruta described fourteen different kinds of bandages (*bandha*) as defined in Table 10.1.[174] On the harmful effects of non-bandaging, Suśruta suggested that flies and gnats will be attracted to the wound. Foreign matters, such as dust and weeds will settle on the wound. Sweat, heat and cold can make the wound malignant.[175]

Suśruta advises doctors to clean wounds first to get rid of dust, hairs, nails, loose pieces of bones, and other foreign objects before suturing to avoid suppuration. If not done, it can cause severe pain and other problems.[176] Suśruta used innovative ways to perform various surgical procedures: horse hairs were used as suture material while black carpenter ants were used in closing incisions in soft tissues. Blood or pus was drawn from body by using leeches.

New doctors practiced incisions first on plants or dead animals. The veins of large leaves were punctured and lanced to master the art of surgery. Cadavers were advocated as an indispensable aid to the practice of surgery. Bandaging, amputations, and plastic surgery was first practiced on flexible objects or dead animals just as is done by medical students today. Suśruta suggested that "a surgeon in seeking a reliable knowledge must duly prepare a dead body and carefully ascertain from his own eyes what he learnt with books."[177]

The amputation of the leg and other body parts was done during the *Ṛgvedic* period. The *Ṛgveda* tells us that the lower limb of Queen Viśpalā was severed in King Khela's battle. Āsvins connected an artificial limb to her

[173] *Suśruta-Saṁhitā*, *Sūtrasthānam*, 5: 3.
[174] *Suśruta-Saṁhitā*, *Sūtrasthānam*, 18: 14 - 18; also read Kutumbiah, 1962, p. 163.
[175] *Suśruta-Saṁhitā*, *Sūtrasthānam*, 18: 23.
[176] *Suśruta-Saṁhitā*, *Sūtrasthānam*, 25: 18.
[177] *Suśruta-Saṁhitā*, *Śarīrsthānam*, 5: 6.

Table 10.1: Types of Bandages in *Suśruta-Saṁhitā*

| | |
|---|---|
| 1. | *Kośa*, egg-shaped, applied to joints of the thumb and fingers |
| 2. | *Dāma*, tail of a quadruped, tied round a part for the relief of pain |
| 3. | *Svastika*, portico shaped, for joints and spaces between the tendons of the great and second toe, eyebrows, breasts, soles, palm, and ears |
| 4. | *Anuvellita*, encircling, applied to limbs |
| 5. | *Pratolī*, broad, for neck and penis |
| 6. | *Maṇḍala*, circular, for round parts |
| 7. | *Sthagikā*, giving firmness, bandage filled with plates, applied to thumb and fingers |
| 8. | *Yamaka*, applied to ulcers |
| 9. | *Khaṭvā*, four tailed, for cheeks, temples, and lower jaw |
| 10. | *Cīna-vibandha*, banner, for the inner angles of the eyes |
| 11. | *Vibhandha*, firm bandage, for back, abdomen, and chest |
| 12. | *Vitāna*, canopy, large bandage for head |
| 13. | *Gophaṇa*, sling for throwing stones, concave bandage for chin, nose, shoulders, and pelvis. |
| 14. | *Pancāṅgi*, bandage with five tails |

at night and she was able to fight the battle again:[178] This indicates the existence of a surgical amputation facility during the *Ṛgvedic* period. This also shows that when the warriors were engaged in a battle, surgeons attended the battlefields and performed surgery to amputate the legs, performed eye surgeries, and extracted the arrow-shafts from the limbs.

A neolithic 4000 year-old skeleton with a multiple-trepanated skull was recently found in Kashmir. The trepanation on this skull was perhaps accomplished using drills of various diameters and could be a result of an elaborate medico-ritual ceremonial procedure.[179] Similarly, the Mehrgarh site, near Baluchistan, provides ample evidences for perfect tiny holes in molar teeth as a dental treatment, that are about 7,500 to 9,000 years old. Eleven drilled molar crowns from nine adults were studied from the Mehrgarh site. The drilled holes are about 1.3 - 3.2 mm in diameter and angled slightly to the occlusal plane, with a depth of between 0.5 to 3.5 mm.[180] Cavity shapes were conical, cylindrical or trapezoidal. In at least one case, "subsequent micro-tool carving of the cavity wall" was performed after the removal of the tooth structure by the drill.[181] In all cases, marginal smoothing confirms that drilling was performed on a living person who later continued to chew on the tooth surfaces.[182] Using the flint drills that are similar to the ones found in Mehrgarh and using a bow-drill, the team of Coppa *et al* could drill a hole in human enamel in less than a minute. According to researcher Andrea Cucina of the University of Missouri-Columbia, she looked the cavities using an electron microscope and found that the sides of the cavities were too perfectly rounded to be caused by bacteria. She could see concentric groves left by the drill. According to her, some plant or other substances were put into the hole to prevent any further decay in the molars. These fillings were decayed during the 9,000 years of seasoning.[183]

---

[178] *Ṛgveda*, 1: 116: 15.
[179] Sankhyan and Weber, 2001.
[180] Coppa *et al*, 2006
[181] Coppa *et al*, 2006
[182] Coppa *et al*, 2006
[183] Coppa *et al*, 2006.

### 10.6.1 Leech Therapy

Suśruta was astute in the use of leeches, carpenter ants and other life forms in various treatments. In the treatment of intestinal surgery to remove obstructing matter, Suśruta suggested to "bring the two ends of intestines that needed to join together. The intestines should be firmly pressed together and large black ants should be applied to grip them quickly with their claws. Then the bodies of the ants having their heads firmly adhering to the spots, as directed, should be severed and the intestines should be gently reintroduced into the original position and sutured up."[184] The use of leeches was to suck pus and blood in the cure of boils, tumors, and other similar diseases. Suśruta defines the types of leeches that should be used and provided details of the process. Leeches worked well when the patients were not fit for operations and bleeding was an issue since, with leeches, almost no quality blood was lost.[185]

For the use of leeches, Suśruta first described the type of leeches based on their dwelling. He described twelve different kinds of leeches and suggested to use only six types for the purpose of surgery. Suśruta suggested that the "leeches should be caught hold of with a piece of wet leather, or by some similar article, and then put in to a large-sized new pitcher filled with the water and ooze or slime of a pool. Pulverized zoophytes and powder of dried meat and aquatic bulbs should be thrown into the pitcher for their food, and blades of grass and leaves of water-plants should be put into it for them to lie upon. The water and the edibles should be changed every second or third day, and the pitchers should be changed every week."[186] "The affected part [of the patient] should be sprinkled over with drops of milk or blood, or slight incisions should be made into it in the event of their refusing to stick to the desired spot. . . while sucking, the leeches should be covered with a piece of wet linen and should be constantly sprinkled over with cold water. A sensation of itching and of a drawing pain at the seat of the application would give rise to the presumption that fresh blood was being sucked, and the leeches should be forthwith removed. Leeches refusing to fall off even after the production of the desired effect, or sticking to the affected part out of their fondness for the smell of blood, should be sprinkled with the dust of

[184] *Suśruta-Saṁhitā*, *Cikitsāsthānam*, 14: 20 - 21.
[185] *Suśruta-Saṁhitā*, *Sūtrasthānam*, Chapter 13
[186] *Suśruta-Saṁhitā*, Sūtra-sthāna, 13: 15.

powered rock salt."[187]

To use these leeches again, Suśruta suggested that the "leeches should be dusted over with rice powder and their mouths should be lubricated with a composition of oil and common salt. Then they should be caught by the tail-end with the thumb and the forefinger of the left hand and their backs should be gently rubbed with the same fingers of right hand from tail upward to the mouth with a view to make them vomit or eject the full quantity of blood they had sucked from the seat of the disease. The process should be continued until they manifest the fullest symptoms of disgorging."[188]

Using leeches for surgical procedures is not just an ancient practice. The Food and Drug Administration in America considers the use of leeches (*Hirudo medicinalis*) as a "medical device" appropriate for certain procedures. The first application was cleared by the Food and Drug Administration in June 2004. Thus, the practices suggested by Suśruta are carrying over in today's medical practices in its original form. Today, surgeons use leeches often in procedures that require skin grafting or regrafting amputated appendages, such as finger or toes. The main problem in such procedures is that the veins do not flow blood to the region due to the damage. Leeches cause the blood to ooze so that the body's own blood supply eventually gets established and the limb survives.[189]

The saliva of leeches contains an anti-clotting agent, called hirudin, which allows the blood to flow freely and avoid congested skin flaps problem. It also has hyaluronidase, histamine-like vasodilators, collagenase, inhibitors of kallikrein and superoxide production, and poorly characterized anaesthetic and analgesic compounds. Thus, the saliva is analgesic, anaesthetic and has histamine-like compounds.[190] The therapy is pretty inexpensive, with each session taking about 40 minutes and leeches worth $7 - 10.[191] For each session, a new group of leeches are used and dumped as infectious waste material after the treatment. In Germany alone, about 350,000 leeches were sold in

[187] *Suśruta-Saṁhitā*, *Sūtrasthānam*, 13: 18.
[188] *Suśruta-Saṁhitā*, *Sūtrasthānam*, 13: 19 - 20.
[189] Michaelsen *et al*, 2001; Rados, 2004.
[190] Michaelsen et al, 2001; Oevi *et al*, 1992.
[191] Rados, 2004.

2001.[192] Their use is even more prevalent now. It led Andreas Michalsen, Manfred Roth and Gustav J. Dobos of Essen-Mite Clinic in Essen, Germany to write a book on *Medicinal Leech Therapy*.[193]

In a study conducted at the Department for Internal and Integrative Medicine, University of Essen, Germany, sixteen people in the test group were recruited who endured persistent knee pain for more than six months and had definite radiographic signs of knee osteoarthritis. Ten out of sixteen proceeded with leech therapy and avoided conventional treatment while the remaining six tried only the conventional treatment. In all patients with leeches therapy, no adverse effect or local infection was observed. Some patients described the initial bite of leeches as painful. It was found that the group with leech therapy did considerably better than the group with conventional therapy. Their pain was reduced significantly after three days and the good effects lasted up to four weeks. Later, a second group with 52 people was was experimented with similar therapy that yielded similar results.[194] Similarly, medical maggots (mostly *Phaenicia sericata*, known as green blowfly) are being promoted for the treatment of wounds and skin ulcers. Maggots have a taste for dead flesh that they eat and avoid live tissues. This helps to clean a wound by taking out dead skin, disinfect the wound by killing bacteria, and recovery is much faster.

## 10.6.2 Plastic Surgery

In ancient India, earlobes or nose were cut off as a punishment for some crimes. For example, in *Rāmāyaṇa*, attractive and voluptuous Sūrpaṇakhā, sister of Rāvaṇa, tried to devour Lord Rāma, a married man. Lord Lakṣmaṇa, younger brother of Lord Rāma, cut off her nose and earlobes as punishment. King Rāvaṇa took care of the problem by asking his surgeon to reconstruct the nose and earlobes for his sister.[195] Nasal amputation has also worked its way into the Indian metaphors and the Hindi term *Nāk kat gai* (nose is chopped) implies that a person is insulted. Also, "saving nose" (*nāk bacā lī*) is a colloquial term meaning to go through difficult circumstances without

[192] Michalsen *et al*, 2001.
[193] Michalsen et al, 2006.
[194] Moore and Harrar, 2003.
[195] Brain, 1993.

any embarrassment.

Suśruta described the technique to graft skin, presently known as "plastic surgery" as a general umbrella term. He is considered as the father of plastic surgery, and the Western world gives credit to India for the method of rhinoplasty surgery. In India, noses or earlobes were repaired using an adjacent skin flap. This procedure is popularly called as "the Indian method of rhinoplasty"[196] although no plastic is used in the process. Live flesh from the thigh, cheek, abdomen, or the forehead was used to make the new artificial parts.

This ancient procedure was not practiced in the West until the second half of the 15th century in Sicily, an empire with considerable contact with Arabia.[197] In England, the first article on the subject appeared in the *Gentleman's Magazine* in 1794 by Colly Lyon Lucas, a British surgeon and member of the Medical Board at Madras, India.[198] He described the process in his letter to the Editor, describing the process as "long known in India" and not known to the British. Colly Lyon Lucas witnessed a case where a local Indian serving the British army, in the war of 1792 A.D., was captured by King Tipu Sultan. Unable to defeat the British outright, the sultan tried to starve his enemies by ambushing the Indian bullock drivers who transported grain to the British. Tipu decided to humiliate the bullock drivers by mutilating their noses and ears. Lucas describes one such victim of this practice, the Mahratta bullock driver Cowasjee, who, on his capture, had his nose and one of his hands amputated by the sultan. After one year, this man decided to get his nose repaired. An operation was performed by the Indian doctors who were fairly well established in the medical community. In this operation, skin was taken from the forehead and placed as a nose. The whole process took about 25 days. The artificial nose looked "as well as the natural one" and the scar on the forehead was not very observable after a "length of time." The picture of the patient was published and the nose looked perfectly normal.[199] The procedure generated great interest among

[196]Sorta-Bilajac and Muzur, 2007.

[197]Pothula, 2001.

[198]Brain, 1993; Lucas, *The Gentleman's Magazine*, 64, pt. 2, no. 4, 891 - 92, October, 1794. In this article, only the initials of the author are provided.

[199]Brain, 1993.

the British surgeons. Lucas writes:[200] "For about 12 months he remained without a nose, when he had a new one put on by a man of the Brickmaker cast near Poonah [Poona]. This operation is not uncommon in India, and has been practiced from time immemorial.. . . A thin plate of wax is fitted to the stump of the nose, so as to make a nose of good appearance. It is then flattened, and laid on the forehead. A line is drawn round the wax, and the operator then dissects off as much skin as it covered, leaving undivided a small slip between the eyes. This slip preserves the circulation till an union has taken place between the new and old parts.

The cicatrix of the stump of the nose is next pared off, and immediately behind this raw part an incision is made through the skin, which passes around both alae, and goes along the upper lip. The skin is now brought down from the forehead, and, being twisted half round, its edge is inserted into this incision, so that a nose is formed with a double hold above, and with its alae and septum below fixed in the incision. A little Terra Japonica is softened with water, and being spread on slips of cloth, five or six of these are placed over each other, to secure the joining. No other dressing but this cement is used for four days. It is then removed and cloths dipped in ghee (a kind of butter) are applied.. . . For five or six days after the operation, the patient is made to lie on his back; and on the tenth day, bits of soft cloth are put into the nostrils, to keep them sufficiently open. The artificial nose is secure, and looks nearly as well as the natural one; nor is the scar on the forehead very observable after a length of time."

In England, The first operation of rhinoplasty was performed by Joseph Constantine Carpue on October 23, 1814, before a large group of surgical colleagues and his students. Carpue performed the second operation on an army officer who lost his nose during the Peninsular War against Napoleon, and later wrote a monograph about it.[201]

Prior to England, plastic surgery was practiced in Italy. Ackernecht, in his book, *A Brief History of Medicine*, firmly stated that "[t]here is little doubt that plastic surgery in Europe, which flourished first in medieval Italy,

[200]taken from Ang, 2005.

[201]Brain, 1993; Carpue, 1816; a brief summary of the history of this procedure in India is provided by Ang, 2005; Rana and Arora, 2002.

is a direct descendant of classic Indian surgery."[202] A well known European surgeon, who restored a lost nose, was Branca de Branca from Sicily, using Suśruta's adjacent flap method. His son, Antonio Branca, used tissue from the upper arm as the reparative flap in his operations (around 1460), and "the Italian method" using a distant flap was born. The method was most extensively described by Gaspare Tagliacozzi in his *Chirurgia curtorum* in 1597.[203] Tycho Brahe, a noted astronomer who lost his nose in a fight, had to live with an artificial nose that used to fall off in the absence of quality glue. Brahe had no financial issue. He could have gone for a rhinoplasty procedure if it had been prevalent in Europe at that time.

In the repair of a nose, the following method was provided by Suśruta: "The portion of the nose to be covered should be measured with a leaf. A piece of skin of the required size should then be dissected from the cheek, and turned back to cover the nose. The part of the nose to which this skin is to be attached or joined, should be made raw, and the physician should join the two parts quickly but evenly and calmly, and keep the skin properly elevated by inserting two tubes in the position of nostrils, so that the new nose may look normal. When the skin has been properly adjusted a powder composed of licorice, red sandal-wood, and extract of barberry should be sprinkled on the part. It should be covered with cotton, and white sesame oil should be constantly applied. The patient should take some clarified butter. When the skin has united and granulated, if the nose is too short or too long, the middle of the flap should be divided and an endeavor made to enlarge or shorten it."[204]

The *Suśruta-Saṁhitā* demonstrated a procedure to mend an earlobe with a patch of skin-flap scraped from the neck or the adjoining parts. "The modes of bringing about an adhesion of the two severed parts of an ear-lobe are innumerable; and a skilled and experienced surgeon should determine the shape and nature of each according to the exigencies of a particular case."[205]

"A surgeon well-versed in the knowledge of surgery should slice off a patch of living flesh from the cheek of a person devoid of ear-lobes in a manner so

[202] Ackernecht, 1982. p. 41.
[203] Sorta-Bilajac and Muzur, 2007.
[204] *Suśruta-Saṁhitā*, *Sūtrasthānam*, 16: 46 - 51.
[205] *Suśruta-Saṁhitā*, *Sūtrasthānam*, 16: 25.

as to have one of its ends attached to its former seat. Thus the part, where the artificial ear-lobe is to be made, should slightly scarified (with a knife), and the living flesh, full of blood and sliced off as previously directed, should be attached to it (so as to resemble a natural ear-lobe in shape)."[206]

After some two thousand years, this operation essentially follows the same procedure. In these operations, "the Indians became masters in a branch of surgery that Europe ignored for another two thousands years," acknowledges Majno, a well-known historian of medicine.[207] Veith and Zimmerman, in their book *Great Ideas in the History of Surgery*, conclude that "[i]t is an established fact that Indian plastic surgery provided basic pattern for Western efforts in this direction."[208] Indians became so skillful in producing artificial nose, ear, and lips without much scarring to show that some westerners, including doctors, wrongly considered *tilak* or *bindī* commonly used on foreheads as attempts to "conceal the scar left by the reparative operation" of the nose.[209]

## 10.7 Hindu Medicine in Other World Cultures

Ayurvedic or Hindu medicine was popular in China, the Middle East, and Europe from the ancient period. Ktesias, a Greek physician who lived in the Persian court during the early fifth-century B.C. for some 17 years, wrote that the Indians did not suffer from headaches, or toothaches, or ophthalmia.[210] They also did not have "mouth sores or ulcers in any part of their body."[211] Claudius Aelianus (175 - 235 A.D.), who lived during the period of Emperor Septimus Severus in Rome, preferred drugs from India in comparison to the Egyptian drugs. He wrote: "So let us compare Indian and Egyptian drug and see which of the two was to be preferred. On the one hand the Egyptian drug repelled and suppressed sorrow for a day, whereas the Indian drug

[206] *Suśruta-Saṁhitā, Sūtrasthānam*, 16: 4.
[207] Majno, 1975, p. 291.
[208] Veith and Zimmerman, 1993, p. 63.
[209] Maltz, 1946, p. 19
[210] McCrindle, 1973, p. 18.
[211] McCrindle, 1973, p. 18.

caused a man to forget his trouble for ever."[212] This statement implies that the Indian herbs and drugs were well received by the Romans.

Jean Filliozat (1906 - 1982), a French Indologist and medical doctor who was well versed in Sanskrit, Pali, Tibetan, and Tamil, found several common theories in Plato's work and the Indian medicinal system.[213] He suggests that Plato perhaps knew about the Indian system of medicine. "It is, therefore, in no way impossible that Plato had knowledge of the Indian Medical theory before he explained the one which is near it and which he does not seem to have borrowed from the known schools of Greek medicine."[214] In a recent book, Thomas McEvilley has also compared Plato's work with the Indian literature. McEvilley also noticed similarities between the two. Platonic monism, Plato's ethics, his theory of soul (lower and higher) has something comparable in the Indian literature. However, McEvilley made no conclusion on how the diffusion of knowledge was made one way or the other, unlike Filliozat. Jean Filliozat believes that the analogy of the human body as a garden with arteries as canals is strikingly similar between Suśruta and Plato. Also, the geometrical shapes of elements in Plato's theory are similar to the *maṇḍalas* in tantra.[215]

Pliny and Galen have mentioned the import of *lycium*, a drug for eye problems found in India.[216] According to Guido Majno, a professor of pathology from the University of Massachusetts, the cataract operation described by Aulus Cornelius Celsus (25 B.C. - 50 A.D.) is perhaps derived from Suśruta.[217] Celsus was a celebrated Roman medical writer and physician. His book *De Medicina* was a standard book of medicine in Rome.

The *Caraka-Saṁhitā* transcended the geographical boundaries of India and became popular in the Middle East. Al-Bīrūnī (973 - 1050 A.D.) mentioned the availability of an Arabic translation of *Caraka-Saṁhitā* in the Middle East.[218] As Arab medicine became popular in Europe, the name of

[212] Aelianus, 1958, 4: 41.
[213] Filliozat, 1964, Chapter 8.
[214] Filliozat, 1964, p. 253.
[215] Filliozat, p. 235.
[216] Majno, 1975, p. 377.
[217] Majno, 1975, p. 378. Majno's book is considered a milestone in the history of medicine.
[218] Sachau, 1964, p. XXVIII.

Caraka is repeatedly mentioned in medieval-Latin, Arabic, and Persian literature on medicine.[219]

Al-Jāḥiz (ca. 776 - 868 A.D.), a Muslim natural philosopher from Arabia, acknowledged that the Hindus possess a good knowledge of medicine and "practice some remarkable forms of treatment."[220] "They were the originators of the science of *fikr*, by which a poison can be counteracted after it has been used."[221]

Ṣā'id al-Andalusī (d. 1070 A.D.) from Spain also acknowledged the vast knowledge of the Hindus in medicine. "They [Hindus] have surpassed all the other peoples in their knowledge of medical science and the strength of various drugs, the characteristics of compounds and the peculiarities of substances."[222]

Nearby in England, Roger Bacon (1214 - 1292 A.D.), a twelfth century natural philosopher, wrote a book for the Pope and noticed that the Indians "are healthy without infirmity and live to a great age."[223] Marco Polo, an Italian traveler, also noticed the long lives of the Hindus. "[T]hese Brahmins live more than any other people in the world, and this comes about through little eating and drinking, and great abstinence which they practise [practice] more than any other people. And they have among them regulars and orders of monks, according to their faith, who serve the churches where their idols are, and these are called yogis, and they certainly live longer than any others in the world, perhaps from 150 years to 200."[224]

Some of the above statements should not be taken in literal sense. They merely imply that the ancient Hindus enjoyed excellent health. The common ailments did not cause much suffering among the Hindus since they had proper treatments for these ailments. Also, Marco Polo's statement about a 150 or 200 year life-span for the Hindus merely indicates that the Hindus were known to live for long periods.

---

[219]Royle, 1837, p. 153.
[220]Pellat, 1969, p. 197.
[221]Pellat, 1969, p. 197.
[222]Salem and Kumar, 1991, p. 12.
[223]Bacon, *Opus Majus*, p. 372.
[224]Needham, 1981, p. 81.

"Hindoo [Hindu] works on medicine having been proved to have existed prior to Arabs, little doubt can be entertained," writes Royle to advocate the antiquity of Hindu medicine and its role in modern medicine.[225] "We can hardly deny to them [Hindu] the early cultivation of medicine; and this so early, as, from internal evidence, to be second, apparently to none with whom we are acquainted. This is further confirmed by the Arabs and Persians early translating of their works; so also the Tamuls [Tamils] and Cingalese [Singhalese, people of Sri-Lanka] in the south; the Tibetans and Chinese in the East; and likewise from our finding, even in the earliest of the Greek writers, Indian drugs mentioned by corrupted Sanscrit [Sanskrit] names. We trace them at still earlier periods in Egypt, and find them alluded to even in the oldest chapters of the Bible."[226]

The Arabs, the Greeks and other Europeans were not the only ones to learn from India. Indian culture and therewith Indian medicine spread just as much in an eastern direction, as is evident from Tibetan, Indo-Chinese, and Indonesian medicine, concludes Erwin Heinz Ackernecht in his book, *A Short History of Medicine*.[227]

Abu Yusuf Ya'aub ibn Ishaq al-Kindī, popularly known as al-Kindī (ca. 800 - 870 A.D.), is considered to be the "philosopher of Arabia." He was born and lived in Baghdad. He wrote a medical formulary, called *Aqrābādhīn*, which was translated by Martin Levey into English.[228] According to Levey, about 13 percent of *materia medica* comes from the Indian region. In his view, "many of the Persian *materia medica* may more properly be considered to be Indian."[229] In that case, the Indian-Persian *materia medica* in al-Kindī's work turns out to be about 31 percent.[230]

Alī B. Rabban Al-Ṭabarī (783 - 858 A.D.), a Persian physician from Tehran, who later lived in Baghdad serving Caliph al-Mutawakkil (reigned

[225]Royle, 1837, p. 62. Royle's book is still a classic on the subject. The strength of this book is in its cross-cultural comparisons between the Middle East and Greece.
[226]Royal, 1837, p. 152.
[227]Ackernecht, 1982, p. 43
[228]al-Kindī, 1966.
[229]al-Kindī, 1966, p. 20.
[230]Levey and al-Khaledy, 1967, p. 33.

847-861 A.D.) as his physician, wrote an encyclopedic book on medicine, *Firdaws al-Ḥikmah* (*Paradise of Wisdom*).[231] It was translated into English by Muḥammad Zubair Siddiqi in 1928.[232] This book contains some thirty-six chapters on Hindu medicine and refers to the works of noted Indian physicians/thinkers such as Caraka, Suśruta, Cāṇakya (Kauṭilaya), Mādhavakara, and Vagbhata II. Al-Ṭabarī mentions three *doṣa* and seven *dhātu* of ayurveda for medical treatments. This is in accordance with the tradition of ayurveda in India.

Max Meyerhof has written excellent accounts of al-Ṭabarī's work and connected it with *Suśruta-Saṁhitā*, *Caraka-Saṁhitā*, the *Nidana* and the *Aṣṭāṅgahṛdava-Saṁhitā*. The *Nidāna* is a work on pathology written by Mādhavakara. This work was translated into Arabic under the patronage of Hārun al-Rashīd. The *Aṣṭāṅgahṛdava-Saṁhitā* is the work of Vagbhata II.[233] Max Meyerhof concluded the existence of Indian drugs in Arabic treatises: "We find indeed, in the earliest Arabic treatises on medicine, the mention of many Indian drugs and plants which were unknown to Greeks, and all of them bearing Sanskrit names which had passed through the New Persian language."[234]

*Triphalā* is a popular drug in India. The word *Triphalā* means "three-fruits" since the drug is made from three ingredients: *haritaki* or simply *har* (*Terminalia chebula*), *bahera* (*Terminalis belerica*), and *āmalaki* (*Phyllanthus emblica*). The Arabs used the term *atrifal* for *triphalā* that is similar in pronunciation, while the Chinese literally translated the term and called it *san-teng*, implying three herbs.[235] These literary evidences are examples of the transmission of medical knowledge from India to Arabia and China.

In studying the transmission of ayurvedic knowledge to China, Chen Ming suggests that "scholars of medical history have long been well aware of the Indian influence on Chinese medicine."[236] He provides examples of such knowledge in China. For example, the division of medicine into eight branches, as

[231]see Meyerhof, 1931; *Dictionary of Scientific Biography*, 13: 230.
[232]Siddiqi, 1928.
[233]Meyerhof, 1931 and 1937.
[234]Meyerhof, 1937.
[235]Mahdihassan, 1978.
[236]Ming, 2006.

done by Caraka and Suśruta, was defined by Dharmakṣema in *Daban nie jing* in 421 A.D., Paramārtha (499 - 569 A.D.) translated a Sāṁkhya text, *jinqishilum*, that refers to eight divisions of medical remedies. Sun Simiao (581-682), a medical writer and physician, wrote about ayurvedic practices in his work *Beiji Qianjin yaofang* (Important medicinal formulae [worth] a thousand [pieces of] gold). He mentioned Jīvaka's story, discussed in Section 10.2, that all plant life has medicinal value. Simiao also attributed several medicinal formulations to Jīvaka: Jīvaka's ball medicine for ten thousand illnesses, Jīvaka's medicines for illnesses caused by evil spirits, Jīvaka's soup, Jīvaka's medicine for prolonging life without getting old. He introduced "*chan*" or "*dhyāna* (meditation) and yoga into China, and also introduced a particular massage technique and called it Brāhmin's method.[237]

Tombs in Turfan (also Turpan, 42°59′N, 89°11′E, near the Silk Road) yield medical fragments in Sanskrit and Tocharian, the local language. Fragments of the ayurvedic texts, *Bhela Saṁhitā*, *Siddhasāra*, and *yoga-śataka*, have been found in these tombs. The epitaph of Lüsbuai (Battalion Commander) Zhang Xianghuan (681 A.D.) mentions two ayurvedic physicians, Jīvaka and Nāgārjuna. In his studies, Chen Ming concludes that "Indian medicine also influenced medical practices in Medieval Turfan."[238]

[237]Deshpande, 2008. This article provides some more examples of the transmission of Indian medicinal knowledge to China.
[238]Ming, 2006.

# Chapter 11

# A World View

The previous chapters defined a wide variety of knowledge that is a part of the ancient Hindu corpus. This includes the issues of the origin of life; the limits of logistic knowledge; the numeral system with its place-value notations; mathematical processes that cover arithmetic, algebra, and trigonometry; astronomical knowledge of eclipses; the shape of the Earth and planetary motions; the constitution of matter; the properties of matter; standards for mass, length and time; and physical and chemical processes in the making of drugs, poisons, and new compounds. This list basically covers most branches of modern science. The ancient Hindus made substantial contributions that played an important role in the evolution of science, including Western science as well as the European Renaissance. Several of these contributions have survived the test of time and are used in the modern world while the others are lost in the annals because they were improved upon by the later scholars.

If we need to mention just one thought that is central to Hinduism and which could play an important role in the future as it did play in the past in India, it is a hymn of *Ṛgveda*: "To what is One, sages call by different names."[1] The underlying meaning is that God is one and people call Him by different names. Though this doctrine is not a part of science, it is perhaps the most timely message in view of the religious strife around the world. Al-Bīrūnī, an Islamic philosopher, understood this. He wrote, "The Hindus believe with regard to God that he is one, eternal, without beginning and end,

[1] *Ṛgveda*, 6: 22: 46.

acting by free will, almighty, all-wise, living, giving life, ruling, preserving, . . ."[2] It is due to the impact of this doctrine that India never subjugated other religions and became a secular country after independence. It happened under the circumstances that the other part of India, now Pakistan, adopted Islam as the state religion. In the history of India, people spread their message by winning the hearts of people. This is the case with China, where not even a single Chinese was killed before China adopted Buddhism as the state religion. Christians, Jews, and Muslims found refuge in India at different times in its history and could flourish there. In contrast, in world history, there have been more wars in the name of religion than anything else. We talk of religion as a way to tranquility and salvation. However, the results of our actions in propagating religion are often the opposite.

For the ancient Hindus, learning science was considered helpful in achieving salvation – the most important goal for all Hindus. Therefore, scientific investigations were encouraged through religious codes, as explained in Chapter 2. The ancient Hindus wrote about logic, philosophy, individual and social ethics, love and hate, astronomy, mathematics, medicine, and countless other topics. Interestingly, all was done in poetic verses to aid in memorization. They used gnomic poetry, similes and metaphors to explain complexities of an individual, in nature, and in society. For the ancient Hindus, knowledge served as a liberator from the cycle of death and rebirth.

Long before the Christian era, the ancient Hindus had established a teaching system which was comparable to the present university system. The intellectual centers of Nālandā (near Patna, India) and Taxila (Takṣaśilā, near Islamabad, Pakistan), Kanchipuram (Tamilnadu, India), Vikramśilā (Bihar, India), and Valabhi (Gujrat, India) were perhaps the most well known during the ancient period. Taxila was visited by Alexander the Great and later supported by Aśoka. It had a large number of *stupas* (pillars) and monasteries. The center of Taxila was destroyed during the fifth century A.D. by the Huns. Nālandā (near Patna, Bihar) is the place where Nāgārjuna and Āryabhaṭa I taught. Nālandā was visited by three famous Chinese visitors: Faxian (or Fa-Hien), Xuan Zang (or Hiuen Tsang), and Yijing (or I-tsing). At one time, it had about 10,000 students, 1,500 teachers, and 300 classrooms. It was completely destroyed in 1193 by the army of Bakhtiyar Khilji. The library

[2]Sachau, 1964, vol. 1, p. 27.

was set on fire and, due to the extensive collection of books, it took over three months for the fire to be extinguished. Other intellectual centers, such as in Varanasi, destroyed in 1194 A.D. by Qutb-ud-din Aibak, and Mathura, destroyed in 1018 A.D. by Mahmud of Ghazni, also suffered.[3] Although the intellectual centers in India were subject to destruction for extended periods by the occupying foreign rulers, not all was lost due to the prevalent oral tradition, which involved memorization of the holy books.

There is a lot more effort needed to know the intellectual treasure that we owe to the ancient Hindus. For some reason, the scholarship of the last century on Indian science is scant and has not received the interest of scholars as it did during the nineteenth century. There could be several reasons for it. First, proper translations of the ancient literature are not commonly available to scholars. Most translations that are in circulation are rather old and are not indexed by subject. A computer search for some keyword is out of question as the *Vedas*, *Upaniṣads*, and *Purāṇas* are not freely available to academic scholars in digital form. There are only a small number of courses on Hinduism in the West. Many of these courses focus on sensational issues, such as *sati*, the caste-system, eroticism in literature, prostitution, etc. At the time of this writing, there was not even a single course in America that dealt with Indian or Hindu science. The same cannot be said about the Greeks whose contributions to science are comparable to those of the Hindus. The main reason is that most academic institutes are shaped by the interests of donors. It may sound ridiculous to some. However, it is the brutal reality of academia. Princeton University offered 13 courses just on Greece during the Spring 2008 semester. This is in part due to Stanley J. Seeger's endowment to the university.[4] Similar endowments from the private sector or from the Indian government are lacking.

---

[3]I have decided to avoid any description of the destruction that the ancient Hindus and their institutions had to deal with. The destruction of Baghdad by Hulagu Khan, the destruction of Spain during the Inquisition, and the destruction of Alexandria by a religious zealot group are some examples of how a country or a culture can be subdued. Most examples of destruction cited here occurred for a short period. In contrast, the Hindus faced subjugation and destruction of their intellectual centers for extended periods. It is almost magical, in contrast to other regions, the way India bounced back in science and technology after the colonial period.

[4]Arenson, 2008.

Most of the knowledge of the ancient Hindus was transferred to the West via the Islamic domination of Europe. In the centers of Baghdad and Damascus in Arabia and the centers of Sicily, Toledo, and Cordoba in Europe during the early medieval period, this knowledge was preserved and disseminated. Several words in the English language are either borrowed from the Sanskrit language or metamorphosed from the Sanskrit language. A list of these words make a strong case for the Hindu science, as they are related to the scholarly tradition of the ancient Hindus. Numbers (two, three, six, seven, eight, and nine), Pundit, guru, read, sine (as a trigonometric function), zero, sulfur, gold, silver, lead, uterus, yoga are some of the examples of the Sanskrit words in English.[5] All cultures have words to define a scholar in their languages. However, it is the Sanskrit word "guru" or "pundit" that has prevailed in English.

One of India's contributions to the world is the orange. This fruit is called *nārangī* in India. First, a bitter variety was introduced in the West. Later, in the fifteenth century, Portuguese traders introduced a much sweeter variety from India. In modern Greek, the bitter variety (*nerantzi*) is distinguished from the sweeter variety. The sweeter variety is called *portokali*, the Greek word for Portuguese. On his second voyage, in 1493, Christopher Columbus brought seeds of oranges and lemons with him and planted them on the American continent. In Rhone, France, Orange was the town (or a feudal state) that was an importing center for oranges. King William III (1650 - 1702 A.D.), who ruled England and Ireland, was from this town. He became seriously involved in the promotion of Protestantism. This is how the orange color became associated with the Protestant religion. Thus, the color of an Indian fruit became a symbol of a religion in the West.

[5]amrita, Aryan, ashram, atman, acharya, ahimsa, bandana, basmati, candy, cheetah, cot, jackal, jungle, karma, kin, kundalini, lemon, mahout, man, Mandarin, mantra, maya, mix, namaste, nirvana, orange, pajamas, puja, samadhi, sapphire, shampoo, sugar, swastika, tantra, yoga, yogi, and zen are non-science representative words that are of Sanskrit in origin. I have chosen the spellings that are used in English dictionaries and not that should be from the system of transliteration. These words clearly indicates India's contribution to nonviolence, and mysticism. However, not all is rosy. The followers of Lord Kṛṣṇa celebrated festivals by carrying images of Him in a huge wagon all over India. He was called Jagan-nāth, the lord of the world. For Europeans, it was a waste of time and money for senseless devotion. Thus, the word juggernaut evolved to define large overpowering and destructive forces or objects. Similarly, thug and dacoit are also Sanskrit in origin. The list is too long to be a part of this book.

## 11.1 Impacts during the Ancient and Medieval Periods

During the ancient and medieval periods, the Hindu corpus attracted the great minds of Pythagoras, Apollonius of Tyana, and Plotinus from the Greek or Roman tradition, al-Bīrūnī, al-Fazārī, Ibn Sīnā, al-Jāḥiz, al-Khwārizmī, al-Mas'udī, Kūshyār Ibn Labbān, and al-Uqlidīsī from the Islamic tradition, Faxian (Fa-Hien), Xuan Zang (or Hiuen Tsang), and Yijing (or I-tsing) from China, and Pope Sylvester, Ṣā'id al-Andalusī, Roger Bacon, Adelard of Bath, and Leonardo Fibonacci from the European tradition.

India was known throughout ancient history for the subtlety of its philosophers, for its rich literature in the arts and sciences, for the keen observations of its people, the diversity of its people and ideas, and its good economy. It was this reputation of the ancient Hindus that prompted Alexander to visit gymnosophists, during his invasion of India, for an interaction and to know their wisdom. There is no counter-view among the ancient civilizations where the ancient Hindus were denigrated for the lack of ethics, prosperity, or intellectual tradition, for about 2,500 years. The literature of the Greeks, the Arabs, the Chinese, and the Europeans demonstrates that the ancient Hindus were respected for their ethics, for their contributions to the sciences and fine arts, for their astute observations, and for their creativity.

The work of al-Khwārizmī in Baghdad was made possible due to the discoveries and inventions of the ancient Hindus. As mentioned in the preceding chapters, his work in mathematics and astronomy was based on the achievements of the ancient Hindus, as he acknowledged. He is the person who is responsible for a number of words in the English dictionary: algebra, zero, algorithm, sine function, etc. His books were read by well-known scholars such as Copernicus, Adelard of Bath, and Leonardo Fibonacci, either in Arabic or in translation. It was the Hindu mathematical tools that allowed Copernicus to figure out his heliocentric solar system. al-Khwārizmī also introduced the trigonometric function "sine," which allowed medieval scholars to measure the height of a distant object, the motions of astronomical objects, and the distance of a ship from the shore.

The role of mathematics in the growth of science needs no introduction here. Many operations of arithmetic, algebra, and trigonometry were invented by the ancient Hindus. These inventions played a major role in the scientific revolution that occurred at *Bayt al Hikma* (House of Wisdom) in Baghdad. These mathematical operations played an equally important role in the European Renaissance to which Pope Sylvester, Roger Bacon, and Leonardo Fibonacci contributed. All these well-known scholars looked to the East for learning.

The ancient Hindus invented the game of chess, called *caturaṅga* in Sanskrit. The word signifies "four members of an army" – elephants, horses, chariots, and foot soldiers. It was called *al-śatranj* in Arabic. In the game, a key move is made to trap the king. This was called *eschec* in French and "check" in English. This led to the word chess for the game. The game epitomizes the science of warfare that helped the ancient and medieval warriors. The role of horse, elephant, foot-soldier, chariots, and king are well defined in a set configuration. The main task was to protect your own king and kill the king of your enemy. Ṣā'id al-Andalusī (died in 1070 A.D.) of Spain considered the discovery of chess as a testimony of the "clear thinking and the marvels of their [Hindu] inventions. . . . While the game is being played and its pieces are being maneuvered, the beauty of structure and greatness of harmony appear. It demonstrates the manifestation of high intentions and noble deeds, as it provides various forms of warnings from enemies and points out ruses as well as ways to avoid dangers. And in this there is considerable gain and useful profit."[6] Today, chess is played all over the world and international tournaments are conducted to select a world champion.

Al-Bīrūnī studied Hinduism from its scriptures and translated a number of Sanskrit works into Arabic, including selections from Patañjali's *Yoga-sūtra* and the *Bhagavad-Gītā*. In his book *India*, he shared the *Bhagavad-Gītā*'s doctrine that God takes human form sometimes to "cut away evil matter which threatens the world." He correctly understood the philosophy behind idol-worship in Hinduism since "the popular mind leans towards the sensible world, and has an aversion to abstract thought which is only understood by

[6]Salem and Kumar, 1991, p. 14.

highly educated people, of whom in every time and every place there are only few. . . This the cause which leads to the manufacture of idols, monuments in honour [honor] of certain much venerated persons, people . . ."[7] Al-Bīrūnī clearly knew that idol worship is for the average person and that the intellectuals believe in one god: "For those who march on the path of liberation, or those who study philosophy and theology, and who desire abstract truth which they call *sâra*, are entirely free from worshipping anything but God alone, and would never dream of worshipping an image manufactured to represent him."[8] Thus, al-Bīrūnī clearly labels the traditions of Hinduism similar to other monotheistic religions.

Dārā Shukōh (1615 - 1659 A.D.) also shared a similar view as al-Bīrūnī on idol worship and considered the Hindu practice of idol worship to play a positive role in the "development of the religious consciousness." Dārā Shukōh believed that "the idols are indispensable for those who are not yet aware of the inner (*bāṭin*), meaning of religion and need therefore a concrete representation of the deity." Once educated, they "dispense with the idols."[9]

Alexander the Great, during his visit to India, received gifts from an Indian king: a girl, a philosopher, a magic cup, and a physician. Aristotle instructed Alexander to not accept any gift, cautioning him of poison. Aristotle wrote:[10] "Remember what happened when the King of India sent thee rich gifts and among them that beautiful maiden whom they had fed on poison until she was of the nature of a snake. Had I not perceived it because of my fear, for I feared the clever men of those countries and their craft, and had I not found by proof that she would be killing thee by her embrace and by her perspiration, she would surely have killed thee." This episode alludes to a practice in India where young and attractive females were given toxic substances (poison) in small amounts regularly. In due course of time, they became so poisonous that, like snakes, their bite or kiss could kill a person. They were called *viṣa-kanyā* (poison girl). These girls were used in warfare and politics to destroy an enemy.

[7]Sachau, 1964, vol. 1, p. 111.

[8]Sachau, 1964, vol. 1, p. 113.

[9]Friedmann, 1975.

[10]Elgood, 1951, p. 31. Aristotle wrote a series of letters as advice to Alexander the Great when the latter was away from Greece. These letters were first translated by Ibn al-Baṭrīq into Arabic and later were translated into Latin.

Copernicus knew many mathematical tools of the ancient Hindus. He learned about other Hindu mathematical tools, in addition to the astronomical knowledge, perhaps from the work of al-Khwārizmī that was popular in the region. Copernicus may have learned these tools during his stay in Italy where people like Pope Sylvester and Leonardo Fibonacci had already documented the mathematics of the Hindus. Copernicus is not the only celebrated scientist who was benefited with the contributions of the ancient Hindus; the table of parallax of the moon that was suggested by Kepler was identical to the one given by Brahmgupta in *Khaṇḍa-Khādyaka* almost a millennium ago, as inferred by Neugebauer in his analysis.[11]

After the Muslim conquest of India, the Muslim rulers Akbar (1542 - 1605 A.D.), Jahāngīr (1569 - 1627 A.D.), and Dārā Shukōh (1615 - 1659 A.D.) were engaged in translating Indian literature into Persian or Arabic. During the period of Akbar, two well-known Indian classic books, *Mahābhārata* and *Bhāgavata Purāṇa*, were translated. Akbar chose the title of *Mahābhārata* as the *Razmnāmah* or *The Book of War*. Another popular Sanskrit text, the *Singhāsan Battīsī* or *Thirty-Two Tales of the Throne*, concerning the fortunes of the ancient Indian king Vikramaditya, was translated into Persian as *Shāhnāmah*, meaning *The Book of Kings*. In 1597 A.D., the Indian classic book *Yogavasiṣṭha* was translated by Nizam al-Din Pānipati at the request of Jahāngīr who considered the book to contain the doctrines of Sufism(*taṣawwuf*) and commentaries on "realities, diverse morals, and remarkable advice." In 1656 A.D., the crown prince Dārā Shukōh, the eldest son of Shāh Jahān and the great-grandson of Akbar, assembled a team of scholars to translate three classic Hindu texts: the *Upaniṣads* (*Sirr-i Akbar*, meaning *The Greatest Mystery*), the *Bhagavad-Gītā* (*Āb-i zindagī*, meaning *The Water of Life*), and the *Yogavasiṣṭha*. Some fifty works of the Indians were translated under the direction of Dārā Shukōh. In his own words, Dārā Shukōh wrote in the translation of the Upaniṣads that the book is "words of God" that can make a person "imperishable, fearless, unsolicitous and eternally liberated." Dārā Shukōh hoped to show that Hinduism and Islam had a common identity.[12]

---

[11]Neugebauer, 1962, p. 124.

[12]Most information in this paragraph is provided in an excellent article by Ernst, 2003.

*Khalīl wa Dimna* is one of the celebrated books in the Islamic world since the early medieval period. This book is based on an Indian book, *the Pañcatantra*, meaning "five-principles." The original Sanskrit book was written before the Christian era by Viṣṇu Sharma to educate the sons of an Indian king on human conduct and the art of governing (*nīti*) using animal fables. Burzuwaih (or Borzuy), a personal physician of Anoushiravan (fl. 550 A.D.) of the Sassanid dynasty, made a trip to India to collect medicinal herbs for the king. He came back with the book and translated it into the Persian language. After the Islamic conquest of Persia, the book was translated by Ibn al-Muqaffa (8th century A.D.) into Arabic as *Khalīl wa Dimna* where Khalīl and Dimna are the two jackal characters in several stories in the book. All stories ended with a question and the next story was the answer of the previous question. These stories have ethical, social, and political wisdom. This book became popular in the Middle East and was known in Europe by the eleventh century. Ṣā'id al-Andalusī praised its contents in his book *Ṭabaqāt al-'Umam* and correctly labeled India as the country of origin. He recognized the role of Persia and Arabia in the promulgation of the book around the world. In his view, the book is useful for the "improvement of morals and the amelioration of upbringing" and "is a book of noble purpose and great practical worth."[13] Like *Khalīl wa Dimna*, *The Arabian Nights* (also known as *A Thousand and One Nights*) and *The Pool of Nectar* are the other two books that became quite popular in the Islamic world and are largely based on Indian books.

Similarly, the legend of Barlaam and Josaphat is the Christianized version of the legend of Buddha. It is accepted even by the Christians, including the *Catholic Encyclopedia*. The name Josaphat is evolved from the name of Lord Buddha, who was known as Budsaif in the Persian language. Josaphat was born as a prince and wanted to be a holy man. His father King Abenner tried everything possible to stop this. However, the prince became a Christian with the help of a monk, named Barlaam. This story was published around 1225 A.D. by Rudolph von Ems, in German. It later appeared in French, Greek, and other European languages.

As mentioned before, *Mandarin* is the word assigned to one of the two

---

[13]Salem and Kumar, 1994, p. 14.

most prominent languages in China.[14] It is the standard language of China and sets norms for aristocratic communication – a language used by the upper class. It was the spoken language of the bureaucrats who were courteous and polite in their behavior. The word itself is derived from the Sanskrit word *mantrin*, defining councilor. Even today in India, all cabinet ministers are called *mantrī* and the Prime-minister is called *Pradhān-mantrī*, meaning the prime councilor. The Sanskrit language influenced the aristocrats of China who adopted the language. Another example of this influence is the fact that Buddhism became the most prominent religion in China without any war or conflict. This is in contrast to several other nations where people had to choose between their lives or their religion. Such was the influence of India on China.

In China, astronomical calendars were designed with the help of Hindu astronomers, and Chinese astronomy prospered with the inclusion of Hindu astronomy. The popular astronomical treatise, *Kaiyuan Zhanjing*, was a translation of a Hindu book. The *Chiu Chih Calendar* was derived from Varāhamihira's work. The nine planets, including nodes (Rāhu and Ketu), were well known in China. During the Tang dynasty, at least two directors of the astronomical bureau were from the Siddhāta school of the Hindus. The trigonometric function "sine" was adapted in China and was called "ming." It is a documented fact that Chinese scholars studied in India, acquired a large number of books during their stay, and carried these books with them on their way back to China.

### 11.1.1 Plotinus

Plotinus (205 - 270 A.D.), the founder of the Neo-Platonic school, attempted to visit India to learn from Hindu Brahmins while he was in Alexandria. He joined the army of Emperor Gordian III on his expedition to Persia in 242 A.D. against the Sassanians king, Sapor. This is obviously a major sacrifice for a philosopher to pick up weapons and fight, just for a possibility to learn from India. However, he was not the first. Before Plotinus, Pyrrhon pursued a similar dream to visit India by joining Alexander's army. Pyrrhon succeeded while Plotinus failed since Gordian III was assassinated

[14] Cantonese is the other prominent Chinese language.

in Mesopotamia, and Plotinus could not fulfill his dream of an Indian expedition.[15] Plotinus moved to Antioch and later to Rome where he spent the rest of his life. Plotinus did not write books, like Pythagoras and Socrates. Similar to the Hindu tradition, he gave lectures in Rome. He lectures were later compiled by his followers under the title *Enneads*.

Plotinus' teacher Ammonius Sakkas was perhaps an Indian since his name is not a typical Greek name, and is similar to a Sanskrit name, where *Śākaya* means a monk.[16] Porphry tells us that Plotinus "acquired such a mastery of philosophy, that he became eager to gain knowledge of the teaching prevailing among the Persians, as also among the Indians."[17] It is pretty clear that association with Ammonius kindled a desire in Plotinus to join Emperor Gordian's military expedition to Persia and India. The influence of the Hindu philosophy on Neo-Platonism has been suggested by scholars over the years. Rawlinson writes: "It certainly appears probable that Neo-Platonism was affected by Oriental [Hindu] philosophy . . . Hence we may suppose that the doctrines it inculcates,–abstinence from flesh, subjection of the body by asceticism, and so on – are derived from Oriental [Hindu] sources."[18] Rawlinson concludes: "the debt of Neo-Platonism to Oriental sources is indisputable . . ."

Plotinus, in the words of Sedlar, "produced a new philosophical synthesis: Greek rationalism in the service of Oriental mysticism."[19] He considered the human soul as part of the world-soul, implying that we all have divinity within us. When Plotinus was close to death, as we know from his student Porphyry, he said, "I am striving to give back the divine which is in me to the divine in the universe."[20] This implies that his individual soul will merge with the cosmic divine soul. This philosophy is an essential part of the ancient Hindu philosophy of *Aham brahmāsmi* (I am the God), which is central to the Upaniṣads[21] and is currently propagated by the New Age movement.

[15]Sedlar, 1980, p. 292; McEvilley, 2002, p. 550.
[16]Sedlar, 1980, p. 292; Halbfass, 1988, p. 17.
[17]Gregorios, 2002, p. 17; McEvilley, 2002, p. 550.
[18]Rawlinson, 1926, p. 175.
[19]Sedlar, 1980, p.292.
[20]MacKenna, 1956, sec. 1, p. 1.
[21]Bṛhadāraṇyaka Upaniṣad, 1: 4: 10.

Thomas McEvilley explored the Greek works and compared with Indian thoughts and noticed quite a few similarities between the two. On the question of exchange, McEvilley suggests that "[i]t can be argued that Plato was already Indianized through Orphic and Pythagoran influences, and on that basis alone some, at least, of his works cannot be regarded as 'purely Greek thought.' Plotinus, then may have received the Indian influence from Gymnosophists in Alexandria, or from the works of Plato, or both; it comes to the same thing: he was philosophizing in an Indianized tradition. It is not just a question of whether Plotinus's philosophy was derived from India *by him*. Its major outlines, in the view presented in this book, had been derived from India almost a thousand years earlier and handed down through what might be called the Indianized, or Indian-influenced, strand of Greek philosophy, to which Plotinus emphatically belonged. He could, then, and perhaps would, have come up with his model of things without any additional Indian input in his lifetime, though it seems clear, in any case, that he had some."[22]

## 11.2 Impacts During the Modern Period

The literature of the ancient Hindus continued to attract modern scientists and philosophers: Ralph Waldo Emerson, Johann Wolfgang von Goethe, Johann Gottfried Herder, Aldous Huxley, Carl Jung, Max Müller, Robert Oppenheimer, Erwin Schrödinger, Henry David Thoreau, and Hideki Yukawa, to name a few. These scholars found the validity of their ideas in Hindu literature.

Voltaire (Actual name, Francois-Marie Arouet; 1694 - 1778 A.D.), a French historian, philosopher, and dramatist, visited India, Egypt, and Arabia during the eighteenth century, and was well versed with the history of these regions. In the view of Voltaire, "As India supplies the wants of all the world but is herself dependent for nothing, she must for that very reason have been the most early civilized of any country. . ."[23] Voltaire considered the origin of "humankind in the East on the banks of the Ganges, as opposed to the account found in Genesis."[24] In Voltaire's view, Indian science was more ancient than Egyptian science. The recent excavations in Margarh is

[22]McEvilley, 2002, p. 550 - 551.
[23]Voltaire, 1901, vol. 29, p. 180.
[24]Patton, 1994, p. 207.

proving him right. Voltaire writes, "If the question was to decide between India and Egypt, I should conclude that the sciences were much more ancient in the former [India],"[25] Voltaire is not alone in assigning priority to the Indians in comparison to Egypt. Benjamin Farrington, a noted historian of science, expressed similar views in his analysis of the Pythagorean Theorem. "The degree of this knowledge, and the possibility of this diffusion from a common centre [center], are questions that may one day be answered with a confidence that is now impossible. But when the answer is given, if it ever is, perhaps neither Babylon nor Egypt will appear as the earliest exponent of civilization. The Nile and the Euphrates may have to yield place to the Indus [India]."[26] This is in reference to the ancient Hindus' *Śulbasūtra* that contains the so-called Pythagorean Theorem long before it was proposed by Pythagoras.

Johann Gottfried Herder (1744 - 1803 A.D.) considered the region of Ganges as "the primordial garden" where human wisdom started and was nourished, and the birth place of all languages, the Sanskrit language being the mother. He also considered Sanskrit poetry as the mother of all other poetry, indicating Vedic literature as the source of most other poetry works elsewhere.[27]

On the literary contribution of Hindu literature, the famous German poet Johann Wolfgang von Goethe (1749 - 1832 A.D.) wrote, "We would be extremely ungrateful if we were not to think of Indian poems, of those which are to be particularly admired because, in their happy unconstrainedness, they have escaped from the conflict with the most abstruse philosophy on one side and the most monstrous religion on the other and have taken over, from both, no more than can help them to inner depth and outward worth."[28] Goethe's interest in India came with his reading of Kālīdāsa's *Śakuntalā* and *Meghduta* and flourished with his interaction with noted philosophers such as Johann Gottfried von Herder.[29] Goethe even tried to learn Sanskrit to read

---

[25]Voltaire, 1901, vol. 24, p. 41.

[26]Farrington, 1969, p. 14.

[27]Patton, 1994, p. 207 - 208.

[28]Taken from Dasgupta, 1973, p. 7.

[29]Herder was influenced by Hinduism and was called German Brahmin by some. Herder wrote *Thoughts of Some Brahmins* in 1792. To read more about Herder and Hinduism, read Ghosh, 1990.

Hindu literature.[30] He tried to visit India, but without success.[31] Goethe's poor health did not allow him to take the arduous sea journey to India. His popular drama, *Faust*, was inspired by his reading of the German translation of *Śakuntalā* by Forster in 1791.[32] He also wrote two ballads dealing with India: "Der Gott und die Bajadere" and "Der Paria."[33]

Carl Jung, one of the leading psychologists of the twentieth century, was greatly influenced by the wisdom of the ancient Hindus. The ancient works of Patañjali influenced Jung greatly. The books of the Hindus not only influenced his thinking but also provided Jung confirming parallels for his independent insights, especially in his early days after his break-up with Sigmund Freud in 1910. "In the absence of like-minded colleagues, Hinduism provided him with evidence that his differences with Freud were founded on experiences shared by other human beings, therefore were not simply the products of a deranged mind," writes Harold Coward.[34] Jung wrote two articles that directly dealt with his impression of India: "The Dreamlike World of India" and "What India Can Teach Us." Both articles were published in the journal *Asia* in the 39th volume in 1939. In the second article, Jung provided a very positive view of the Indian civilization. In his view, India was "more balanced psychologically than the West and hence less prone to the outbreaks of barbarism which at that time were only too evident" in Europe.[35] It is suggested that the concept of "self," as developed by Jung was largely based on the Upaniṣadic notion of *ātman.*[36] As Jung writes, "The East [India] teaches us another broader, more profound, and higher understanding - understanding through life."[37] According to Jung, the influence of Hindu literature is nothing new; such influences "may be found in the works of Meister Ekhart, Leibniz, Kant, Hegel, Schopenhauer, and E. von Hartmann."[38] In a study conducted by Nicholson, Hanson, and Stewart, they found that "[t]here is no question that India had a major influence upon Jung's thought. Although

[30]Dasgupta, 1973, p. 21.
[31]Steiner, 1950, p. 1.
[32]Remy, 1901, p. 20.
[33]Remy, 1901, p. 20.
[34]Coward, 1984.
[35]Clarke, 1994, p. 62.
[36]Nicholson, Hanson, Stewart, 2002, p. 116.
[37]Jung, *Collected Works*, vol. 13, p. 7.
[38]Jung, *Letters*, vol. 1, p. 87.

many Hindu and Buddhist concepts fascinated him, it was the idea of karma and rebirth that continued to play a central role in the development of Jung's thinking to the end of his life."[39]

On the question of the impact of the ancient Hindus on European intellectual renaissance, Halbfass writes that "India has had a significant impact upon the manner in which Europe has articulated, defined, and questioned itself and its fundamental and symptomatic concepts of theory, science and philosophy."[40] This view is supported by the fact that many European philosophers, including Goethe, Jung, Herder, and Humboldt, studied Hindu books and introduced Hindu concepts in their own works.[41]

J. Robert Oppenheimer (1904 - 1967 A.D.), an atomic physicist, director of the Manhattan Project, and the so-called father of the atomic bomb, not only read books of the ancient Hindus but even tried to learn the Sanskrit language to get a firsthand experience of its profoundness. While working with Ernest Lawrence in Berkeley, after he got his B.S. degree from Johns Hopkins University in 1933, he wrote to his brother Frank: "Lawrence is going to the Solvay conference on nuclei, and I shall have double chores in his absence . . . I have been reading the Bhagwad Gita [*Bhagavad-Gītā*] with Ryder and two other Sanskrists. It is very easy and quite marvelous. I have read it twice but not enough. . ."[42] It is interesting that the great mind of Oppenheimer could not feel satisfied with two readings of *Bhagavad-Gītā*. In another letter to his brother Frank, he wrote, "Benevolences starting with the precious Meghduta [a book by *Kālīdāsa*] and rather too learned *Vedas* . . ."[43] Within a year of study on the side, Oppenheimer became well versed in reading classics in Sanskrit on his own.[44] It is obvious that reading of the sacred books of the ancient Hindus shaped Oppenheimer's world view and helped him in his researches in physics. As described in Chapter 3, he tried to explain the reality of an electron using the concept of *neti-neti*. Oppenheimer wrote about the parallels in atomic physics and Eastern philosophies

[39] Nicholson, Hanson, and Stewart, 2002, p. 116.

[40] Halbfass, 1988, p. 159.

[41] Sedlar, 1982 and 1980.

[42] Smith and Weiner, 1980, p. 162. The reference alludes to Arthur Rider who was an excellent poet in Sanskrit.

[43] Smith and Weiner, 1980, p. 180.

[44] Nisbet, p. 136.

of India: "the general notions about human understanding and community which are illustrated by discoveries in atomic physics are not in the nature of things wholly unfamiliar, wholly unheard of, or new. Even in our own culture they have a history, and in Buddhist and Hindu thought a more considerable and central place."[45]

In his condolence message upon the death of President Roosevelt, he asked people to have courage and continue with the task of Manhattan Project at Los Alamos: "In the Hindu Scripture, in the Bhagwad Gita [*Bhagavad-Gītā*], it says, 'Man is a creature whose substance is faith. What his faith is, he is.' "[46] Most of his speech was taken from *Bhagavad-Gītā*. It is also well known that he started chanting the verses of *Bhagavad-Gītā* when he witnessed the test explosion of the atom bomb as director of the Manhattan Project. In the main verse, the aura of God is described as brighter than a thousand suns. It has led to a book an engaging and popular book on the atomic bomb with the same title.[47]

Brian Josephson received the Nobel Prize in physics for his discovery of the Josephson tunneling effect in superconductivity. His discovery has led to changes in technology as well as in quantum physics. SQUID (Superconducting quantum interference device), a magnetometer to measure low magnetic fields, is a device that is based on the Josephson effect. Many companies are trying to come up with faster computers and switches that will be based on the Josephson junctions. Josephson suffered from insomnia after he received Nobel Prize and was hooked on tranquilizers. In his attempt to protect his health, he started practicing transcendental meditation, as propagated by Maharshi Mahesh Yogi. These regular practices gave him "inner peace" and sleep. He is currently the director of the Mind-Matter Unification Project at the Cavendish Laboratory in England and a regular practitioner of yoga and meditation. His current researches mostly deal with the uncommon subjects of science: consciousness, the role of the observer and mind in quantum mechanics, analogies between quantum mechanics and Eastern mysticism,

---

[45] Oppenheimer, 1954, p. 40.

[46] Smith and Weiner, 1980, p. 288.

[47] *Brighter than a Thousand Suns* by Robert Jungk (1913 - 1994 A.D.). He was an Austrian writer who wrote extensively on nuclear weapons. He even ran for the presidency of Austria in 1992 for the Green Party. The book has appeared in several editions with several publishers.

physics and spirituality, and yoga. Here Eastern mysticism implies the practices of the ancient Hindus and their world view. Josephson believes that scientists "can enhance their abilities through meditation."[48] Josephson is currently working on issues that bridge the gap between science and spirituality.[49]

### 11.2.1 Erwin Schrödinger: the Guru of Dual Manifestations

Erwin Schrödinger, a Nobel Laureate in physics and one of the prominent architects of modern physics, wrote on the connectedness of the various events in nature in his book, *My View of the World*: "Looking and thinking in that manner you may suddenly come to see, in a flash, the profound rightness of the basic conviction in Vedanta . . . Hence this life of yours is not merely a piece of the entire existence, but in a certain sense the whole. This, as we know, is what the Brahmins express in that sacred, mystical formula which is yet really so simple and so clear: *Tat tvam asi* (that is you). Or, again in such words as 'I am in the east and in the west, I am below and above, I am this whole world."'[50] It is interesting to see some similarities of Schrödinger's philosophy of life that he derived from the books of the Hindus and the quantum mechanical wavefunction he created to define the microscopic reality. Though anecdotal, it is interesting to note the similarities of quantum mechanical waves and wavefunctions and its all pervading nature, as defined by Schrödinger, with the all pervading (omnipresent) description of God. The wavefunction, though abstract, materializes itself in a region of space, in either its particle or its wave aspect, when squared (probability density). This is similar to the *Nirguṇa-svarūpa* (*amūrta*, without form) of God that manifests itself in human form (*saguṇa-svarūpa*) from time to time in different regions. It is important to mention that these analogies are not real connections. However, they do play an important role in the thinking of the inventors, as mentioned in Chapter 1. Schrödinger's worldview helped him in "hatching" scientific and mathematical ideas.

---

[48] Horgan, 1995

[49] Josephson, 1987.

[50] Schrödinger, 1964, p. 21.

Schrödinger was influenced with the philosophy of Schopenhauer (1788 - 1860 A.D.). He was not the only one. Many noted European intellectuals and philosophers were influenced by Schopenhauer, including Immanual Kant, Friedrich Nietzsche, Thomas Mann, Sigmund Freud, Albert Einstein, Carl Jung, and Leo Tolstoy.[51] Schopenhauer's philosophy was greatly influenced by the teachings of Hinduism and Buddhism. He suggested a life style of negating desires, similar to Pythagoras and the Hindu philosophy. His book, *The World as Will and Representation*, emphasizes the Upaniṣadic teaching: *Tat tvam asi* (that is you). Schopenhauer had a dog named Atman (meaning soul in Sanskrit). Though Schopenhauer was popular in Europe, his philosophy was not popular among physicists, the kind of intellectuals Schrödinger was dealing with. Schrödinger realized that his views may not be appreciated by his physicist colleagues: "I know very well that most of my readers, despite Schopenhauer and the Upaniṣads, while perhaps admitting the validity of what is said here as a pleasing and appropriate metaphor, will withhold their agreement from any *literal* application of the proposition that all consciousness is essentially *one*."[52]

Observation is central to the growth of science. However, it provides a plurality of realities as the observation may differ from person to person, depending on the aspects these observers are interested in. This creates multifarious realities. This plurality does not get much attention by scientists. Schrödinger did consider the issue of plurality in life. "For philosophy, then, the real difficulty lies in the spatial and temporal multiplicity of observing and thinking individuals. . . the plurality that we perceive is only *an appearance; it is not real.* Vedantic philosophy, in which this is fundamental dogma, has sought to clarify it by a number of analogies, one of the most attractive being the many faceted crystal which, while showing hundreds of little pictures of what is in reality a single existent object, does not really multiply that object."[53] Schrödinger believed in the advocacy of the Hindu philosophy on oneness: "In all the world, there is no kind of framework within which we can find consciousness in the plural; this is simply something we construct because of the spatio-temporal plurality of individuals, but it is a false construction. . . The only solution to this conflict, in so far

[51]Moore, 1992, p. 112
[52]Schrödinger, 1964, p. 29.
[53]Schrödinger, 1964, p. 18.

as any is available to us at all, is in the ancient wisdom of the Upanishads."[54]

In his essay on *The Diamond Cutter*, Schrödinger wrote: "The ego is only an aggregate of countless illusions, a phantom shell, a bubble sure to break. It is *Karma*. Acts and thoughts are forces integrating themselves into material and mental phenomena – into what we call objective and subjective appearances . . . The universe is the integration of acts and thoughts. Even swords and things of metal are manifestations of spirit. There is no birth and death but the birth and death of Karma in some form or condition."[55] These statements led Walter Moore, who has written a book on the life of Schrödinger, to infer that Schrödinger "thought deeply about the teachings of Hindu Scriptures, reworked them into his own words, and ultimately came to believe in them."[56] Also, his worldview helped him in creating the wave-mechanics which is counterintuitive to Newtonian thinking: "[p]erhaps these thoughts recurred to Erwin when he made his great discovery of wave mechanics and found the reality of physics in wave motions, and also later when he found that this reality was part of an underlying unity of mind."[57] Thus, it is an example of how the worldview of a person, such as Schrödinger, played an important role in the science that he produced.

The philosophy of multiple representations of truth, thin boundaries between abstract thoughts (*amūrtā*), actual realities (*mūrta*) and universal consciousness played an important role in the science that Erwin Schrödinger created. Moore concludes, "Vedanta and gnosticism are beliefs likely to appeal to a mathematical physicist, a brilliant only child, tempted on occasion by intellectual pride. Such factors may help to explain why Schrödinger became a believer in Vedanta, but they do not detract from the importance of his belief as a foundation for his life and work. It would be simplistic to suggest that there is a direct causal link between his religious beliefs and his discoveries in theoretical physics, yet the unity and continuity of Vedanta are reflected in the unity and continuity of wave mechanics."[58] In the view of Moore, "Schrödinger and Heisenberg and their followers created a universe based on superimposed inseparable waves of probability amplitudes. This

[54]Schrödinger, 1964, p. 31.
[55]taken from Moore, 1992, p. 114.
[56]Moore, 1992, p. 113.
[57]Moore, 1992, p. 114.
[58]Moore, 1992, p. 173.

view would be entirely consistent with the Vedantic concept of the All in One."[59]

### 11.2.2 Emerson and Thoreau – Two Celebrated American Scholars

Ralph Waldo Emerson (1803 - 1882 A.D.) and Henry David Thoreau (1817 - 1862 A.D.) were two of the most celebrated American scholars of the nineteenth century. With their writings on philosophy and poetry, they proposed transcendentalism as a philosophy and thereby influenced many American lives. Their worldview was influenced by the sacred books of the Hindus and other classics of India. Emerson was fond of reading the *Bhagavad-Gītā* and popularized the book among his friends by loaning his personal copy. This copy was quite worn due to its extensive use.[60] Emerson shared his appreciation for the Hindu philosophy in the following language: "This belief that the higher use of the material world is to furnish us types or pictures to express the thoughts of the mind, is carried to its extreme by the Hindoos [Hindus] . . . I thinks Hindoo [Hindu] books are the best gymnastics for the mind."[61]

Russell B. Goodman, in his analysis of Emerson's work, concludes that "Emerson's philosophy, from his college days onward, grew up together with his knowledge of and interest in Hindu philosophical writings."[62] "His relationship with Hinduism, as with many other systems of thought, was transformative and respectful at once – reconstructive rather than deconstructive." At the youthful age of seventeen, Emerson wrote that during the ancient period in India "fair science pondered" and "sages mused."[63] Emerson wrote to John Chapman in May 1845 that he "very much want[ed]" the *Bhagavad-Gītā* to read the dialogs of Lord Kṛṣṇa and Arjun.[64] In August 1845, Emerson went to the mountains of Vermont and read the *Viṣṇu-Purāṇa*. He commented that "[n]othing in theology is so subtle as this & the Bhagwat

---

[59]Moore, 1992, p. 173.

[60]Horton and Edwards, 1952, p. 118; taken from Gangadharan, Sarma and Sarma, 2000, p. 311. For more information on Emerson, read Acharya, 2001.

[61]Emerson, 1904, vol. 8, p. 14.

[62]Goodman, 1990.

[63]Goodman, 1990.

[64]Rusk, 1939, vol. III, p. 288; taken from Goodman, 1990.

[*Bhagavad-Gītā*]."[65]

Thoreau praised Hindu literature in his writing: "What extracts from the *Vedas* I have read fall on me like the light of a higher and purer luminary . . . It rises on me like the full moon after the stars have come out, wading through some far summer stratum of the sky."[66] A. K. B. Pillai, in his book *Transcendental Self: A Comparative Study of Thoreau and the Psycho-Philosophy of Hinduism and Buddhism*, evaluated the connections between the tenets of Hinduism and transcendentalism and concluded that "*Walden* is the closest to the Yogic system of all major American writings."[67] *Walden* is perhaps the most popular book of Thoreau.

In Walden, Thoreau writes, "It appears that the sweltering inhabitants of Charleston and New Orleans, of Madras and Bombay and Calcutta, drink at my well. In the morning I bathe my intellect in the stupendous and cosmogonal philosophy of the Bhagvat-Geeta [*Bhagavad-Gītā*], since whose composition years of the gods have elapsed, and in comparison with which our modern world and its literature seem puny and trivial. . . I lay down the book and go to my well for water . . . There I meet the servant of the Brahmin, priest of Brahma [Brahmā] and Vishnu [Viṣṇu] and Indra, who still sits in his temple on the Ganges reading the *Vedas*, or dwells at the root of a tree with his crust and water jug. . . The pure Walden water is mingled with the sacred water of the Ganges."[68] Thoreau wrote quite approvingly about the Hindus: "the Hindoos [Hindus] . . . possess in a wonderful degree the faculty of contemplation," "their religious books describe the first inquisitive & contemplative access to God," or "their method is pure wisdom or contemplation."[69] Thoreau promoted vegetarianism: "I believe that every man who has ever been earnest to preserve his higher or poetic faculties in the best condition has been particularly inclined to abstain from animal food, and from much food of any kind."[70]

Thoreau and Emerson were the leaders of the American Transcenden-

[65]Rusk, 1939, vol. III, p. 288; taken from Goodman, 1990.
[66]Thoreau, 1906, 2: 4.
[67]Pillai, 1985, p. 4, 88.
[68]Taken from Scott, 2007
[69]Scott, 2007.
[70]Henry David Thoueau, *Walden; or, Life in the Woods*, Dover, 1995, p. 139.

talism movement of the nineteenth century.[71] They were drawn to Hindu texts since they received support to their own way of thinking in these texts. "There is little doubt that both Emerson and Thoreau shared an overwhelming and sustained enthusiasm for Vedic literature and Vedāntic philosophy and became their ardent votaries," concludes Sundararaman.[72] This support for Hinduism by Emerson and Thoreau has also been noticed by others. Raj Kumar Gupta shares on the enthusiasm and struggle Emerson and Thoreau dealt with: "[i]n their fervent and enthusiastic admiration for Hindu idealism and spiritualism, Emerson and Thoreau lost sight of the fact that they were contrasting American *practice* with Indian *theory*."[73] In his analysis, Gupta made the following judgment about Emerson and Thoreau: "Hindu ideas and ideals are used to bring out, explicitly or by implication, the failures and deficiencies of nineteenth century American life and thought, and in an attempt to fill the void and supply the deficiency. Thus Hinduism represents not only ideals against which American values are judged and found wanting, but also a corrective and an antidote to those values."[74]

* * * * * * * *

Yoga is of immense value in remolding the human personality as a whole. The practice of yoga opens up a completely new realm of existence that is beyond any specific religion. This is the principal reason why yoga transcended the boundaries of the Hindu religion and has permeated the whole world. Yoga is practiced by devoted Christians, Muslims, and atheists, without discrimination. It is currently prescribed in many hospitals in the Western world for pain management and cures of various diseases. It is the effectiveness of yoga that prompted Bill Moyer to have one segment on yoga in his five-part series, in his well known documentary *Healing and the Mind*. This series, in Part 2, shows how people with incurable diseases and pain could function normally with minimal medicine with the practice of yoga. When the medical conditions arise for a person, the usual approach is to consult the family doctor. These visits are usually covered by medical insurance. In case of

---

71Christy, 1932; Riepe, 1970.

72Sundararaman, David Thoreau: The Walden-Ṛṣi, in the book by Gangadharan, Sarma, and Sarma, 2000, p. 311.

73Gupta, 1986, p. 81.

74Gupta, 1986, p. 81.

need, medical insurances will pay for Prozac for depression, chemotherapy, or heart surgery. However, medical insurances mostly do not pay for preventive measures which will keep a person healthy. There is little support for an average person with medical insurance to remain healthy by practicing yoga or ayurveda to avoid future expenses in cancer, heart disease, or other ailments. The British phrase "penny-wise and pound-foolish" fits so well to describe this situation.

Many schools, universities, and health clubs offer classes in yoga. It is a useful tool to help at-risk adults and kids to concentrate on their studies; it attracts a large number of people who want to "look good" and reach their fullest potential. The practice of yoga is as relevant in today's society as it was during the period of Patañjali. It is the best Rx (prescription) that a person can find to do better physically and emotionally.

Most cities in America have doctors that are trained in the ayurvedic system of medicine. Deepak Chopra, Vasant Lad, and Maharishi Mahesh Yogi have trained a fairly large number of doctors who are scattered all over the country. In a simple Google search on ayurveda, one can find a multitude of sites that are devoted to ayurveda in America. Though ayurveda is multifaceted, in the U.S., it mostly focuses on massages, dietary and herbal advice, and yoga. The body-mind approach of ayurveda is effective in the cure of various diseases. It "is the ultimate feel-good, do-it-yourself healthcare system."[75]

It does not matter whether you were schooled in Africa, Asia, or Europe. Everywhere in the civilized world, kids learn to count even before they learn the alphabet of their own language. In different languages, the numbers may be pronounced differently. However, they all use the same mathematical principles in counting and in arithmetic operations that were invented long ago by the ancient Hindus. These mathematical methods are so effective and important, most kids learn them without ever inquiring about the alternatives. "The Indian system of counting has been the most successful intellectual innovation ever made on our planet. It has spread and been adopted almost universally, far more extensively even than the letters of the Phoenician alphabet which we now employ. It constitutes the nearest thing

[75]Doheny, 2006.

we have to a universal language,"[76] writes John D. Barrow in his book, *Pi in the Sky: Counting, Thinking, and Being.*

In dealing with numbers, the size of the atom that was recommended by the ancient Hindus comes close to $10^{-10}$m. For the age of the universe, they came up with a number that is of the order of a billion years. Both of these speculations were ridiculed in history by foreign scientists during the early medieval period. However, these numbers are in tune with modern science now. Is this a coincidence? How can one reach such striking conclusions that were considered ridiculous just a century ago and stick by them for thousands of years by documenting them in their most sacred books? The answer is perhaps in the role of the mind in learning the mysteries of nature. This is what Brian Josephson is investigating these days. This is what Robert Oppenheimer and Erwin Schrödinger wondered about when they became interested in the sacred books of the ancient Hindus.

The social attitudes toward accepting new scientific theories were in contrast between India and Europe. Unlike Galileo and Copernicus, Āryabhaṭa never faced any hostile emotions of the priestly class or the possibility of any kind of persecution from the king. In contrast, evolution of such ideas led a person to the ultimate goal of salvation, as suggested by Āryabhaṭa I. Thus, becoming an astronomer, medical doctor, or a musician was as holy as becoming a priest. This is the reason Caraka and Suśruta got involved in the medical profession and Āryabhaṭa in astronomy and mathematics. They were all concerned with achieving salvation for themselves.

Many intellectual centers in our history used multicultural approaches to learning. Damascus and Baghdad became leading centers of learning when they started attracting Greek, Indian, Persian, and Egyptian scholars. Similarly, Spain became a center of learning from the eighth century to the eleventh century, because it attracted scholars from all over the world.[77] However, these centers eventually declined when the scholars were judged on ill-conceived grounds (race, gender, religion, or culture), rather than on the basis of merit. Such declines occurred in Spain, Egypt, the Middle East, and elsewhere in Europe.

[76]Barrow, 1992, p. 92.

[77]Salem and Kumar, 1991.

Modern science is international in character. Its ideas are studied and developed in many languages from all over the world. Thus, science education with a multicultural approach can be a broadening and humanizing experience. It teaches us how human beings are alike. It allows people to learn and appreciate each other. It allows us to function as a cohesive society. However, so far, science education has not utilized this opportunity to achieve its full potential.

Preserving knowledge is a process in which all generations must participate or else become susceptible to external pressures and lose the knowledge. Museums are created to slow down or stop this tide of entropy. However, they mostly secure artifacts and have limitations. Perhaps, libraries are the places where knowledge can be documented and preserved. However, these libraries are prone to destruction as we know from the history of Alexandria, Nālandā, and Taxila. Therefore, dissemination of knowledge to the following generation is a duty that each generation must involve itself in. This is a *guru-ṛṇa* (debt to the guru) that each disciple owes to his or her teacher. To take care of this debt, I have written this book and tried to preserve the knowledge of the ancient Hindus and to recognize their appropriate place in world history. Of course, it is a small effort and much more is to be done on a continual basis to preserve the knowledge. Since supporting documents to the text are provided, this should promote more scholarship that is much needed on this topic. Some of my assertions will survive the test of time while others may fail. After all, it took about 300 years of intense scholarship by the Western world to bring Greek literature to its current status.

As George Sarton pointed out, a teacher of political history is not expected to cover all aspects of the history. However, a historian of science is expected to "know the whole history plus the whole of science – a pretty big order."[78] No person can be an expert in so many disciplines. Yet, it is what the readers expect from an author. It is this "pretty big order" that I decided to take on. It has been a challenging and satisfying experience. Only the readers can judge the validity of this endeavor.

---

[78]Sarton, 1955

# Appendix A

# The Travels of Pythagoras

While dealing with the Pythagorean theorem, the transmigration of the soul, establishing connections between plants, animals and humans, orality in preserving knowledge, and vegetarianism in this book, striking similarities between the ancient Hindu perspectives and the doctrines of Pythagoras were noticed. Was it a mere coincidence? This appendix provides a possibility of connection between Pythagoras and the ancient Hindus. This scholarship is not as conclusive as I would like. However, it is scholarly, interesting, and is based on historical documents. More work is needed to ascertain this possibility. Not much will change about the conclusions drawn in this book if the readers disagree with the possibility presented here about the travels of Pythagoras.

The connections of Pythagoras with India have been a matter of speculations from the ancient period. Several scholars during the ancient period have written about his travels to Egypt, Babylon and India. These documents are highly credible, as they are written by the followers of Pythagoras during the ancient period or by authors of great repute. The ancient Hindus did not document these visits since these visitors were there for learning, not for teaching. Such visits are documented in Greek, Syrian, British, Spanish, and French literature.

The ancient world is considered, in the popular opinion, to be primitive and isolated into small nations or tribes. However, the ancient world was not as segregated and isolated as some may think. Exchanges of ideas between cultures and civilizations were possible and indeed occurred. Several natural philosophers migrated from the regions of their birth to far lands in order to acquire knowledge. This is particularly true for the Greeks during the ancient period. This trend continues even now as scientists travel abroad to teach and learn from their colleagues. This has always resulted in the blending of ideas.[1]

Thales (ca. 640 - 546 B.C.), Anaximinder (ca. 611 - 547 B.C.), Pythagoras (ca. 582 - 500 B.C.), Isocrates (ca. 436 - 338 B.C.), Democritus (460 - 370 B.C.), Plato (427 - 347 B.C.), Eudoxus (ca. 400 - 350 B.C.) from Greece, and Pliny the Elder (23 - 79 A.D.), Apollonius of Tyana (fl. 1st century A.D.), Plotinus (205 - 270 A.D.), Porphyry (233 - 305 A.D.), and Galen (130 - 200 A.D.) from the Roman world tried to visit Egypt or other Eastern lands, including Persia and India.[2]

India has a unique geographical location as it is on sea as well as land routes of the West and the Far East. Land and sea were the only two available modes of long-distance travel in the ancient world. Though the modes of transportation were certainly primitive where horses, mules, or elephants were used as carriages on land, and sailboats or masted boats by sea, such exchanges between various civilizations did occur. The travels of Alexander the Great and his large army are well known. He invaded and captured Egypt, Persia, and parts of northern India in the fourth century B.C. under extremely hostile conditions. When an army of that size, which had to overcome the hostility of powerful Egyptian, Persian, and Indian empires, could reach India, what would stop a lone yet determined Greek traveler from performing a similar journey in the absence of any hostility? Indeed such exchanges between India, Greece and Egypt had often taken place since antiquity.

The first extended contact between India and Greece probably must have happened during the reign of Cyrus the Great of Persia (558 - 535 B.C.) whose kingdom controlled the Greek cities of Asia Minor to the West and the northwestern region of India to the East. Darius, Cyrus' successor, consolidated his predecessor's conquests and extended his kingdom to the Indus-Sarasvatī

[1]Casson, 1994; Halbfass, 1988, Pococke, 1856; Prasad, 1977; Rawlinson, 1926; Sedlar, 1980; Warmington, 1974; and West, 1971.

[2]Casson, 1994; Halbfass, 1988; Rawlinson, 1926; Sedlar, 1980; and West, 1971.

regions of Punjab and Sindh. Because the Achaemenid Empire covered a large region, the transportation of merchandise for trade, and military troops to suppress local unrest, was necessary. It was accomplished by constructing "royal roads," which were maintained with the utmost priority. Thus, the Achaemenid Empire became a meeting ground of the Hindus and the Greeks.[3] Panini, a Hindu grammarian, used the word *yavana* in his work that is definitely anterior to the period of Alexander the Great.[4] The Sanskrit term *yavana* perhaps originated from the Persian word *yauna*, meaning Ionian.

Xenophon (born 429 B.C.), an historian and philosopher from Greece and a mentor and disciple of Socrates, wrote that an Indian king was considered to be an arbitrator in a war between the Medes [ancient country in NW Iran] and Assyrians [in modern Iraq].[5] Elsewhere he wrote on the frequent visits of Chaldaeans to India.[6] The Persian King Cyrus hired Indians to work for him as spies.[7] All this is reasonable due to the proximity of Persia and Chaldea to India. There was a land route between the countries which was inhabited by people.

The Achaemenid dynasty of Persia, which extended as far as Greece, was destroyed by a young man who later became known as Alexander the Great. In 336 B.C., at the age of 20, he became the king of Macedon and inherited, from his father Phillipus, the army and master plan with which to invade Asia. For the Greeks, it was a plan of revenge, to annex back the territory that was lost some 150 years earlier.

The prosperity of the Indus-Sarasvatī region and good climatic and social conditions attracted many Greeks who preferred to remain in the region after their services to the Achaemenid empire, in view of the arduous journey back home to Greece. Small colonies of the Greeks were established in northern India where the settlers practiced their own customs and language. Alexander encountered such settlements during his invasion of the Indus-Sarasvatī region. According to the history books, Alexander and his army were helped

---

[3]Casson, 1994, p. 53.

[4]Karttunen, 1989, p. 57.

[5]Xenophon, *Cyropaedia*, II: IV: 1-9

[6]Xenophon, *Cyropaedia*, III: II: 26

[7]Xenophon, *Cyropaedia*, VI: 2: 1 - 4

by these settlers and enjoyed a few days of festivities and drinking at Nyasa in the Swat valley.[8]

Megasthenes (350 - 290 B.C.), a Greek traveler to India, mentioned that India was invaded by Dionysos, in earlier times, from the West.[9] Later, due to the excessive heat of summers in the region, the soldiers moved from the plains to the hills[10] and ruled the area until the invasion of Alexander.[11] They founded a new city, Nyasa.[12] Arrian (flourished second century B.C.) tells us that the residents of Nyasa were not of Indian races; they were the descendants of Dionysos, an ancient Greek god of wine, vegetation, and theater.[13]

Elsewhere, King Solomon, a king of Tyre and a figure in the Old Testament of the *Bible*, built a fleet of ships around 975 B.C. to visit India with the assistance of King Eiromos of Tyros (Phoenicia) who provided experienced sailors. The purpose of the visit was to collect gold, ivory, and peacocks.[14] The fleet sailed to Ophir, a seaport in India. Ophir, Sopheir, or Sophara were the different names used in history to describe this location. The word is perhaps derived from the name of a tribe, Abhir, that lived near the seaport near Kathawar, a city in Gujrat near the coastal area: ". . . Solomon ordered to sail along with his own stewards to the land anciently called Sopheir, but now the Land of Gold; it belongs to India. And when they had amassed a sum of four hundred talents they returned again to the king," wrote Josephus (37 - 100 A.D.), a well-known Jewish historian and geographer.[15] Such exchanges with India were not restricted to gold, ivory or peacocks but were also extended to culture, rituals, sciences, language, and ideas.

In the procession of Ptolemaios Philadelphus II (285 - 246 B.C.) (or

[8] Doshi, 1985, p. 1; Karttunen, 1989, p. 56 and references therein.

[9] McCrindle, 1926, p. 35 and 109.

[10] McCrindle, 1926, p. 35.

[11] McCrindle, 1926, p. 38.

[12] McCrindle, 1926, p. 110.

[13] McCrindle, 1926, p. 183.

[14] Rawlinson, 1926, p. 10; Sedlar, 1980, p. 266; Josephus, *Jewish Antiquities*, vol. 8, p. 163 - 164.

[15] Josephus, *Jewish Antiquities*, vol. 8, p. 164.

Ptolemy II), a Greek ruler in Egypt, Indian women, hunting dogs, cows, and Indian spices on camel were displayed.[16] Also, King Ptolemy IV (r. 221 - 203 B.C.), in Egypt, made a dining room on his boat where the columns were made from Indian marble.[17].

Apollonius of Tyana (fl. 1st century A.D.), in the words of Philostratus, wrote on the presence of exported cotton in the shrines of Egypt.[18] On a possible exchange of Indian philosophy to Egypt, Philostratus wrote that "they [Egyptians] have such a high opinion of the Indian's doctrines about the creator of the universe that they teach it to others, although it is Indian."[19]

### A.0.3 About Pythagoras

Pythagoras advocated vegetarianism, metempsychosis (transmigration of the soul, where the soul migrated from one physical body to another just like we change our clothes), fasting as a way of purification, chastity as a virtue, and the sanctity of animal life that must be honored. He promoted orality in the preservation of knowledge, monism as an ideology, and believed on the existence of a soul in plants, animals, and humans. He founded the Pythagorean Brotherhood. This was a school of thought that influenced the ideas of Socrates, Plato, Aristotle, and many others. The above-mentioned doctrines of Pythagoras were essential to his school, and were practiced by his followers in the West for almost a millennium. These doctrines were new to Greece during the period of Pythagoras. Elsewhere, there is only one place in the world where all the above-mentioned doctrines existed prior to Pythagoras. This is the Indus-Sarasvatī region which was populated by the ancient Hindus.

In tune with the Eastern tradition, Pythagoras did not write any book since he contended that knowledge should be veiled from undesirable people; only oral communication was used to transfer knowledge. His disciple, Hippasus (470 B.C.), was drowned at sea for revealing some mathematical result to the general public.[20] For example, he divulged the way to construct

[16] Majumdar, 1964, p. 212; Rawlinson, 1926, p. 93.
[17] Sedlar, 1980, p. 88.
[18] Philostratus, 1970, p. 56.
[19] Philostratus, 1970, p. 216.
[20] Rouse Ball, 1908, p. 20.

a dodecahedron inside a sphere. The number of similar ideas is so large, it is highly improbable that the ideas evolved independently. Historical accounts indicate that Pythagoras did come in direct contact with the Indian philosophers.

It is known that Pythagoras left Samos at a young age and traveled extensively for the purpose of learning. He later came back to Greece as an established philosopher. After a short period in Greece, following his return, he again had to leave Greece due to the hostile political climate there. Such accounts are documented in the history books that were written all the way from the ancient to the medieval period.

Pythagoras believed that the human soul is a part of Divinity [God] and that it merges with the Divinity after completing the cycle of transmigrations – thus propagating the immortality of a soul. Herodotus (ca. 484 - 425 B.C.) mentioned that certain Greek writers had learned the concept of transmigration in Egypt.[21] Unfortunately, he did not provide the writers' names. It is plausible that Herodotus had Pythagoras in mind when he mentioned the Greek philosophers who had studied in Egypt. Pythagoras was the leading proponent of the concept of metempsychosis at that time.

Iamblichus (250 - 330 A.D.), a Neo-Platonic Syrian philosopher, wrote an encyclopedia on Pythagorean doctrines in nine books. Many of these books are now lost; only some have survived in small fragments. According to Iamblichus, Pythagoras, on the advice of Thales, went to Egypt for some twenty years to learn philosophy, astronomy, and geometry. Later on he was captured by the soldiers of Cambyses, a king of Persia and the son of Cyrus the Great, and was brought to Babylon, where he lived for another twelve years. At the age of fifty-six years, he went back to Greece and then started his school of thought.

Thales suggested to Pythagoras that he would become "the wisest and most divine of all men" by his association with the Egyptian priests. "Indeed, after Thales had gladly admitted him [Pythagoras] to his intimate confidence, he admired the great difference between him [Pythagoras] and other young men, whom Pythagoras left far behind in every accomplishment. And be-

[21]Herodotus, 1928, vol. 2, p. 123; vol. 1, p. 425.

sides this, Thales increased the reputation Pythagoras already acquired, by communicating to him such disciplines as he was able to impart: and, apologizing for his old age, and the imbecility of his body, he extorted him to sail into Egypt, and associate with the Memphian and Diospolitan priests. For he confessed that his own reputation for wisdom was derived from the instructions of these priests; but that he was neither naturally, nor by exercise, educated with those excellent prerogatives, which were so visibly displayed in the person of Pythagoras. Thales, therefore, gladly announced to him, from all these circumstances, that he [Pythagoras] would become the wisest and most divine of all men, if he [Pythagoras] associated with these Egyptian priests."[22]

Isocrates (436 - 338 B.C.), a Greek philosopher, wrote of the Egyptian roots in Pythagorean philosophy.[23] Isocrates was a disciple of Socrates and was mentioned by Plato in *Pheadrus*, as a companion of Socrates. Isocrates wrote: "If one were not determined to make haste, one might cite many admirable instances of the piety of the Egyptians, that piety which I am neither the first nor the only one to have observed; on the contrary, many contemporaries and predecessors have remarked it, of whom Pythagoras of Samos is one. On a visit to Egypt he became the student of the religion of the people, and was first to bring to the Greeks all philosophy, and more conspicuously than others he seriously interested himself in sacrifices and in ceremonial purity, since he believed that even if he should gain thereby no greater reward from the gods, among men, at any rate, his reputation would be greatly enhanced. And this indeed happened to him."[24] This remark is from a person who lived with Socrates, a Pythagorean at the time when Pythagorean teachings were relatively new and untouched. Isocrates is clear in his view that the Pythagorean doctrines were not his own but learned from philosophers on his visit to Egypt, and which were later taught in Greece.

The popularity of the Pythagorean doctrines among the Greek intellectuals is evident from the following statement of Isocrates: "And these reports we cannot disbelieve; for even now persons who profess to be followers of his [Pythagoras'] teachings are more admired when silent than are those who

[22]Taylor, 1976, p. 8.
[23]Sedlar, 1980, p. 31; Guthrie, 1962, vol. 1, p. 163.
[24]*Busiris*, 28, 29; Norlin and Van Hook, 1961, vol. 3, p. 119.

have the greatest renown for eloquence."[25] The above-mentioned citations establish Pythagoras' journey out of Greece, his presence in Egypt, and the influence of Pythagorean philosophy in Greece.

Did Pythagoras visit only Egypt? Or, did he visit other nations too? Once you move out from your own surroundings to a far place for learning for decades, what stops you from continuing with such travels? This indeed happened with Pythagoras, as written by Apollonius of Tyana (fl. 1st century A.D.), a well-respected Pythagorean philosopher during the Roman period. Apollonius was revered enough that the Roman Emperor Septimius Severus (146 - 211 A.D.) erected a statue of Apollonius in his private shrine, along with Christ and Abraham. Philostratus (170 - 244 A.D.), a Greek philosopher, taught in Athens and then served in the court of Julia Domna in Rome, the wife of emperor Septimius Severus, and compiled the travel accounts of Apollonius. Apollonius learned of the doctrine of nonviolence or the sanctity of human and animal life in the tradition of Pythagoras who, in Apollonius' view, was himself taught by Indian philosophers. Philostratus quotes the words of Apollonius: "I did not sacrifice, I do not: I do not touch blood, even on the alter, since that was the doctrines of Pythagoras. . . it was the doctrines of the Naked Philosophers of Egypt and the Wise Men of India, from whom Pythagoras and his sect derived the seeds of their philosophy."[26]

Apollonius correctly considered the Indian civilization to be much more ancient than the period of Pythagoras. "Pythagoras was anticipated by the Indians, lasts not for brief time, but for an endless and incalculable period," opines Apollonius.[27] This is a clear statement from someone who was a Pythagorean, a person of repute, who visited India, and connected the Pythagorean philosophy directly to Egypt and India.

Titus Flavius Clemens (c.150 - 215 A.D.), popularly known as Clement of Alexandria, was a Christian theologian who was familiar with classical Greek philosophy and literature. Similar to Apollonius, Clement, in his book *Stromata* (*The Miscellanies*), writes that Pythagoras went to Persia and came in

---

[25] *Busiris* 29; Norlin and Van Hook, 1961, vol. 3, p. 119.
[26] Book 8, Chapter 7, 12; Philostratus, 1960, vol. 2, p. 339.
[27] Philostratus, 1960, vol. 2, p. 49.

contact with the Brahmins. In his view, philosophy flourished in other cultures first and then from those cultures came to Greece. Clement used the term "barbarians" to define people from these other cultures. In his view, Pythagoras learned philosophy from his interactions with these brahmins: "Pythagoras was a hearer of . . . the Brahmins."[28]

Eusebius (263 - 339 A.D.) (also known as Eusebius of Caesarea), a Roman Christian historian, in his book *Praeparatio Evangelica* (*Preparation for the Gospel*) mentions Pythagoras' visit to Babylon, Egypt and Persia. Similar to Apollonius of Tyana and Clement of Alexandria, Eusebius also shares the opinion that Pythagoras "studied under the Brahmans [or Brahmins]" (these are Indian philosophers), and learned geometry, arithmetic, and music from these foreign lands. Eusebius is also quite clear in his statement that Pythagoras learned "nothing" from the Greek philosophers and "became the author of instruction to the Greeks in the learning which he had procured from abroad."[29]

Apuleius (Lucius Apuleius; 125 - 180 A.D.), a Platonist scholar from Algeria who was trained in Athens, tells us that Pythagoras lived in Egypt, Babylon, Persia and later learned from the "Brahmins." Pythagoras "derived the greater part of his philosophy, the arts of teaching the mind and exercising the body, the doctrines as to the parts of the soul and its various transmigrations" from the Brahmins.[30]

Although much later in history, Ṣāʻid al-Andalusī, an Islamic philosopher of eleventh-century Spain, wrote, "He [Pythagoras] studied philosophy in Egypt under the disciples of Solomon, son of David . . . Prior to that he studied geometry under the tutelage of Egyptian masters."[31]

The visit of Pythagoras to India was conclusive in the mind of Voltaire (1694 - 1778) (pen name of Francois Marie Arouet) as he wrote that "Pythagoras, the gymnosophist, may alone serve an incontestable proof that true science was cultivated in India. . . . It is even more probable that Pythagoras learnt the properties of the right-angled triangle from the Indians, the inven-

[28] *Stromata*, Book 1, Chapter 15.
[29] *Praeparatio Evangelica*, Book 10, Chapter 4.
[30] *The Florida*, chapter 15.
[31] Salem and Kumar, 1991, Chapter VII.

tion of which was afterward ascribed to him."[32] According to Voltaire, "the Greeks, before Alexander, traveled into India in quest of science . . . The Orientals, and particularly the Indians, treated all subjects under the veil of fable and allegory: for that reason Pythagoras, who studied among them, expresses himself always in parables."[33] In the view of Voltaire, the issue of Pythagoras' visit to India was more or less settled. "All the world knows that Pythagoras, while he resided in India, attended the school of Gymnosophists and learned the language of beasts and plants."[34]

D. E. Smith, a noted historian who is known for his classic book, *History of Mathematics*, points out a resemblance between the Hindu and Pythagorean philosophies: "In spite of the assertions of various writers to the contrary, the evidence derived for the philosophy of Pythagoras points to his contact with the Orient. The mystery of the East appears in all his teaching . . . indeed his [Pythagoras'] whole philosophy savors much more of the Indian than of the Greek civilization in which he was born."[35]

Jean W. Sedlar concludes in his studies on possible exchanges between India and Greece that "the contribution of India to Hellenistic civilization was unquestionably significant, although not precisely measurable."[36] Her book, *India and the Greek World*, provides details of the impacts of India on the West. In another book, *India in the Mind of Germany*, she provided the way Indian philosophy and poetry influenced the great thinkers from Europe, particularly from Germany. Both books are ground-breaking in their approach and provide valuable information on the way Europe and the West assimilated Indian philosophy in their mainstream thinking.

William Enfield published the *History of Philosophy* during the early part of the nineteenth century. In this book, he concluded that Pythagoras visited Egypt, Babylon, Persia, and India. The whole journey took about 34 years. He returned to Samos at about the age of sixty. On the similarities of Pythagorean philosophy and the philosophies of the ancient Hindus, Royle concludes: "In the same manner Pythagoras, who we have seen trav-

[32]Voltaire, 1901, vol. 29, p. 174.
[33]Voltaire, 1901, vol. 24, p. 39.
[34]Voltaire, vol. 4, p. 47.
[35]Smith, 1925, vol. I, p. 72.
[36]Sedlar, 1980, p. 302.

eled in Egypt and the East, likewise resembles the Hindoos [Hindus] in his metaphysical doctrines. It is remarkable also, that he should be thought to have discovered the celebrated properties of the right-angled triangle, as these have long been known to the Hindoos [Hindus], and which it is proper to add, they 'demonstrate in a very singular way, which partakes more of the nature of algebraic reasoning than of pure geometry.' The properties of numbers, always a favorite speculation with the Brahmans [Brahmins], was also entered into by Pythagoras; and as he likewise treated of music, so we find it one of the subjects included in the ancient Vedas."[37]

Edward Pococke, an anthropologist, wrote a detailed book on the interactions between India and Greece in 1852 and summed up his view about the visit of Pythagoras to India in the following sentence: "Certain it is, that he [Pythagoras] visited India, which I trust I shall make it self-evident."[38] Pococke showed connections between the philosophies in India and their similarities with those of the Pythagoreans and Platonists. In the view of Pococke, that Pythagoras "derived his doctrines from an Indian source is very generally admitted [by the scholars]." He used the words of H. T. Colebrooke, a noted Indologist from the nineteenth century, to summarize the Indian and Greek connection: "I should be disposed to conclude that the Indians were in this instance teachers, rather than learners."[39]

On a possible Indian influence on Pythagoras in comparison to Egyptian influence, H. W. Rawlinson (1810 - 1895 A.D.) concludes: "It is more likely that Pythagoras was influenced by India than by Egypt. Almost all the theories, religious, philosophical, and mathematical, taught by the Pythagoreans were known in India in the sixth century B.C."[40] Rawlinson is the person who first decoded cuneiform language of Babylon after discovering the Darius Behistun inscriptions. He also served the British empire and lived in India.

As evident, Pythagoras' visit to Egypt, Babylon, Persia, and India is documented from the pre-Christian period to the nineteenth century. These authors include respectable scholars of history or philosophy and also some Pythagoreans. These accounts make it quite clear that Pythagoras went to

[37]Royle, 1837, p. 185.
[38]Pococke, 1856, p. 353.
[39]Pococke, 1856, p. 363.
[40]Rawlinson, 1938, p. 5.

Egypt as a young man and lived most of his life away from Greece. Historians of the later periods have supported the idea. Still, most current textbooks on Greek philosophy and Pythagoras simply ignore this issue.

# Appendix B

# Timeline of the Hindu Manuscripts

A lack of clear-cut chronology is the weakest point of the Hindu manuscripts. There are several reasons for it: first, the documents are from a period where chronology is not available. Second, the Hindu culture practiced orality to preserve their ideas unlike the Egyptians (papyrus) and Babylonians (clay tablets). Third, the Hindu culture basically followed in the *same tradition* from the earliest periods. As you may remember, events are essential to define time. Similarly, events of atrocity, changes in value systems, changes in prosperity, etc. are the landmarks of a culture. However, in the case of the Hindus, a similar pattern of prosperity and cultural traits continued for so long, it is difficult to assign events and a time frame.

All we know is that the *Ṛgveda* is the earliest book of the Hindu literature, and is certainly one of the earliest known manuscripts in human history. Some Western historians look for a "hard-copy" or "hard-evidence" to accept a particular date for the antiquity of the Hindu manuscripts. Due to the prevalent oral tradition, such evidences are not possible with the ancient Hindus. The ancient Hindus did not write books on tablets or tree leaves, and did not erect monuments, until the period of Emperor Aśoka. They did not even mention the name of the authors for their books. Jessica Frazier, who in her quest to define history and historiography in Hinduism, tried to create a timeline for various Hindu events, shared her experience in the following language: "The search for an Indian historiography is like the work of a geologist who must sift innumerable layers of compacted material, or perhaps better, a zoologist who can never stop seeking new species of history, and never assume that any given specimen will remain long in a single place or form." She criticizes too much reliance on the accounts of travelers such as Herodotus and others who visited India and wrote their accounts. In her view, "[t]he ideal Western historian has classically been seen as an itinerant Godlike traveller." These travellers had a limited exposure of the culture and wrote their view from their bird's eye view. Even the astronomical observations of the heavenly bodies which should provide a linear scale of time fails because of "multiple parallel chronologies." She concludes that, "'while India's Hindu past was ever-present in divine reality and recursive myth, it was neverthelss unreachable by accepted historiographical methods."[1]

Al-Bīrūnī (973 - 1050 A.D.) noticed the problem of historiography in assigning actual dates to various events in the eleventh century and criticized the ancient Hindus: "Unfortunately the Hindus do not pay much attention to the historical order of things, they are very careless in relating the chronological succession of their kings."[2]

Carl Jung (1875 - 1961 A.D.), a psychoanalyst and philosopher, aptly answered such criticisms of the absence of a chronological history in India, by attributing it to the antiquity of the Hindu civilization. "After all why should there be recorded history [in India]? In a country like India one does not really miss it. All her native greatness is in any case anonymous and impersonal, like the greatness of Babylon and Egypt. History makes sense in European countries where, in a relatively recent, barbarous, and inhistorical past, things began to shape up . . . But in India there seems to be nothing that has not lived a hundred thousand times before."[3] Even the ancient writers such as Caraka (around 600 B.C.), Suśruta (around 1000 B.C.), and Ārybhaṭa I (5th century) do not represent the first efforts for most theories in their books; they all claim to be merely reproducing old ideas in new forms.

The first book in the modern era on the Hindu sciences was published by Brajendranath Seal in 1915.[4] Seal opted to deal with the absence of a proper chronology right in the first paragraph of his Foreword. He basically

[1]Frazier, 2009.
[2]Sachau, 1964, vol. 2, p. 10.
[3]Jung, *Collected Works*, vol. 10, p. 517.
[4]Seal, 1915

assigned an umbrella period of 500 B.C. to 500 A.D. to all Hindu manuscripts considered in his book. Bose *et al* in 1971 and Chattopadhyaya in 1986 encountered the same issue. This situation has not improved much in the last century; we still cannot assign accurate dates for ancient manuscripts.

The issue of antiquity can only be solved either from the artifacts of the ancient Hindus or from the astronomical maps provided in their oral literature. Previously, the maps provided in the Hindu oral literature were ignored by the scholars as they defined the period of *Ṛgveda* to be more ancient than 1500 B.C., an arbitrary date assigned by Max Müller. Since the recent excavations unearthed artifacts that are some 10,000 years old, a need for the revision of history is warranted.[5] A massive effort is needed to resolve this issue.

Though the dates are not certain, there is a general acceptance about the chronological order of these books and periods can be assigned to these books, as listed in Table B.1. The content of my book will not change much with corrections in this table. Most conclusions will remain valid even with a different chronology.

Table B.1: The Antiquity of Hindu Books

| Manuscript | Date or Period |
|---|---|
| *Āryabhaṭīya* | 5th century A.D. |
| *Bakhshālī Manuscript* | Seventh Century A.D. |
| *Brāhmaṇas* | Between 2000 - 1500 B.C. |
| *Mahābhārata* Battle | Between 2449 B.C. to 1424 B.C. |
| *Pingala's Chandah Sutra* | Between 480 - 410 B.C. |
| *Purāṇa* | Between the Christian era and 11th century |
| *Manu-Saṁhitā* , | Between 500 - 100 B.C. |
| *Caraka-Saṁhitā* | Around 600 B.C. |
| *Suśruta-Saṁhitā* | Around 1000 B.C. |
| *Śulbasūtra* | Between 1700 B.C. to 1000 B.C. |
| *Upanishads* | 800 B.C. to the start of the Christian era |
| *Vaiśesika-Sūtra* | Between 900 - 600 B.C. |

[5]Kenoyar, 1998.

| | |
|---|---|
| *Vedas* | Around 1500 B.C. or earlier |

# A Glossary of Sanskrit Terms

*ācāra-vicāra*, eating and thinking
*ahiṁsā*, nonviolence
*ākāśa*, space
*alaṅkāra*, metaphorical description
*amāvāsya*, new moon
*amṛta*, divine nector
*amūrta*, formless or void
*ānanda*, bliss
*ananta*, infinite
*aṅga*, limb
*aṅgula*, finger breadth
*aṅka*, number
*aṇu*, small particle, molecule
*anubhava*, experience
*anubhūti*, intuition
*ānuvīkṣikī*, investigation through reasoning
*āpaḥ* or *Ap*, water
*apāna*, exhaled air
*aparā-vidyā*, secular knowledge
*āratī*, prayer
*ariṣṭa*, fermented liquor
*Āṣāḍha*, a month in Hindu Calendar
*āsana*, posture or seat
*āśrama*, stage of life, hermitage.
*aṣṭa*, eight
*astra*, weapon
*aṣṭāṅga-yoga*, the eight-limbed yoga
*ātma-śakti*, self-power
*avasthā*, state
*avyakta*, unknown

*Bhāgavad-Gītā*, a holy book
*Bhakti-yoga*, The path of devotion or love
*bhasma*, oxidized or reduced compound in powder form for consumption
*bhīṣak*, physician
*bhūmi*, earth
*bhūta*, element
*brahmacarya*, learning stage of life or celibacy
*brahm-muhūrta*, sunrise

*cakṣu*, eye
*caṇḍāla*, outcast
*capāti*, a flat bread
*carṇāmṛta*, nectar from the feet of god

*dakṣiṇā*, gift to honor guru or gift for his services
*dakṣiṇāyaṇa*, apparent motion of the Sun to the South
*daṇḍa*, stick
*daṇḍanīti*, jurisprudence
*darśana*, philosophy
*daśaharā*, a festival to celebrate the victory of Lord Rāma over Rāvaṇa
Dasaratha, father of Lord Rāma
*dhanuṣ-āsana*, bow posture in yoga

*dhanuṣ-vidyā*, knowledge of the bow
*dharma*, duty
*dhāraṇa*, concentration
*dhātu*, the three constituents of the body in ayurveda, element
Dhruva, north star or a mythical figure
*dhyāna*, meditation
*dīkṣa*, initiation
*dinacaryā*, daily routine
*doṣa*, disorder
Draupadī, wife of Pāṇḍava
*dravya*, matter

*ekādaśī*, eleventh day (from the full Moon)
*ekāgratā*, concentration

*Gaṇga*, a sacred river
*Garuḍa*, eagle like mythical bird
*gṛah-praveśa*, entrance to the house
*guṇa*, property
*Gurukula* , teacher's home
*guru-ṛṇa*, debt to the guru
*gṛhastha*, family stage in life

*hala-āsana*, plow posture in yoga
*hṛdaya*, heart

*indriya*, sensory activity
*itihāsa*, history

*jāti*, caste
*Jñāna-yoga*, the path of knowledge [to achieve salvation]
*jyotiṣa*, science of stars

*kāl*, time
*kalpa*, cosmic age
*kalpavṛkṣa*, a tree that fulfils all human desires
*Kaṇāda*, name of a person who propagated Vaiśeṣika system
*Karma-yoga*, The path of action [to achieve salvation]
*kāraṇā*, half of *tithi*
*khagola-śāstri*, astronomer
*khicaḍī*, a popular rice-lentil preparation
*krounc-āsana*, heron-posture in yoga
*kumbhaka*, retention

Lakṣmī, Goddess of wealth and prosperity

*Mahābhārata*, an epic compiled by sage Vyāsa
*mahābhūta*, the great element
*makar-āsana*, crocodile posture in yoga
*māṃsa*, flesh
*maṇdūka-āsana*, frog-posture in yoga
*manana*, meditation
*mānasa*, mind, inner emotion, thought
Maṅgala, Mars
*mantra*, formula, a sacred phrase or word
*manvantara*, epoch or period of Manu
*mayūr-āsana*, peacock posture in yoga
*mokṣa*, salvation
*muhūrta*, a unit of time, thirtieth part of a day, about 48 minutes

*nakṣatra*, star or constellation
*nakṣatra-vidyā*, astronomy
*nirguṇa*, formless or with no attribute
*niṣkāma-karma*, selfless action

*niyama*, observance

*oṣadhi*, medicine

*padārtha*, matter
*pakṣa*, waxing or waning preiod of the moon
*Pañcāṅga*, five limbs, Hindu almanac
*Pañcā-bhūta*, five elements
*Pañcā-karma*, five actions
*paramāṇu*, atom
*parā-vidyā*, spiritual knowledge
*paryāvaraṇa*, ecology or environment
*paśu*, animal
*paśupati*, Lord Śiva, Lord of Animals
*prakṛti*, primal nature
*pramāṇa*, proof
*prāṇa*, breath or life force
*prāṇāyāma*, breath-control
*pratyāhāra*, sense-withdrawal
*pṛthivī*, Earth
*pūraka*, inhalation
*pūrṇimā*, the full Moon
*puruṣa*, man or spirit

*Raja-yoga*, The mystical path
*Rākṣasa*, demon or unlawful people
*Rāmāyaṇa*, an epic written by Vāmīki
*rāsi-vidyā*, mathematics
*recaka*, exhalation
*roti*, a flat bread
*ṛṣi*, Vedic sage
*ṛtu*, season
*rudrākṣa*, bead of Rudra (Śiva)
*rūpa*, form or color

*saguṇa*, with attribtes
*śalya-vaidya*, surgeon
*samādhi*, tranquil state
*sāmānya*, generality
*saṁhitā*, a division of Veda
*samīkaraṇa*, equation
*saṁkṛta*, coming together
*saṁskāra*, sublimial impression, purification rites, or faculty
*samudra*, ocean
*saṁyoga*, conjunction
*sandhyā*, sunset period
*sannyāsa*, stage of renunciation in life
*sāṅkhya*, the philosophical school
*śānti*, tranquality, peace
*śarīra*, body
*śāstrārtha*, debate or discussion
*satyam*, truth
*siddhānta*, principle or conclusion
*śikṣak*, teacher
*śiṣya*, disciple
*śivam*, benovalent
*smṛti*, memory
*śrotriya*, teacher
*sṛṣṭi*, creation
*śulba-sūtra*, rope measurement or geometry
*sundar*, beautiful
*śūnyatā*, the theory of void
*śūnya*, void
*sūkṣma*, subtle
*Suśruta-Saṁhitā*, the book written by Suśruta
*svābhāvika*, innate or natural
*svarūpa*, attribute
*svayaṁvara*, one's own choice [to choose a husband]

*tantra*, rule
*tark*, logic

*tejas*, fire
*tithī*, date
*trayī*, vedic learning
*trikoṇa-āsana*, triangle posture in yoga

*Upādhāya*, teacher
*Uttarāyana*, apparent motion of the sun to the north

*vāda*, debate
*Vaiśeṣika-sūtra*, a book
Valmīki, author of *Rāmāyaṇa*
*vamana*, vomit
*vānaprastha*, stage of contemplation in life, forest dweller
*vār*, day
*varṇa*, caste
*vārttā*, economics
*varṣā*, rain
Vaśiṣṭha, a Vedic sage
*vasti*, enema
*vasudha*, earth
*vasundharā*, earth
*vāyu*, air
*vidyā*, knowledge
*vijñāna*, wisdom, science
*virechana*, purging
*viśeṣa*, particularity or peculliarity.
*vṛkṣa-devatā*, tree deity

*yama*, moral principle
yojna, a measure of length

# Bibliography

Most primary sources were written about 2000 years ago and are difficult to read due to their style. Even commentaries from the translators do not make the reading easy for a beginner. The last major work on Hindu science was done by Seal (1915). However, the book is not easy to read for a beginner and is outdated. Bose, Sen, and Subbarayappa (1971), Chattopadhyaya (1986) Sarsvati (1986) are perhaps better books on this topic. However, these books are written for scholars and not beginners. It is much better to learn about the Hindu science by reading books on specific topics: Bag (1979), Datta and Singh (1938), Joseph (1991), and Srinivasiengar (1967) for mathematics; Subbarayappa and Sarma (1985) for astronomy; Clark (1930) and Shukla and Serma (1976) for mathematics and astronomy; Ray (1956) for chemistry; Coulter (2001) and Feuerstein (1989) for yoga; Royle (1837), Kutumbiah (1962), and Mukhopādhyāya (1979) for medicine.

### Primary Sources


*Atharvaveda: The Hymns of the Atharvaveda*, Ralph T. H. Griffith, Chowkhamba Sanskrit Series, Varanasi, 1968.

*Bhagavad-Gītā*, S. Radhakrishnan, Unwin Paperbacks, London, 1948.

*Caraka Saṁhitā: The Caraka Saṁhitā*, Satyanārāyaṇa Śāstrī, Chowkhamba Sanskrit Series, Varanasi, 1992, 2 volumes, in Hindi.

*Kauṭilya's Arthaśāstra: Kauṭilya's Arthaśāstra*, Shamasastry, R. Mysore Printing and Publishing House, Mysore, 1960. Fourth Edition. (first published in 1915)

*Purāṇa*, Translator: Pandit Shreeram Ji Sharma, Sanskrit Sansthan, Bareiley, 1988. In Hindi and Sanskrit. All eighteen *Purāṇas* have been translated by the same author and published from the same publications.

*Ṛgveda: The Hymns of the Ṛgveda*, Ralph T. H. Griffith, Chowkhamba Sanskrit Series, Varanasi, 1963.

*Suśruta-Saṁhitā: The Suśruta-Saṁhitā*, Bhishagratna, Kaviraj Kunjalal [Translator], Chaukhambha Sanskrit Series, Varanasi, 1963, 3 vols.

*Upaniṣads*, Translator: Patrick Olivelle, Oxford University Press, Madras, 1996.

*Upaniṣads:The Thirteen Principal Upanishads*, Translator: Robert Ernst Hume, Oxford University Press, Madras, 1965.

*Upaniṣads:108 Upanishads*, Translator: Pandit Shreeram Ji Sharma, Sanskrit Sansthan, Bareiley, 1990. In Hindi and Sanskrit.

*Vedas*, Translator: Pandit Shreeram Ji Sharma, Sanskrit Sansthan, Bareiley, 1988. In Hindi and Sanskrit. All four *Vedas* have been translated by the same author and published from the same publications.

*Viṣṇu Purāṇa*, Wilson, H. H. [Translator] , Punthi Pustak, Calcutta, 1961.

*Yajurveda (Vājasaneyisamhitā): The texts of the White Yajurveda*, Ralph T. H. Griffith, Chowkhamba Sanskrit Series, Varanasi, 1976.

**Secondary Sources**

Abhyankar, K. D., A Search for the earliest Vedic Calendar, *Indian Journal of History of Science*, 28 (1), 1 - 14, 1993.

Abrahamson, Warren G. [Editor], *Plant-Animal Interactions*, McGraw Hill, New York, 1989.

Achar, B. N. Narahari, Āryabhaṭa and the Tables of Rsines, *Indian Journal of History of Science*, 37 (2), 95 - 99, 2002.

Acharya, Shanta, *The Influence of Indian Thought on Ralph Waldo Emerson*, The Edwin Mellen Press, Lewiston, 2001.

Achaya, K. T., Alcoholic Fermentation and its Products in Ancient India, *Indian Journal of History of Science*, 26 (2), 123 - 129, 1991.

Ackernecht, Erwin Heinz, *A Short History of medicine*, Johns Hopkins University Press, Maryland, 1982.

Aelianus, Claudius, *On the characteristics of Animals*, A. F. Scholfield [Translator], Harvard University Press, Cambridge, 1958.

Agarwal, D. P. *Ancient Metal Technology and Archaeology of South Asia: A Pan-Asian Perspective*, Aryan Books International, New Delhi, 2000.

Agarwala, V. S., 1969, *Pāninikālīna Bhāratvarṣa*, The Chowkhamba Vidyabhawan, Varanasi, 1969.

Alborn, H. T., T. C. J. Turlings, T. H. Jones, G. Stenhagen, and J. H. Tunilinson, An elicitor of plant volatiles from beet armyworm oral secretion, *Science*, 276 (5314), 945 - 950, 1997.

Algra, Keimpe, *Concepts of Space in Greek Thought*, Brill, Leiden, 1994.

al-Hasan, A. Y. and Hill, D. R., *Islamic Technology*, Cambridge University, New York, 1986.

al-Kindī, *The Medical Formulary Or Aqrābādhīn of Al-Kindī*, Martin Levey [Translator], The University of Wisconsin Press, Madison, 1966.

Allen, James, *The Art of Medicine in Ancient Egypt*, The Metropolitan Museum of Art, New York, 2005.

Allen, Richard Hinckley, *Star Names: Their Lore and Meaning*, Dover Publications, Inc., New York, 1963 (first published in 1899).

Allen, Robert [Editor], *Bulletproof Feathers: How Science Uses Nature's Secrets to Design Cutting-Edge Technology*, The University of Chicago Press, Chicago, 2010.

Alter, Joseph S., Heaps of Health, Metaphysical Fitness, *Current Anthropology*, 40, S43 - S66, 1999.

Anderson, C. N., *The Fertile Crescent*, Sylvester Press, Florida, 1972.

Ang, Gina C., History of skin transplantation, *Clinics in Dermatology*, 23, 320 - 324, 2005.

Aristotle, *Aristotle Minor Works*, W. S. Hett [Translator], Harvard University Press, Cambridge, 1963 (first published 1936).

Arenson, Karen W., When Strings Are Attached, Quirky Gifts Can Limit Universities, *The New York Times*, April 13, 2008.

Arivazhagan, S., B. Velmurugan, V. Bhuvaneswari, and S. Nagini, Effects of aqueous extracts of garlic (Allium sativum) and neem (Azadirachta indica) leaf on hepatic and blood oxidant-antioxident status during experimental gastric carcinogenesis, *Journal of Medical Food*, 7, 334 - 339, 2004.

Asimov, Isaac, *Asimov's Biographical Encyclopedia of Science and Technology*, Doubleday & Company, Inc., New York, 1982 (first published in 1964).

Atkinson, Nancy L. and Rachel Permuth-Levine, Benefits, Barriers, and Cues to Action of Yoga Practice: A Focus Group Approach, *American Journal of Health Behavior*, 33(1), 3 - 14, 2003.

Baber, Zaheer, *The Science of Empire: Scientific Knowledge, Civilization, and Colonial Rule in India*, State University of New York, Albany, 1996.

Bacon, Roger, *The Opus Majus of Roger Bacon*, Translator: Robert Belle Burke, University of Pennsylvania Press, Philadelphia, 1928. (Originally published with Oxford University Press)

Bag, A. K., *Mathematics in Ancient and Medieval India*, Chaukhambha Orientalia, Delhi, 1979.

Bagchi, P. C., *India and China*, Hind-Kitab, Bombay, 1950.

Bailey, Cyril, *The Greek atomists and Epicurus*, Clarendon Press, Oxford, 1928.

Balasubramaniam, R., V. N. Prabhakar and Manish Shankar, On Technical Analysis of Cannon Shot Crater on Delhi Iron Pillar, *Indian Journal of History of Science*, 44 (1), 29 - 46, 2009.

Balasubramaniam, R. and A. V. Ramesh Kumar, Corrosion resistance of the Dhar iron pillar, *Corrosion Science*, 45, 2451 2465, 2003.

Balasubramaniam, B., A New Study of Dhar Iron Pillar, *Indian Journal of history of Science*, 37 (2), 115 - 151, 2002.

Balasubramaniam, B., New insights on the 1600 year-old corrosion resistant Delhi iron pillar, *Indian Journal of history of Science*, 36 (1 - 2), 1 - 49, 2001.

Balle, David W., Elemental Etymology: What's in a Name?, *Journal of Chemical Education*, 62 (9), 787 - 788, 1985.

Balslev, A. N., *A Study of Time in Indian Philosophy*, Otto Harrassowitz, Wiesbaden, 1983.

Barnwart, G. J., *Basic Food Microbiology*, Van Nostrand-Reinhold, New York, 1989.

Barua, Beni Madhab, *Aśoka and His Inscriptions*, New Age Publishers, Calcutta, 1946.

Barrow John D., *Pi in the Sky: Counting, Thinking, and Being*, Oxford University Press, New York, 1992.

Bays, G., *The Voice of Buddha*, Dharma Publishing, California, 1983. (original title, *Lalitvistara*).

Beal, Samuel, *Si-Yu-Ki Buddhist Records of the Western World*, Paragon Book Reprint Corp., New York, 1968.

Beckman, Johann, A History of Inventions, Discoveries, and Origins, Translated by William Johnston, William Francis, and J. W. Griffith, Henry G. Bohn, London, 1846, Volume 2.

Bender, R. C., M. Spraul Trautnet, and M. Berger, Assessment of excess mortality in obesity, *American Journal of Epidemiol.*, 147 (1), 42 - 48, 1998.

Benyus, Janine M, *Biomimicry*, William Morrow and Company, New York, 1997.

Bernstein, Richard, *Ultimate Journey*, Alfred A. Knopf, New York, 2001.

Bhagvat, R. N., Knowledge of the Metals in Ancient India, *Journal of Chemical Education*, 10, 659 - 666, 1933.

Bhardwaj, H. C., *Aspects of ancient Indian technology: a research based on scientific methods*, Motilal Banarsidas, Delhi, 1979.

Bhatla, Neeraj, Tapan Mukherjee, and Gian Singh, Plants: Traditional Worshipping, *Indian Journal of History of Science*, 19(1), 37 - 42, 1984.

Bigwood, J. M. Ctesias, his royal patrons and Indian swords, *Journal of Hellenic Studies*, 115, 135 - 140, 1995.

Billard, Roger, Aryabhaṭa and Indian Astronomy, *Indian Journal History of Science*, 12 (2), 207 - 224, 1977,

Billing, Jennifer and Paul W. Shermon, Antimicrobial functions of spices: Why some like it hot, *The Quarterly Review of Biology*, 73, 3 - 49, 1998.

Birdee, Gurjeet S., Gloria Y. Yeh, Peter M. Wayne, Russell S. Phillips, Roger B. Davis,and Paula Gardiner, Clinical Applications of Yoga for the

Pediatric Population: A Systematic Review, *Academic Pediatrics*, 9 (4), 212 - 220, July/August 2009.

Biswas, Arun Kumar, Brass and Zinc Metallurgy in the Ancient & Medieval World: India's Primacy and the Technology Transfer to the West, *Indian Journal of History of Science*, 41 (2), 159 - 174, 2006.

Biswas, Arun Kumar and Sulekha Biswas, *Minerals and Metals in Ancient India*, D. K. Printworld, New Delhi, 1996, three volumes.

Biswas, Arun Kumar and Sulekha Biswas, *Minerals and Metals in Pre-Modern India*, D. K. Printworld, New Delhi, 2001.

Boesche, Roger, Kautilya's *Arthaśāstra* on War and Diplomacy in Ancient India, *The Journal of Military History*, 67, 9 - 38, January 2003.

Bohm, David, *Wholeness and the Implicate Order*, Routledge, London, 1980.

Boncompagni, B. [Editor], *Scritti di Leonardo Pisano*, Rome, 1857 - 1862. (2 volumes) [In Italian]

Bond, John David, The Development of Trigonometric Methods down to the Close of the XVth Century, *Isis*, 4(2), 295 - 323, 1921.

Bose, D. M., S. N. Sen, and B. V. Subbarayappa, *A concise history of science in India*, Indian National Science Academy, New Delhi, 1971.

Boulting, William, *Giordano Bruno: His Life, thought, and Martyrdom*, Books for Libraries Press, New York, 1972 (first published 1914)

Boyer, Carl B., *A History of Mathematics*, revised by Uta C. Merzbach, John Wiley & Sons, New York, 1991. (II Edition)

Brain, David J., The Early History of Rhinoplasty, *Facial Plastic Surgery*, 9 (2), 91 - 88, April 1993.

Brennard, W., *Hindu Astronomy*, Caxton Publications, Delhi, 1988.

Brier, Bob, Napoleon in Egypt, *Archaeology*, 52 (3), 44 - 53, May/June 1999.

Bronkhorst, Johannes, A Note on Zero and the Numberican Place-Value System in Ancient India, *Asiatische Studien Études Asiatiques*, 48 (4) 1039 - 1042, 1994.

Brown, Ronald A. and Alok Kumar, A New Perspective on Eratosthenes' Measurement of the Earth, The Physics Teacher, 19, 445 - 447, October 2011.

Bryan, Cyril, *The Papyrus Ebers*, Appleton, New York, 1931.

Bubnov, Nikolai Michajlovic, *Gerberti, Opera Mathematica*, Berlin, 1899. [In Latin]

Burgess, J., Notes on Hindu Astronomy and the History of Our knowledge of It, *Journal of Royal Asiatic Society*, p. 717 - 761, 1893.

Burke, Robert Belle, *The Opus Majus of Roger Bacon*, University of Pennsylvania Press, Philadelphia, 1928.

Burnett, Charles, *The Introduction of Arabic Learning into England*, The British Library, London, 1997.

Burnett, Charles, The Semantics of Indian Numerals in Arabic, Greek and Latin, *Journal of Indian Philosophy*, 34, 15 - 30, 2006.

Cajori, Florian, *A History of Mathematics*, Chelsea Publishing Company, New York, 1980.

Calinger, Ronald, *A Contextual History of Mathematics*, Prentice Hall, New Jersey, 1999. (first published in 1893)

Capra, Fritjof, *The Tao of Physics*, Bantam New Age Books, New York, 1980.

Carpue, J. C., *An account of two successful operations for restoring lost nose, from the integuments of the forehead*, Longman, London, 1816.

Carson, James W., Kimberly M. Carson, Kim D.Jones, Robert M. Bennett, Cheryl L. Wright , and Scott D. Mist, A pilot randomized controlled trial of the Yoga of Awareness program in the management of fibromyalgia, *Pain*, 151, 530 - 539, 2010.

Casson, Lionel, *The Periplus Maris Erythraei*, Princeton University Press, 1989.

Casson, Lionel, *Travel in the Ancient World*, The Johns Hopkins University Press, Baltimore, 1994.

Castillo-Richmond, Amparo, Robert H. Schneider, Charles N. Alexander, Robert Cook, Hector Myers, Sanford Nidich, Chinelo Haney, Maxwell Rainforth, John Salerno, Effects of Stress Reduction on Carotid Atherosclerosis in Hypertensive African Americans, *Stroke*, 31, 568 - 573, 2000.

Chabás, José, The Diffusion of the Alfonsine Tables: The case of the Tabulae resolutae, *Perspectives on Science*, 10 (2), 168 - 178, 2002.

Chapple, Christopher Key and Mary Evelyn Tucker [Editors], *Hinduism and Ecology: The Intersection of Earth, Sky and Water*, Harvard Divinity School, Cambridge, 2000.

Chaudhuri, Ratan K., Emblica cascading antioxidant: a novel natural skin care ingredient. *Skin Pharmacology and Applied Skin Physiology*, 15 (5), 374 - 380, 2002.

Chattopadhyaya, D. *Science and Society in Ancient India*, Research India Publication, Calcutta, 1977.

Chattopadhyaya, D., *History of Science and Technology in Ancient India*, Firma KLM Pvt. Ltd., Calcutta, 1986.

Chaurasia, S. S., S. Panda, and A. Kar, *Withania somnifera* root extract in the regulation of lead-induced oxidative damage in male mouse, *Pharma-*

*col. Research* 41(6), 663 - 6, 2000.

Chikara, Sasaki, Mitsuo Sugiura, and Joseph W. Dauben [Editors], *The Intersection of History and Mathematics*, Birkhäuser Verlag, Basel, 1994.

Chopra, R. N., S. I. Nayar, and I. C. Chopra, *Glossary of Indian Medical Plants*, CSIR Publication, New Delhi, 1956.

Christy, Arthur, *The Orient in American Transcedentalism: A Study of Emerson, Thoreau and Alcott*, Columbia University Press, New York, 1932.

Chuntharapata, Songporn, Wongchan Petpichetchian, and Urai Hatthakit, Yoga during pregnancy: Effects on maternal comfort, labor pain and birth outcomes, *Complementary Therapies in Clinical Practice* 14,105 115, 2008.

Clark, W. E., *Indian Studies in the honor of Charles Rockwell Lanman*, Harvard University Press, Cambridge, 1929.

Clark, W. E., *The Āryabhaṭīya of Ārybhaṭa I*, The University of Chicago Press, Illinois, 1930.

Clarke, J. J., *Jung and Eastern Thought: A Dialogue With the Orient*, Routeledge, New York, 1994.

Cobb, Cathy and Harold Goldwhite, *Creations of Fire*, Plenum Press, New York, 1995.

Colebrooke, Henry Thomas[Translator], *Algebra with Arithmetic and Mensuration from the Sanscrit of Brahemegupta and Bhascara*, John Murray, London, 1817.

Conboy, Lisa. A., Ingrid Edshteyn, and Hilary Garivsaltis, Ayurveda and Panchakarma: Measuring the Effects of a holistic Health Intervention, *The Scientific World Journal*, 9, 272 - 280, 2009.

Cooke, Roger, *The History of Mathematics*, John Wiley & Sons, Inc., New York, 1997.

Copernicus, Nicholas, *On the Revolutions*, E. Rosen [translator], The Johns Hopkins University Press, Maryland, 1978.

Coppa, A., L. Bondioli, A. Cucina, D. W. Frayer, C. Jarrige, J. -F. Jarrige, G. Quivron, M. Rossi, M. Vidale, and R. Macchiarelli, Palaeontology: Early Neolithic tradition of dentistry, *Nature*, 440, p. 755 - 6, April 6, 2006.

Coulter, H. David, *Anatomy of Hatha Yoga*, Body and Breath, Pennsylvania, 2001.

Coward, H, Jung and Hinduism, *The Scottish J. Religions*, 5, 65 - 68, 1984. also read *Jung and Eastern Thought*, State University of New York Press, 1985.

Cowen, Virginia, Functional fitness improvements after a worksite-based yoga initiative, *Journal of Bodywork & Movement Therapies*, 14, 50 - 54, 2010.

Cowen, Virginia S. and Troy B. Adams, Heart rate in yoga asana practice: A comparison of styles, *Journal of Bodywork and Movement Therapies* 11, 91 - 95, 2007.

Cowen, Virginia S. and Troy B. Adams, Physical and perceptual benefits of yoga asana practice: results of a pilot study, *Journal of Bodywork and Movement Therapies*, 9, 211 - 219, 2005.

Crawford, S. Cromwell, *Dilemmas of Life and Death: Hindu Ethics in North American Context*, SUNY Press, Albany, 1995.

Cromer, Alan, *Uncommon Sense: The Heretical Nature of Science*, Oxford University Press, New York, 1993.

Crosby, Alfred W., *The Measure of Reality*, Cambridge University Press, New York, 1997.

Crosby, Alfred, *Throwing Fire*, Cambridge University Press, Cambridge, 2002.

Crossley, J. N. and A. S. Henry, Thus Spake al-Khwarizmi, *Historia Mathematica*, 17, 103 - 131, 1990.

Crossley, John N. and Anthony W. -C. Lun, *The Nine Chapters on the Mathematical Art*, Oxford University Press, New York, 1999.

Curtze, Max, Üer eine Algorimusschrift des 12. Jahrhundert, *Geschichte der Mathematik*, 8, 3 - 27, 1898.

da Silva, Tricia L., Lakshmi N. Ravindran, and Arun V. Ravindran, Yoga in the treatment of mood and anxiety disorders: A review, *Asian Journal of Psychiatry* 2, 6 16, 2009.

Dalby, A., *Dangerous tastes: the story of spices*, University of California Press, Berkeley, 2000.

Danucalov, Marcello A., Roberto S. Simões, Elisa H. Kozasa, and José Roberto Leite, Cardiorespiratory and Metabolic Changes During Yoga Sessions: The Effects of Respiratory Exercises and Meditation Practices, *Applied Psychophysiology and Biofeedback*, 32(2), 77-81, 2008.

Darlington, Oscar, G., Gerbert, 'Obscuro loco natus', *Speculum*, 11 (4), 509 - 520, 1936.

Darlington, Oscar, G., Gerbert, the Teacher, *The American Historical Review*, 52 (3), 456 - 476, 1947.

Dasgupta, A., *Goethe and Tagore*, South Asia Institute, Heidelberg, 1973.

Dasgupta, S. N., *History of Indian Philosophy*, Cambridge University Press, Cambridge, 1922 - 1955, 5 volumes.

Dasgupta, Surendra, *A History of Indian Philosophy*, Cambridge University Press, Cambridge, 1963.

Datta, B. B., Early Literary Evidence of the Use of Zero in India, *American Mathematical Monthly*, 33, 449 - 454, 1926.

Datta, B. B., *The Science of Sulba*, University of Calcutta, Calcutta, 1932.

Datta, Bibhutibhusan and Avadhesh Narayan Singh, *History of Hindu Mathematics*, Asia Publishing House, New York, 2 volumes, 1961. [First published in 1938]

Deshpande, Vijaya Jayant, Glimpses of Ayurveda in Medieval Chinese Medicine, *Indian Journal of History of Science*, 43 (2), 137 - 161, 2008.

DeYoung, Donald B. and Derrik Hobbs, *Discovery of Design; Searching Out the Creator's Secrets*, Master Books, Green Forest, 2009.

Dikshit, Moreshwar G., *History of Indian Glass*, University of Bombay, Bombay, 1969.

Diogenes Laertius, *Lives of Eminent Philosophers*, English translation, R. D. Hicks, Harvard University Press, 1958, 2 volumes.

Doberer, K. K., *The Goldmakers: 10,000 Years of Alchemy*, W. W. Dickes [Translator] Nicholson & Watson, London, 1948.

Doheny, Kathleen, Calm Mind, Heal Your Body, *Natural Health*, 36 (5), 56 - 63, May 2006.

Dollemore, Doug, Mark Giuliucci, Jennifer Haigh, Sid Kirchheimer, Jean Callahan, *New Choices in Natural Healing*, Rodale Press, Inc., Emmaus, Pennsylvania, 1995.

Dombrowski, D. A., Was Plato a Vegetarian, *Apeiron*, 18, 1- 9, 1984a.

Dombrowski, D. A., *The Philosophy of Vegetarianism*, University of Massachusetts Press, Amherst: l984b.

Doshi, Saryu, *India and Greece: Connections and Parallels*, Marg Publications, Bombay, 1985.

Dubs, H. H., The Beginnings of Alchemy, *Isis*, 38 , 62, 1947.

Durant, Will, *Our Oriental Heritage*, Simon and Schuster, New York, 1954.

Dwivedi, O. P., Vedic Heritage for Environmental Stewardship, *Worldviews: Environment, Culture, Religion*, 1, 25 - 36, 1997.

Ebeid, Nabil I., *Egyptian Medicine in the Days of the Pharaohs*, General Egyptian Book Organization, Cairo, 1999.

Elgood, Cyril, *A Medical History of Persian and the Easter Caliphate from the Earliest Times until the year A.D. 1932*, Cambridge University Press, London, 1951.

Eliade, Mircea, *Yoga: Immortality and Freedom*, Translated by Willard R. Trask, Princeton University Press, New Jersey, 1969.

Eliot, Charles, *Hinduism and Buddhism*, Routledge and Kegan Paul Ltd., London, 1954. (3 volumes)

Elizarenkova, Tatyana J., *Language and Style of the Vedic Ṛṣis*, State University of New York Press, Albany, 1995.

Emerson, Waldo, *The Complete Works of Ralph Waldo Emerson*, Houghton, Miffin and Company, Cambridge, 1904, ten volumes.

Enfield, William, *The History of Philosophy: From the Earliest Times to the Beginning of the Present Century*, London, 1819, 2 volumes.

Ernst, Carl W., Muslim Studies of Hinduism? A Reconsideration of Arabic and Persian Translations from Indian Languages, *Iranian Studies* 36 (2), 173 - 195, June 2003.

Eshel, Gidon and P. A. Martin, Diet, Energy, and Global Warming, *Earth Interactions*, 10 (9), 1 - 17, 2006.

Estes, J. Worth, *The Medical Skills of Ancient Egypt*, Science History Publications, Canton, 1989.

Farrington, B., *Science in Antiquity*, Oxford University Press, London, 1969. (first published in 1936)

Frazier, Jessica, History and Historiography in Hinduism, *The Journal of Hindu Studies*, 2, 1 - 16, 2009.

Feuerstein, Georg, *The Yoga-sutra of Patañjali*, Inner Traditions International, Rochester, Vermont, 1989.

Fiala, Nathan, Meeting the demand: An estimation of potential future greenhouse gas emissions from meat production, *Ecological Economics*, 67, 412 - 419, 2008.

Fields, R. Douglas, Making memories stick, *Scientific American*, 292 (2), 75 - 81, Feb. 2005.

Figiel, L. S., *On Damascus Steel*, Atlantas Arts Press, Atlanta, 1991.

Filliozat, J., *The Classical Doctrine of Indian Medicine: Its Origins and its Greek Parallels*, [Translator] Dev Raj Chanana, Munshiram Manoharlal, Delhi, 1964.

Filliozat, Pierre-Sylvain, Making something out of nothing, *UNESCO Courier*, 30 - 34, November 1993.

Findly, Ellison B., Jahāngīr's Vow of Non-Violence, *Journal of the American Oriental Society*, 107 (2), 245 - 256, 1987.

Finocchiaro, M., *The Galileo Affair: A Documentary History*, University of California Press, Los Angeles, 1989.

Fitzgerald, James L., Sanskrit *Pita* and *Śaikya/Saikya* Two Terms of Iron and Steel Technology in Mahabharata, *The Journal of the American Oriental Society*, 120 (1), 44 - 61, 2000.

Folkerts, Menso, Early Texts on Hindu-Arabic Calculation, *Science in Context*, 14 (1/2), 13 - 38, 2001.

Fraser, G. E., Associations between diet and cancer, ischemic heart disease, and all-cause mortality in non-Hispanic white California Seventh-day Adventists, *American Journal of Clinical Nutrition*, 70, 532 - 538, 1999.

Friedmann, Yohanan, Medieval Muslim Views of Indian Religions, *Journal of the American Oriental Society*, 95 (2), 214 - 221, 1975.

Gangadharan, N., S. A. S. Sarma, and S. S. R. Sarma, *Studies on Indian Culture, Science, and Literature*, Sree Sarada Education Society, Chennai, 2000.

Garbe, R., *The Philosophy of Ancient India*, The Open Court Publishing Company, Chicago, 1897.

Garfinkel, M. S., A. Singhal, W. A. Katz, A. A. David, and R. Reshetar, Yoga-based inervention for carpal tunnel syndrome: a radomized trial, *JAMA*, 280, 1601 - 03, 1998.

Gazalé, Midhat, *Number: From Ahmes to Cantor*, Princeton University Press, New Jersey, 2000.

Geoghegan, D., A Licence of Henry VI to Practice Alchemy, *Ambix*, 6, 10 - 17, 1957.

Ghosh, Pranabendra Nath, *Johann Gottfried Herder's Image of India*, Visva-Bharati Research Publication, Calcutta, 1990.

Gies, Joseph and Frances Gies, *Leonardo of Pisa and the New Mathematics of the Middle Ages*, Thomas Y. Crowell Company, New York, 1969.

Gilbert, Christopher, Yoga and Breathing, *Journal of Bodywork and Movement Therapies*, 3 (1), 44 - 54, 1999a.

Gilbert, Christopher, Breathing and the cardiovascular system, *Journal of Bodywork and Movement Therapies*, 3 (4), 215 - 224, 1999b.

Gill, D. and L. Levidow, *Anti-Racist Science Teaching*, Free Association Books, London, 1987.

Gingerich, Owen [Editor], *The Eye of Heaven: Ptolemy, Copernicus, and Kepler*, American Institute of Physics, New York, 1993.

Gohlman, William E., *The Life of Ibn Sina*, State University of New York Press, Albany, 1974.

Goldstein, B. R., Astronomy and Astrology in the Works of Abraham Ibn Ezra, *Arabic Sciences and Philosophy*, 6(1), 9 - 21, 1996.

Goldstein, B. R., *Ibn al-Muthannâ's Commentary on the Astronomical Tables of al-Khwârizmī*, Yale University, New Haven, 1967.

Goodman, Russell B., East-West Philosophy in Nineteenth-Century America: Emerson and Hinduism, *Journal of the History of Ideas*, 51 (4), 635 - 645, 1990.

Goonatilake, S., *Aborted Discovery: Science and Creativity in the Third World*, Zed Books, London, 1984.

Goonatilake, S., The Voyages of Discovery and the Loss and Rediscovery of 'Others' Knowledge, *Impact of Science on Society*, no. 167, 241 - 264, 1992.

Gorini, Catherine A. [Editor], *Geometry at Work*, The Mathematical Association of America, 2000.

Gosling, David L., *Religion and Ecology in India and Southeast Asia*, Routledge, New York, 2001.

Govindarajan, R., M. Vijayakumar, and P. Pushpangadan, Antioxidant approach to disease management and the role of 'Rasayana' herbs of Ayurveda, *Journal of Ethnopharmacology*, 99, 165 - 178, 2005.

Greener, William Wellington, *The Gun and Its Development*, Cassell, London, 1907.

Gregorios, Paulos Mar [Editor], *Neoplatonism and Indian Philosophy*, State University of New York Press, 2002.

Grimm, R. E. The Autobiography of Leonardo Pisano, *Fibonacci Quarterly*, 11 (1), 99 - 104, 1973.

Grombrich, E. H., Eastern Inventions and Western Response, *Daedalus* 127 (1), 193 - 205, 1998.

Gupta, Mradu, Pharmacological Properties and Traditional Therapeutic Uses of Important Indian Spices, *International Journal of Food Properties*, 13, 1092 - 1116, 2010.

Gupta, R. C., Indian astronomy in China during ancient times, *Vishveshvaranand Indological Journal* (India), 9, 266 - 276, 1981.

Gupta, R. C., Spread and Triumph of Indian Numerals, *Indian Journal of the History of Science*, 18 23 - 38, 1983a.

Gupta, R. C., Decimal Denomination Terms in Ancient and Medieval India, *Gaṇita Bhāratī*, 5 (1 - 4), 8 - 15, 1983b.

Gupta, R. C., Who Invented the Zero?, *Gaṇita Bhāratī*, 17 (1 - 4) 45 - 61, 1995.

Gupta, Raj Kumar, *Great Encounter: A Study of India-American Literature and Cultural Relations*, Abhinav Publications, New Delhi, 1986.

Gupta, S. K., A. Dua, B. P. Vohra, *Withania somnifera* (*Ashwagandha*) attenuates antioxidant defense in aged spinal cord and inhibits copper induced lipid peroxidation and protein oxidative modifications, *Drug Metabol Drug Interact*, 19(3), 211 - 22, 2003.

Gupta, Shakti M., *Plants Myths and Traditions in India*, E. J. Brill, Leiden, 1971.

Guthrie, W. K. C., *A History of Greek Philosophy*, Cambridge University Press, Cambridge 1962 - 1981.

Hairetdinova, N. G., On Spherical Trigonometry in the Medieval Near

East and in Europe, *Historia Mathematica*, 13, 136 - 146, 1986.

Halbfass, W., *India in Europe*, State University of New York, Albany, 1988.

Halstead, George Bruce, *On the Foundation and Technic of Arithmetic*, The Open Court Publishing Company, Chicago, 1912.

Hammett, Frederik S., Agricultural and Botanic Knowledge of Ancient India, *Osiris*, 9, 211 - 226, 1950.

Hammett, Frederik S., The Ideas of the Ancient Hindus Concerning Man, *Isis*, 28 (1), 57 - 72, 1938.

Hanachi, P., O. Fauziah, L. T. Peng, L. C.Wei, L. L.Nam, T. S. Tian, The effect of Azadirachta indica on distribution of antioxidant elements and glutathione S-transferase activity in the liver of rats during hepatocarcinogenesis. *Asia Pacific Journal of Clinical Nutrition*, 13, S170, 2004.

Harding, Sandra, *Whose Science? Whose Knowledge?*, Cornell University Press, Ithaca, 1991.

Harding, Sandra, Is Science Multicultural? Challanges, Resources, Opportunity, Uncertainties, *Configuration*, 2, 301 - 330, 1994.

Hayasi, T., Āryabhaṭa's Rule and Table for sine-differences, *Historia Mathematica*, 24, 396 - 406, 1997.

Hayashi, T, Takanori Kusuba, and Michio Yano, Indian Values for $\pi$ from Āryabhaṭa's Value, *Historia Scientiarum*, no. 37, 1 - 16, 1989.

Hayasi, T., *The Bakhshālī Manuscript: An ancient Indian mathematical treatise*, Egbert Forsten, Groningen, 1995.

Headrick, Daniel R., *The Tools of Empire: Technology and European Imperialism in the Nineteenth Century*, Oxford University Press, New York, 1981.

Helms, Michael, Swaroop S. Vattam and Ashok K. Goel, Biologically inspired design: process and products, *Design Studies* 30, 606 - 622, 2009.

Helson, Julie E., Todd L. Capson, Timothy Johns, Annette Aiello, and Donald M. Windsor, Ecological and evolutionary bioprospecting: using aposematic insects as guides to rainforest plants active against disease, *Frontiers Ecology Environment*, 7, 1 - 10, 2008.

Herodotus, *The History of Herodotus*, H.G. Rawlinson [Translator], Tudor Publishing Company, New York, 1928.

Herur, Anita, Sanjeev Kolagi, Surekharani Chinagudi, Effect of yoga on body mass index and gender on the cardiovascular and mental response to yoga, *Biomedical Research*, 22 (4), 499 - 505, 2011.

Higbee, Kenneth L., *Your Memory: How It Works and How to Improve it*, Paragon House, New York, 1993.

Hirasa, K. and M. Takemasa, *Spice Science and Technology*, Marcel Dekker Inc., New York, 1998.

Hitti, P. K., *History of the Arabs*, Macmillian & Company Ltd., New York, 1963.

Hoernle, A. F. R., The Bakhshālī Manuscript, *Indian Antiquary*, 17, 33 - 48, 1888; 17, 275 - 274, 1888.

Hoffman, Michelle, The Enemy of My Enemy is My Friend, *American Scientist*, 80, 535 - 536, November-December 1992.

Hoffman, Roald, Storied Theory, *American Scientist*, 93 (4), 308, 2005.

Hogendijk, Jan P. and Abdelhamid I. Sabra, *The Enterprise of Science in Islam*, The MIT Press, Cambridge, 2003.

Hoggatt, V. E., *Fibonacci and Lucas Numbers*, Fibonacci Association, San Jose, 1973.

Holford, Patrick, and Jerome Burne, *Food Is Better Medicine Than Drugs: Your Prescription for Drug-free Health*, Piatkus, London, 2007.

Hooda, D. S. and J. N. Kapur, *Āryabhaṭa: Life and Contributions*, New Age International, New Delhi, II Edition, 2001.

Hopkins, E. Washburn, *Ethics of India*, Kennikat Press, New York, 1968 (first published 1924)

Horadam, A. F., Eight Hundred Years Young, *The Australian Mathematics Teacher*, 31, 123 - 134, 1975.

Horgan, John, Josephson's Inner Juction, *Scientific American*, 272 (5), 40 - 42, May 1995.

Horne, R. A., Atomism in Ancient Greece and India, *Ambix*, 8, 98 - 110, 1960.

Horne, R. A., Atomism in Ancient Medical History, *Medical History*, 7(4), 317 - 329, 1963.

Horton, Rod W. and Herbert W. Edwards, Backgrounds of American Literary Thought, New York, 1952. First self published, later by Prentice Hall.

Høyrup, Jens, *In Measure, Number, and Weight: Studies in Mathematics and Culture*, State University of New York, Albany, 1994.

Høyrup, Jens, A New Art in Ancient Clothes, *Physis*, 35, 11 - 50, 1999.

Hughes, Barnabas, *Robert of Chester's Latin Translation of al-Khwarizmi's al-Jabr*, Franz Steiner Verlag, Stuttgart, 1989.

Hsu, M. C., H. Schubiner, M. A. Lumley, J. S. Stracks, D.J. Clauw, D.A. Williams, Sustained pain reduction through affective self-awareness in fibromyalgia: a randomized controlled trial, *Journal of General Internal Medicine*, 25 (10), 1064 - 70, Oct. 2010.

Ifrah, Georges, *From One to Zero*, [Translator] Lowell Bair, Viking Penguin Inc., New York, 1985.(original in French)

Ilyas, M., *Astronomy of Islamic Prayer Times for the Twenty-first Century*, Mansell, London, 1989.

Iyengar, B. K. S., *Light on Yoga*, Schocken Books, New York, 1966.

Jain, G. R., *Cosmology Old and New*, Bhartiya Jnanpith Publications, New Delhi, 1975.

Jain, S. K., *Dictionary of Indian Folk Medicine and Ethnobotany*, Deep Publication, New Delhi, 1991.

Jha, L. K. and K. N. Jha, Chanakya: the pioneer economist of the world, *International Journal of Social Economics*, 25 (2-4), 267 - 282, 1998.

Johns, T., *With Bitter Herbs They Shall Eat it: Chemical Ecology and the Origins of Human Diet and Medicine*, University of Arizona Press, Tucson, 1990.

Joseph, George Gheverghese, *The Crest of the Peacock*, Penguin Books, London, 1991.

Josephson, Brian, Physics and Spirituality: the Next Grand Unification?, *Physics Education*, 22, 15 - 19, 1987.

Josephus, Flavious, *Josephus*, H. St. J. Thackeray [Translator], Harvard University Press, Cambridge, 1956.

Joshi, M. C. and S. K. Gupta [Editors], *King Chandra and the Meharauli Pillar*, Kusumanjali Prakashan, Meerut, 1989.

Joshi, Rasik Vihari, Noes on Guru, Dīkṣā, and Mantra, *Ethnos*, 37, 103 - 112, 1972.

Jung, C. G., *The Collected Works of C. G. Jung*, [Editors]. H. Read, M. Fordham, and G. Alder, Princeton University Press, Princeton, 1954 - 1979.

20 volumes.

Jung, C. G., *Letters*, edited by G. Adler, Princeton University Press, 1973 - 75. 2 volumes.

Jungk, Robert, *Brighter Than a Thousand Suns*, HarCourt Brace, New York, 1958.

Kabat-Zinn, J., S. Santorelli, W. Lenderking, A. Massion, and L. Kristeller, Effectiveness of a meditation-based stress reduction program in the treatment of anxiety disorders, *The American Journal of Psychiatry*, 149, 936 - 943, 1992.

Kadvany, John, Positional Value and Linguistic Recursion, *Journal of Indian Philosophy*, 35, 487 - 520, 2007.

Kak, S. C., Some Early Codes and Ciphers, *Indian Journal of science*, 24, 1 - 7, 1989.

Kak, S. C., The Sign for Zero, *The Mankind Quarterly*, 30, 199 - 204, 1990.

Kane, P. V., *History of Sanskrit Poetics*, Motilal Banarsidass, Delhi, 1961.

Kapil, R. N., Biological Sciences: Biology in Ancient and Medieval India, *Indian Journal History of Science*, 5 (1), 119 - 140, 1970.

Karpinski, L. C., *The History of Arithmetic*, Russell & Russell, New York, 1965 (first published 1925).

Karttunen, Klaus, *India in Early Greek Literature*, Studia Orientalia, Helsinki, 1989.

Katz, Victor J., *History of Mathematics*, HarperCollins College Publishers, New York, 1993.

Kaza, Stephanie, Western Buddhist Motivations for Vegetarianism, *Worldviews*, 9(3), 211 - 227, 2005.

Keith, A. B., *The Religion and Philosophy of the Veda and Upanishads*, Greenwood Press, Connecticut, 1971 (first published in 1925).

Kennedy, E. S. and W. Ukashah, Al-Khwārizmī's Planetary Latitude Tables, *Centaurus*, 14, 89 - 96, 1969.

Kennedy, E. S., The Arabic heritage in the exact sciences, *Quarterly Journal of Arab Studies*, 23, 327 - 344, 1970.

Kenoyar, Jonathan Mark, *Ancient Cities of the Indus Valley Civilization*, Oxford University Press, New York, 1998.

Khalidi, T., *Islamic Historiogaphy*, State University of New York Press, Albany, 1975.

Khalsa, Dharma Singh. *Food as Medicine*, Atria books, New York, 2003.

King, David, *Al-Khwārizmī and New Trends in Mathematical Astronomy in the Ninth Century*, Hagop Kevorkian Center for Near Eastern Studies, New York, 1983.

King, D.A. and G. Saliba [Editors], *From Deferent to Equant*, The New York Academy of Science, New York, 1987.

King, D. A., *Astronomy in the Service of Islam*, Variorum, Aldenshot 1993.

King, Richard, *Indian Philosophy: An Introduction to Hindu and Buddhist Thought*, Georgetown University Press, Washington D.C., 1999.

Kissen, Morton and Debra A Kissen-Kohn, Reducing addictions via the self-soothing effects of yoga, *Bulletin of the Menniger Clinic*, 73 (1), 34 - 43, Winter 2009.

Kokatnur, Vaman R., Chemical Warfare in Ancient India, *Journal of Chemical Education*, 25, 268 - 272, 1948.

Kolb, Vera M., Biomimicry as a basis for drug discovery, *Progress in Drug*

*Research*, 51, 181 - 213, 1998.

Krishnamurthy, K. H., *The Wealth of Suśruta*, International Institute of Ayurveda, Coimbatore, 1991.

Krupp, E. C., Arrows in Flight, *Sky and Telescope*, 92 (4), 60 - 62, October 1996.

Kulkarni, R. P., The Value of $\pi$ Known to Śulbasūtrakarās, *Indian Journal of History of Science*, 13, 32 - 41, 1978a.

Kulkarni, R. P., Geometry as Known to the People of Indus Civilization, *Indian Journal of History of Science*, 13, 117 - 124, 1978b.

Kulkarni, R. P., *Geometry According to Śulba Sūtra*, Vaidika Samsodhana Mandala, Pune, 1983.

Kulkarni, T. R., *Upanishads and Yoga*, Bhartiya Vidya Bhavan, Bombay, 1972.

Kumar, Alok, Improving Science Instruction by Utilizing Historical Ethnic Contributions, *Physics Education*, 11 (2), 154 - 163, 1994.

Kumar, Alok and Ronald A. Brown, Teaching Science from a World-Cultural Point of View, *Science as Culture*, 8(3), 357 - 370, 1999.

Kunitzsch, P., How we got our Arabic star names, *Sky and Telescope*, 65, 20 - 22, 1983.

Kunitzsch, P., *The Arabs and the Stars: Texts and Traditions on the Fixed Stars and Their Influence in Medieval Europe*, Variorum, Northampton, 1989.

Kutumbiah, P., *Ancient Indian Medicine*, Orient Longmans, New Delhi, 1962.

Lad, Vasant, *Ayurveda: the science of self-healing, a practical guide*, Lotus Press, Santa Fe, 1984.

Lahiri, Nayanjot, Indian Metal and Metal-Related Artefacts as Cultural Signifiers: An Ethnographic Perspective, *World Archaeology*, 27 (1), 116 - 132, 1995.

Landes, David S., *Revolution in Time: Clocks and the Making of the Modern World*, Harvard University Press, Cambridge, 1983.

Larson, G. J., Ayurveda and the Hindu philosophical systems, *Philosophy of East and West*, 37, 245 - 259, 1997.

Lattin, Harriet Pratt [Translator], *The Letters of Gerbert: With His Papal Privileges as Sylvester II*, Columbia University Press, New York, 1961.

Le Coze, J., About the Signification of Wootz and Other Names Given to Steel, *Indian Journal of History of Science*, 38 (2), 117 - 127, 2003.

Leeds, Anthony and Andrew Vayda (Editors) *The Role of Animals in Human Ecological Adjustments*, American Association for the Advancement of Science, Washington D.C., 1965.

Levey, Martin and Marvin Petruck [Translator], *Principles of Hindu Reckoning*, by Kushyar Ibn Labban, University of Wisconsin Press, Milwaukee, 1965.

Levey, Martin and Noury al-Khaledy, *The Medical Formulary of Al-Samarqandi*, University of Pennsylvania Press, Philadelphia, 1967.

Lishk, S. S., *Jaina Astronomy*, Vidya Publications, Delhi, 1987.

Little, A. G. [Editor], *Roger Bacon*, Clarendon Press, Oxford, 1914.

Lochan, Kanjiv, *Medicines of Early India*, Chaukhambha Sanskrit Bhawan, Varanasi, 2003.

Loosen, Penate and Franz Vonnessen [Editor], *Gottfried Wilhelm Leibniz. Zwei Briefe über das Binäre Zahlensystem und die Chinesische Philosophie*, Belser-Presse, Stuttgart, 1968. [In German]

Mabilleau, L., *Histoire de la philosophie atomistique*, Félix Alcan, Paris, 1895.

Macdonell, A. A., *A History of Sanskrit Literature*, Haskell House, New York, 1968 (first published 1900).

MacKenna, Stephen, *Porphery on the life of Plotinus* in Plotinus' Enneads, Faber and Faber Limited, London, 1956.

Macrina, Francis L., *Scientific Integrity*, ASM Press, Washington D.C., 1995.

Magnus, Albert, *Book of Minerals*, Dorothy Wyckoff [Translator], Claredon Press, Oxford, 1967.

Mahdihassan, S., *Triphalā* and Its Arabic and Chinese Synonyms, *Indian Journal of History of Science*, 13 (1), 50 - 55, 1978.

Mahdihassan, S., *Indian Alchemy or Rasayana*, Vikas Publishing House Pvt. Ltd., New Delhi, 1979.

Majno, Guido, *The Healing hand: man and wound in the ancient world*, Harvard University Press, Cambridge, 1975.

Majumdar, R. C., *Ancient India*, Motilal Banarsidas, New Delhi, 1964.

Maltz, Maxwell, *Evolution of Plastic Surgery*, Froben Press, New York, 1946.

Margenau, H., *Physics and Philosophy*, D. Reidel Publishing Company, London, 1978.

Marlow, Harold J., William K. Hayes, Samuel Soret, Ronald L. Carter, Ernest R. Schwab, Joan Sabaté, Diet and the environment: does what you eat matter? *The American Journal of Clinical Nutrition*, 89 (5), 1699S, 2009.

Martzloff, Jean-Claude, *A History of Chinese Mathematics*, Springer-Verlag, New York, 1997.

Marx, Karl, The Difference Between the Democritean and Epicurean Philosophy of Nature, 1841, Doctoral Thesis, available online www.marxists.org/archive/marx/works/1841/dr-theses/.

McCall, Timothy B., *Yoga as Medicine: the Yogic Prescription for Health & Healing*, Bantam Books, New York, 2007.

McCrindle, J. W., *The Invasion of India by Alexander the Great as Described by Arrian, Q. Curtis, Diodoros, Plutarch and Justin*, Westminister, 1896.

McCrindle, J. W., *Ancient India as described by Ktesias the Knidian*, Trubner and Company, London, 1882. Reprint by Manohar Reprints, Delhi, 1973.

McCrindle, J. W., *Ancient India as Described by Megasthenes and Arrian*, Chuckervertty, Chatterjee & Company, Calcutta, 1926.

McDonell, John, J., *The Concept of an Atom From Democritus to John Dalton*, The Edwin Mellen Press, Lewiston, 1991, p. 11 - 12.

McDonald, Robert J., Bryan D. Devan, and Nancy S. Hong, Multiple memory systems: The power of interactions, *Neurobiology of Learning and Memory*, 82, 333 - 346, 2004.

McDougall, Walter A., Merits and Perils of Teaching Cultures, *Orbis*, 43 (4), 599 - 604, 1999.

McEvilley, Thomas, *The Shape of Ancient Thought*, Allworth Press, New York, 2002.

Menninger, Karl, *Number Words and Number Symbols: A Cultural History of Numbers.* MIT Press, Massachusetts, 1970.

Meyerhof, Max, Al at-Tabarî's "Paradise of Wisdom", one of the oldest

Arabic Compendiums of Medicine, *Isis*, 16 (1), 6 - 54, 1931.

Meyerhof, M., On the Transmission of Greek and Indian Science to the Arabs, *Islamic Culture*, 11, 17 - 29, 1937.

Michalsen, Andreas, Gustav J. Dobos, and Manfred Roth, *Medicinal Leech Therapy*, Thieme Publishing, 2006.

Michalsen, A., U. Deuse, T. Esch, G. Dobos, and S. Moebus, Effect of leech therapy (*Hirudo medicinalis*) in painful osteoarthritis of the knee: a pilot study, *Annals of the Rheumatic Diseases*, 60 (6), 986, Oct. 2001.

Mikami, Yoshio, *The Development of Mathematics in China and Japan*, Chelsea Publishing Company, New York, 1913.

Milius, Susan, Aspirin works on plants, too, *Science News*, 154 (7), 107, 1998a.

Milius, Susan, Parental care seen in mountain plants, *Science News*, 154 (2), 20, 1998b.

Miller, Jeanine, *The Vision of Cosmic Order in the Vedas*, Routledge and Kegan Paul, Boston, 1985.

Ming, Chen, The Transmission of Indian Ayurvedic Doctrines in Medieval China: A Study of Aṣṭâṅga and Tridoṣa Fragments from the Silk Road, *Annual Report of the International Research Institute for Advanced Buddhology at Soka University for the Academic Year 2005*, 9, 201 - 230, March 2006.

Mishra, V. and S. L. Singh, Height and Distance Problems in Ancient Indian Mathematics, *Gaṇita Bhāratī*, 18, 25 - 30, 1996.

Montgomery, Scott L. and Alok Kumar, Telling Stories: Some Reflections on Orality in Science, *Science as Culture*, 9(3), 391 - 404, 2000.

Mookerji, Radha Kumud, *Ancient Indian Education: Brahmanical and Buddhist*, Motilal Banarsidass, New Delhi, 1989.

Moore, Arden and Sari Harrar, Leeches Cut Knee Pain, *Prevention*, 55 (8), 161, 2003.

Moore, Walter, *Schrödinger: Life and Thought*, Cambridge Univesity Press, London, 1992.

More, Louis Trenchard, Boyle as Alchemist, *Journal of the History of Ideas*, 2, 61 - 76, 1941.

Mukherjee, P. K., Rai, S., Bhattacharya, S., Wahile, A., Saha, B.P., . Marker analysis of polyherbal formulation, Triphala – a well known Indian traditional medicine, *Indian Journal of Traditional Knowledge*, 7, 379 - 383, 2008.

Mukherjee, Pulok K. and Atul Wahile, Integrated approaches towards drug development from Ayurveda and other Indian system of medicines, *Journal of Ethnopharmacology*, 103, 25 - 35, 2006.

Mukherjee, Pulok K., Kuntal Maiti, Kakali Mukherjee, Peter J. Houghton, Leads from Indian medicinal plants with hypoglycemic potentials, *Journal of Ethnopharmacology* 106 1 - 28, 2006a.

Mukherjee, P. K., S. Rai, S. Bhattacharya, P. K. Debnath, T. K. Biswas, U. Jana, S. Pandit, B.P. Saha, P. K. Paul, Clinical study of Triphala–a well known phytomedicine, from India, *Iranian Journal of Pharmacology and Therapeutics*, 5, 51 - 54, 2006b.

Mukherjee, P. K., *Indian Literature Abroad*, Calcutta Oriental Press, Calcutta, 1928.

Mukhopādhyāya, Girinath, *Ancient Hindu Surgery*, Cosmo Publications, New Delhi, 1994, 2 volumes.

Müller, M., *India: what can it teach us?* Funk and Wagnalls Publishers, London, 1883.

Nakatani, N., Antioxidative and antimicrobial constituents of herbs and spices, *In Spices, herbs and edible fungi*, G. Charalambous [Editor], p. 251 -

272, Elsevier, Amsterdam, 1994.

Nakayama, S. and N. Sivin [Editors], *Chinese Science*, MIT Press, Massachusetts, 1973.

Narayana, A., Medical Science in Ancient Indian Culture with Special Reference to Atharvaveda, *Bulletin of the Indian Institute of History of Medicine*, 25 (1 - 2), 100 - 110, 1995.

Narayanan, Vasudha, Water, Wood, and Wisdom: Ecological Perspectives from the Hindu Traditions, *Daedalus*, vol. 130, no. 4, p. 179 -197, 2001.

Needham, J., *Science and Civilization in China*, Oxford University Press, London, 1954 - 1999. 6 volumes.

Needham, Joseph, *Science in Traditional China: A Comparative Perspective*, Harvard University Press, Cambridge, 1981.

Needham, J. and W. Ling, *Science and Civilization in China*, Cambridge University Press, Chicago, 1959.

Nelson, Lance [Editor], *Purifying the Earthly Body of God: Religion and Ecology in Hindu India*, State University of New York Press, Albany, 1998.

Neugebauer, O., *The Astronomical Tables of al-Khwārizmī*, Hist. Filos. Skr. Dan. Vid. Selsk, Copenhagen, 1962.

Newman, William, Technology and Alchemical Debate in the Late Middle Ages, *Isis*, 80, 423 - 445, 1989.

Nicholson, Shirley J., Virginia Hanson, and Rosemarie Stewart, *Karma: Rhythmic Return to Harmony*, Motilal Banarsidass, New Delhi, 2002.

Ninivaggi, Frank John, *Ayurveda: A Comprehensive Guide to Traditional Medicine for the West*, Praeger, Connecticut, 2008.

Nisbet, Robert A., *Teachers and Scholars: A Memoir of Berkely in Depression and War*, Transaction Publishers, 1992.

Nishida, R., Sequestration of defensive substances from plants b Lepidoptera, *Annual Review entomology*, 47, 57 - 92, 2002.

Norlin, G. and L. R. Van Hook, *Isocrates*, Harvard University Press, Cambridge, 1961 - 62.

Nunn, J. F., *Ancient Egyptian Medicine*, University of Oklahoma Press, Norman, 1996.

Oevi M., M. Rigbi, E. Hy-Am, Y. Matzner, and A. Eldor, A potent inhibitor of platelet activating factor from the saliva of the leech Hirudo medicinalis, *Prostaglandins*, 43, 483 - 95, 1992.

Ogawa, M., Science Education in a Multiscience Perspective, *Science Education*, 79, 583 - 593, 1995.

Ōhashi, Yukio, Astronomical Instruments in Classical Siddhāntas, *Indian Journal of History of Science*, 29(2), 155 - 313, 1994.

O'Leary, D. L., *How the Greek Science Passed to the Arabs*, Routledge and Kegan Paul, London, 1949.

Oppenheimer, J. R., *Science and the Common Understanding*, Simon and Schuster, New York, 1954.

Oppert, Gustav Salomon, Lakshmikanta Varma, *On the Weapons, Army Organization, and Political Maxims of the Ancient Hindus*, New Order Book Company, Ahmedabad, 1967 (first pbulished in 1880 by Oppert).

Pacey, Arnold, *Technology in World Civilizations*, The MIT Press, Cambridge, Massachusetts, 1990.

Panikkar, R., Time and History in the Tradition of India: Kala and Karma, in the book *Culture and Time*, The UNESCO Press, Paris, 1976.

Paramhans, A. A., Astronomy in Ancient India–its importance, insight and prevalence, *Indian Journal of History of Science*, 26 (1), 63 - 70, 1991.

Parry J. W., *The Story of Spices*, Chemical Publishers, New York, 1953.

Parsons, Edmund, *Time Devoured: A Materialistic Discussion of Duration*, George Allen & Unwin, London, 1964.

Partington, J. R., The Origins of the Atomic Theory, *Annals of Science*, 4, 245 - 283, 1939.

Patel, P. G., Linguistic and Cognitive Aspects of the Orality-Literacy Complex in Ancient India, *Language & Communication*, 16 (4), 315 - 329, 1996.

Patton, Laurie L., *Authority, Anxiety, and Canon: Essays in Vedic Interpretation*, SUNY Press, Albany, 1994.

Pederson, O., *A Survey of the Almagest*, Odense University Press, Odense, 1974.

Pellat, C., *The Life and Works of al-Jahiz*, University of California, Los Angeles, 1969.

Perrett, Roy W., History, Time, and Knowledge in Ancient India, *History and Theory*, 38(3), 307 - 321, 1999.

Petitjean, P., *Science and Empires: Historical Studies about Scientific Development and European Expansion*, Kluwer, Dordrecht, 1992.

Philostratus, *Life of Apollonius*, C.P. Jones [Translator], Penguins Books, New York, 1970; *The Life of Apollonius*, Translated by F. C. Conybeare, Harvard University Press, Cambridge, 1960 (first published in 1912).

Pichersky, Eran, Plant Scents, *American Scientist*, 92, 514 - 539, 2004.

Pilkington, Karen, Graham Kirkwood, Hagen Rampes, and Janet Richardson, Yoga for depression: The research evidence, *Journal of Affective Disorders*, 89, 13 - 24, 2005.

Pillai, A. K. B., *Transcendental Self: A Comparative Study of Thoreau and the Psycho-Philosophy of Hinduism and Buddhism*, University Press of America, Lanham, 1985.

Pingree, D., *The Thousands of Abu Masher*, Warburg Institute, London, 1968.

Pococke, E., *India in Greece*, Richard Griffin and Company, London, 1852.

Ponnusankar, S., Subrata Pandit, Ramesh Babu, Arun Bandyopadhyay, and Pulok K. Mukherjee, Cytochrome P450 inhibitory potential of Triphala: A Rasayana from Ayurveda, *Journal of Ethnopharmacology*, 133, 120 - 125, 2011.

Pothula, V. B., T. M. Jones, and T. H. J. Lesser, Ontology in Ancient India, *The Journal of Laryngology & Otology*, 115, 179 - 183, March 2001.

Prakash, B and K. Igaki, Ancient Iron Making in Bastar District, *Indian Journal History of Science*, 19, 172 - 185, 1984.

Prakash, Om, *Food and Drinks in Ancient India*, Munshi Ram Manohar Lal, Delhi, 1961.

Prasad, P. C., *Foreign Trade and Commerce in Ancient India*, Abhinav Publications, New Delhi, 1977.

Prasad, Hari Shanker [Editor], *Time in Indian Philosophy*, Sri Satguru Publications, Delhi, 1992.

Pullman, Bernard, *The Atom in the History of Human Thought*, Oxford University Press, 1998.

Puskar-Pasewicz, Margaret [Editor], *Cultural Encyclopedia of Vegetarianism*, Greenwood, Santa Barbara, 2010.

Radhakrishnan, Sarvepalli, *Indian Philosophy*, Macmillan, New York, 1958. 2 volumes.

Rados, Carol, Beyond Bloodletting: FDA Gives Leeches a Medical Makeover, *FDA Consumer*, p. 9, September-October, 2004..

Raghavan, Ellen M. and Barry Wood, Thoreau's Hindu Quotation in "A Week", *American Literature*, 51 (1), 94 - 98, 1979.

Rajgopal, L., G. N. Hoskeri, G. S. Seth, P. S. Bhulyan, and K. Shyamkishore, History of Anatomy in India, *Journal of Postgraduate Medicine*, 48 (3), 243 - 5, 2002.

Ramakrishnappa, K., *Impact of Cultivation and gathering of medicinal plants on biodiversity: case studies from India*, FAO, Rome, 2002. This report can be accessed at
http://www.fao.org/DOCREP/005/AA021E/AA021e00.htm.

Rana, R. E. and B. S. Arora, History of Plastic Surgery in India, *Journal of Postgrad. Med.*, 48 76 - 78, 2002.

Rao, G. Hanumantha, The Basis of Hindu Ethics, *International Journal of Ethics*, 37 (1), 19 - 35, 1926.

Rao, T. R. N. and Subash Kak, *Computing Science in Ancient India*, The Center for Advanced Computer Studies, University of Southwestern Louisiana, Lafayette, 1998.

Rashed, Roshdi [editor], *Encyclopedia of the History of Arabic Science*, Routledge, New York, 1996.

Ray, P. C., *A History of Hindu Chemistry*, Indian Chemical Society, 1956. It was first published in 1902. There are several editions of this book with different publishers.

Ray, Priyada Ranjan, Chemistry in Ancient India, *Journal of Chemical Education*, 25, 327 - 335, 1948.

Rawlinson, H. G., *Intercourse between India and the Western World*, Cambridge University Press, Cambridge, 1926.

Reed, Philip A., A paradigm shift: biomimicry: biomimicry is a new way of linking the human-made world to the natural world, *The Technology Teacher*, 63 (4), 23 - 27, Dec. 2003.

Regan, Tom and Peter Singer [Editor], *Animal Rights and Human Obligations*, Prentice-Hall, Englewood Cliffs, 1976.

Reijnders, Lucas and Sam Soret, Quantification of the environmental impact of different dietary protein choices, *American Journal of Clinical Nutrition*, 78, 664S - 668S, 2003.

Remy, A. F. J., *The Influence of India and Persia on the Poetry of Germany*, Columbia University Press, New York, 1901.

Restivo, Sal P., Parallels and Paradoxes in Modern Physics and Eastern Mysticism: I-A Critical Reconnaissance, *Social Studies of Science*, 8(2), 143 - 181, 1978; Parallels and Paradoxes in Modern Physics and Eastern Mysticism: II–A Sociological Perspective on Parallelism, *Social Studies of Science*, 12 (1), 37 - 71, 1982.

Riché, Pierre, *Gerbert d'Aurillac, le pape de l'an mil*, Fayard, Paris, 1987. [In French]

Richer of Saint-Rémy, *Histoire de France*, Robert Latouche [Editor], Paris, 2 volumes, 1964 - 67.

Ridpath, I., *Star Tables*, Universe Books, New York, 1988.

Riepe, Dale, *The Philosophy of India and Its Impact on American Thought*, Thomas Press, Springfield, 1970.

Ringnes, Vivi, Origin of the Names of Chemical Elements, *Journal of Chemical Education*, 66 (9), 731 - 737, 1989.

Riley, D., Hatha yoga and the treatment of illness (commentary). *Alternative Therapies in Health Medicine*, 10 (2), 20 - 21, 2004.

Robinson, Catherine and Denise Cush, The Sacred Cow; Hinduism and Ecology, *Journal of Beliefs & Values*, 18 (1), 25 - 37, 1997.

Rohini, G., K. E. Sabitha, C. S. Devi, *Bacopa monniera Linn.* extract modulates antioxidant and marker enzyme status in fibrosarcoma bearing rats. *Indian Journal of Experimental Biology*, 42, 776 - 780, 2004.

Ronan, C. A., *Science: Its History and Development Among the World's Cultures*, Facts on the File Publications, New York, 1982.

Roodenrys, Steven, Dianne Booth, Sonia Bulzomi, Andrew Phipps, Caroline Micallef, and Jaclyn Smoker, Chronic Effects of Brahmi (*Bacopa monnieri*) on Human Memory, *Neuropsychopharmacology* 27, 279 - 281, 2002.

Rosaen, Cara and Rita Benn, The Experience of Transcendental Meditation in Middle School Students: A Qualitative Report, *Explore*, 2 (5), 422 - 425, September/October 2006.

Rosemond, A.D., C.B. Anderson, Engineering role models: do non-human species have the answers?, *Ecological Engineering* 20, 379 - 387, 2003.

Rosen, Frederic [Translator], *The Algebra of Mohammed Ben Musa*, by al-Khwārizmī, Georg Olms Verlag, New York, 1986. (first published in 1831)

Rouse Ball, W. W, *A Short Account of the History of Mathematics*, Dover Publications, Inc., New York, 1908.

Roy, Mira, Ecological Concept in Ayurveda: Nature-Man Relations, *Indian Journal of History of Science*, 44 (1), 1 - 19, 2009.

Roy, Mira, Environment and Ecology in the *Rāmāyaṇa*, *Indian Journal of History of Science*, 40 (1), 9 - 29, 2005.

Roy, S. B., *Prehistoric Lunar Astronomy*, Institute of Chronology, New Delhi, 1976.

Royle, J. F., *Antiquity of Hindoo Medicine*, W. H. Allen and Company, London, 1837.

Rubia, Katya, The neurobiology of Meditation and its clinical effectiveness in psychiatric disorders, *Biological Psychology* 82, 1- 11, 2009.

Ruegg, D. Seyfort, Mathematical and Linguistic Models in Indian Thought: The Case of Zero and Śûnatā, *Wiener Zeitschrift für die Kunde Südasiens und Archiv für Indische Philosophie*, 22, 171 - 181, 1978.

Rusk, Ralph L. [Editor], *The Letters of Ralph Waldo Emerson*, Columbia University Press, New York, 1939.

Sabu, M. C. and R. Kuttan, Antidiabetic activity of medicinal plants and its relationship with their antioxidant property, *Journal of Ethnopharmacology*, 81, 155 - 160, 2002.

Sachau, E. C., *Alberuni's India*, S. Chand & Company, Delhi, 1964.

Sache, M., *Damascus Steel, Myth, History, Technology Applications*, Stahleisen, Düsseldorf, 1994.

Saggs, H. W. F., *Civilization Before Greece and Rome*, Yale University, New Haven, 1989.

Saha, M. N. and N. C. Lahri, *History of the Calendar*, Council of Scientific & Industrial Research, New Delhi, 1992. (First published in 1955)

Said, Edward W., *Orientalism*, Pantheon Books, New York, 1978.

Said, Edward, *Culture and Imperialism*, Vintage Books, New York, 1993.

Saidan, A. S., *The Arithmetic of al-Uqlīdisī*, D. Reidel Publishing Company, Dordrecht, 1978.

Saidan, A. S., The Development of Hindu-Arabic Arithmetic, *Islamic Culture*, 39, 209 - 221, 1965.

Salem, S. I. and A. Kumar, *Science in the Medieval World*, University of Texas Press, Austin, 1991.

Saliba, George, *A History of Arabic Astronomy: Planetary Theories During the Golden Age of Islam*, New York University Press, New York, 1994.

Samsó, J., *Islamic Astronomy and Medieval Spain*, Variorum, Aldenshot, 1994.

Sankhyan, Anek R. and George H. J. Weber, Evidence of surgery in Ancient India: Trepanation at Burzahom (Kashmir) over 4000 Years Ago, *International Journal of Osteoachaelogy*, 11, 375 - 380, 2001.

Sarasvati Amma, T. A., *Geometry in Ancient and Medieval India*, Motilal Banarsidas, Delhi, 1979.

Sardar, Z., *Exploration in Islamic Science*, Mansell, London, 1989.

Sarma, Nataraja, Diffusion of astronomy in the ancient world, *Endeavour*, 24(4), 157 - 154, 2000.

Sarsvati, Svami Satya Prakash, *Founders of Sciences in Ancient India*, Govindram Hasaram, Delhi, 2 volumes, 1986.

Sarton, G., *Introduction to the History of Science*, Carnegie Institute, Washington, 1927 - 1947. 3 volumes.

Sarton, George, Decimal Systems Early and Late, *Osiris*, 9, 581 - 601, 1950.

Sarton, George, Chaldaean Astronomy of the Last Three Centuries B.C., *Journal of the American Oriental Society*, 75 (3), 166 - 173, 1955.

Sarton, George, *Ancient Science Through the Golden Age of Greece*, Dover Publications, New York, 1993.

Schoff, Wilfred H., The Eastern Iron Trade of the Roman Empire, *Journal of the American Oriental Society*, 35, 224 - 239, 1915.

Schrödinger, Erwin, *My View of the World*, Cambridge University Press,

England, 1964.

Scott, David, Rewalking Thoreau and Asia: 'Light From the East' for 'a Ver Yankee Sort of Oriental', *Philosophy East & West*, 57 (1), 14 - 39, 2007.

Seal, Brajendranath, *The positive sciences of the ancient Hindus*, Longmans, Green and Company, New York, 1915. This book has been reprinted several times, most recently in 1985 by Motilal Banarsidass, Delhi.

Sedlar, J. W., *India and the Greek World*, Rowman and Littlefield, New Jersey, 1980.

Sedlar, J. W., *India in the Mind of Germany*, University Press of America, Washington, D.C., 1982.

Seeger, Elizabeth, *The Pageant of Chinese History*, David McKay Company, New York, 1962.

Seidenberg, A., The Ritual Origin of Geometry, *Archive for History of Exact Sciences*, 1, 488, 1962.

Selin, H. [Editor], *Encyclopedia of the History of Science, Technology, and Medicine in Non-Western Cultures*, Kluwer Academic Publishers, 1997.

Selin, Helaine, *Astronomy Across Cultures: The History of Non-Western Astronomy*, Kluwer Academic Publishers, Boston, 2000.

Sen, S. N., Influence of India Science on Other Culture Areas, *Indian Journal History of Science*, 5 (2), 332 - 346, 1970.

Sen, S. N. and A. K. Bag, *The Śulbasūtras of Baudhāyana, Āpastamba, Kātyana, and Mānava*, Indian National Science Academy, New Delhi, 1983.

Sequeira, W., Yoga in treatment of carpal-tunnel syndrome, *Lancet*, 353 (9154) 689 - 690, February 27, 1999.

Seymour, Roger S., Plants That Warm Themselves, *Scientific American*, 104 - 109, March 1997.

Seymore, R. S. and P. Schultze-Motel, Thermoregulating Lotus Flowers, *Nature*, 383, 305, Sept. 26, 1996.

Seymore, R. S., G. A. Bartholomew, and M. C. Barnhart, Respiration and Heat Production by the Inflorescence of *Philodendron Selloum*, *Planta*, 157 (4), 336 - 343, April 1983.

Shamasastry, R., *Kauṭilya's Arthśāstra*, Mysore Printing and Publishing House, Mysore, 1960. Fourth Edition. (first published in 1915)

Sharma, Bhu Dev and Nabarun Ghose, *Revisiting Indus-Sarasvati Age and Ancient India*, World Association for Vedic Studies, 1998.

Sheppard, H. J., Alchemy: Origin or Origins, *Ambix*, 17 (2), 69 - 84, 1972.

Sherby, Oleg D. and Jeffrey Wadsworth, Damascus Steel, *Scientific American*, 112 - 120, February 1985.

Sherby, Oleg D. and Jeffrey Wadsworth, Ancient blacksmiths, the Iron Age, Damascus Steels, and modern metallurgy, *Journal of Materials Processing Technology*, 117, 347 - 353, 2001.

Sherman, Paul W. and Jennifer Billings, Darwinian Gastronomy: Why We Use Spices, *BioScience*, 49 (6), 453 - 463, 1999.

Sherman, P. W. and S. M. Flaxman, Protecting ourselves from food, *American Scientist*, 89, 142 - 151, 2001.

Sherman, P. W. and G. A. Hash, Why vegetable recipes are not very spicy, *Evolution and Human Beavior*, 22, 147 - 163, 2001.

Sherman, Paul W., Why We Use Spices. Foods and Food Ingredients, *Journal of Japan*, 198, 56 - 69, 2002.

Shoul Ahmad, M. H., *al-Mas'ūdī and His World*, Ithaca Press, London, 1979.

Shukla, Kripa Shankar and K.V. Sarma, *Āryabhaṭīya of Āryabhaṭa*,, Indian National Science Academy, New Delhi, 1976.

Siddiqi, M. Z., *Paradise of Wisdom* by Ali B. Rabban at-Ṭabarî, Berlin, 1928; reprinted by Hamdard Press, Karachi, 1981.

Sigler, Laurence [Translator], *Liber Abaci by Leonardo Fibonacci*, Springer-Verlag, New York, 2002.

Simoons, Frederick J., *Plants of Life, Plants of Death*, The University of Wisconsin Press, Madison, 1998.

Singer, Charles, *A History of Scientific Ideas*, Dorset Press, New York, 1990 (first published with Oxford University Press, 1959)

Singer, Peter, Utilitarianism and Vegetarianism, *Philosophy and Public Affairs*, 9 (4), 325 -337, 1980.

Singer, Peter, *Animal Liberation*, Avon Books, New York, 1990.

Singh, L. M., K. K. Thakral, and P. J. Deshpande, Suśruta's Contributions to the Fundamentals of Surgery, *Indian Journal History of Science*, 5 (1), 36 - 50, 1970.

Singh, Nand Lal, Ramprasad, P.K. Mishra, S.K. Shukla, Jitendra Kumar, and Ramvijay Singh, Alcoholic Fermentation Techniques in Early Indian Tradition, *Indian Journal of History of Science*, 45 (2), 163 - 173, 2010.

Singh, Permanand, The So-called Fibonacci Numbers in Ancient and Medieval India, *Historia Mathematica*, 12, 229 - 244, 1985.

Singh, Pramil N., Joan Sabate, Gary E. Fraser, Does low meat consumption increase life expectancy in humans?, *The American Journal of Clinical Nutrition*, 78 (3S), S526, Sept. 2003.

Singh S. S., S. C. Pandey, S. Srivastava, V. S. Gupta, B. Patro, and A. C.Ghosh, Chemistry and medicinal properties of Tinospora cordifolia (Guduchi). *Indian Journal of Pharmacology*, 35, 83 - 91, 2003.

Sinha, Nandlal, *Vaisesika Sutras of Kanada*, Bhuvaneswari Ashram, Allahabad, 1911.

Sinha, Braj M., *Time and Temporality in Sāṁkya-Yoga and Abhidharma Buddhism*, Munshiram Manoharlal Publishers, New Delhi, 1983.

Sircar, D. C., *Inscriptions of Aśoka*, Ministry of Information & Broadcasting, Government of India, New Delhi, 1957.

Sivin, Nathan, *Traditional Medicine in Contemporary China*, The University of Michigan Center for Chinese Studies, Ann Arbor, 1987.

Smith, A. K. and C. Weiner, [Editors], *Robert Oppenheimer*, Harvard University Press, 1980.

Smith, Brian K., Classifying Animals and Humans in Ancient India, *Man*, 26 (3), 527 - 548, 1991.

Smith, C. S., Damascus Steel, *Science*, 216, 242-244, 1982.

Smith, David Eugene and Louis Charles Karpinski, *The Hindu-Arabic Numerals*, Ginn and Company, Boston, 1911.

Smith, David Eugene, *History of Mathematics*, Ginn and Company, New York, 1925. 2 volumes.

Smith, Vincent A., *Aśoka, The Buddhist Emperor of India*, S. Chand & Company, Delhi, 1964.

Somayaji, Dhulipala Arka , *A Critical Study of the Ancient Hindu Astronomy in the light and language of the modern*, Karnatak University, Dharwar, 1971.

Sourdel, Dominique, Bayt al-Hikma, in the book *Encyclopaedia of Islam*, Brill, Leiden, 1960, volume 1.

Sorta-Bilajac, Iva and Amir Muzur, The nose between ethics and aesthetics: Sushrutas legacy, *Otolaryngology-Head and Neck Surgery*, 137, 707 - 710, 2007.

Speight, James G., *The Chemistry and Technology of Petroleum*, Marcel Dekker, New York, 1999.

Srinivasiengar, C. N., *The History of Ancient Indian Mathematics*, World Press, Calcutta, 1967.

Staal, Fritz, Greek and Vedic Geometry, *Journal of Indian Philosophy*, 27, 105 - 127, 1999.

Steiner, Rudolf., *Goethe the Scientist*, O. D.Wannamaker [Translator], Anthroposophic Press, New York, 1950.

Steinfeld, Henning, Pierre Gerber, Tom Wassenaar, Vincent Castel, Mauricio Rosales, and Cees de Haan, *Livestocks Long Shadow: Environmental Issues and Options*, Food and Agriculture Organization of the United Nations, Rome, Italy, 2006.

Stillman, John Maxson, *The Story of Alchemy and Early Chemistry*, Dover Publications, Inc., New York, 1960.

Stix, Gary, Spice Healer, *Scientific American*, 296 (2), Feb. 2007.

Strabo, *The Geography of Strabo*, Horace Leonard Jones [Translator], Harvard University Press, Cambridge, 1959.

Stux, Gabriel, Brian Berman, Bruce Pomeranz, P. Kofen, and K. A. Sahm, *Basics of Acupuncture*, Springer-Verlag, New York, 2003.

Subak, S., Global environmental costs of beef production, *Ecological Economics*, 30, 79 - 91, 1999;

Subbarayappa, B.V. and K. V. Sarma, *Indian Astronomy*, Nehru Centre, Bombay, 1985.

Subbarayappa, B. V., An Estimate of the Vaiśesika Sūtra in the History of Science, *Indian Journal of History of Science*, 2(1), 22-34, 1967.

Suzuki, Jeff, *A History of Mathematics*, Prentice Hall, New Jersey, 2002.

Swerdlow, N. M. and O. Neugebauer, *Mathematical Astronomy in Copernicus's De Revolutionibus*, Springer, New York, 1984.

Takakusu, J. [Translator], *The Buddhist Religion by I-tsing*, Munshiram Manoharlal, Delhi, 1966 (first published 1896).

Talbot, Michael, *Mysticism and New Physics*, Routledge, London, 1981.

Tan, Leong T., Margret Y. C. Tan, and I. Veith, *Acupuncture Therapy; Current Chinese Practice*, Temple University Press, Philadelphia, 1973.

Tangley, Laura, By watching What Animals Eat, Experts May Find New Medicine for People, *National Wildlife*, 38 (6), 14 - 15, 2000.

Tattvabhushan, Sitanath, Ethical Science Among the Hindus, *International Journal of Ethics*, 22 (3), 287 - 298, 1912.

Taylor, Thomas [Translator], *Iamblichus' Life of Pythagoras*, A. J. Valpy, London, 1976.

Tekur, Padmini; Singphow, Chametcha; Nagendra, Hongasandra Ramarao; Raghuram, Nagarathna. Effect of Short-Term Intensive Yoga Program on Pain, Functional Disability, and Spinal Flexibility in Chronic Low Back Pain: A Randomized Control Study, *Journal of Alternative & Complementary Medicine*, 14 (6), 637 - 644, 2008.

Teller, Edward, *Energy from Heaven and Earth*, W. H. Freeman and Company, San Francisco, 1979.

Tersi, Dick, *Lost Discoveries*, Simon and Schuster, New York, 2002.

Tewari, Ashok K., Wisdom of Ayurveda in perceiving diabetes: Enigma of therapeutic recognition, *Current Science*, 88 (7), 1043 - 1051, 2005.

Thibaut, G., On the Śulva-Sūtra, *Journal Asiatic Society*, Bengal, India, 1875, p. 227.

Thomas, A. P., Yoga and cardiovascular function, *Journal of the International Association of Yoga Therapists*, 4, 39 - 41, 1993.

Thompson, C. J., *The Lure and Romance of Alchemy*, George G. Harrap and Company Ltd., London, 1932.

Thoreau, H. D., *The Writings of Henry David Thoreau*, AMS Press, New York, 1906.

Thoreau, Henry David, *Walden; or, Life in the Woods*, Dover, 1995.

Thurston, Hugh, Planetary Revolutions in Indian Astronomy, *Indian Journal of History of Science*, 35 (4), 311 - 318, 2000.

Thurston, Hugh, *Early Astronomy*, Springer Verlag, New York, 1994.

Timbs, John, *Stories of Inventors and Discoverers in Science and the Useful Arts: A book for Old and Young*, Harper and Brothers, London, 1860.

Toomer, G. J., A survey of the Toledan Tables, *Osiris*, 15, 5 - 174, 1968.

Toussaint, Friedrich, Was the first steel from ancient India?, *Steel Times*, 228 (5), 194 - 198, 2000.

Tucker, Mary Evelyn and John A. Grim [Editors], *Worldviews and Ecology: Religion, philosophy, and the Environment*, Orbis Books, New York, 1994.

Ülger, Özlem, and Naciye Yağl. Effects of Yoga on the Quality of Life in Cancer Patients, *Complementary Therapies in Clinical Practice*, 6, 60 - 63, 2010.

Underwood, B. J., Forgetting, *Scientific American*, 210, 91 - 99, 1964.

Van Horn, Gavin, Hindu Tradition and Nature, *Worldviews*, 10 (1), 5 - 39, 2006.

Van Nooten, B., Binary numbers in Indian Antiquity, *Journal of Indian Philsophy*, 21, 31 - 50, 1993.

Van Sertima, Ivan, *Blacks in Science*, Transaction Books, London, 1992.

Vartak, Padmakar Vishnu, *Scientific Knowledge in the Vedas*, Nag Publishers, Delhi, 1995.

Vasudevan, Mani and Milind Parle, Memory enhancing activity of Anwala churna (*Emblica officinalis* Gaertn.): An Ayurvedic Preparation, *Physiology & Behavior*, 91, 46 - 54, 2007.

Veith, Ilza and Leo M. Zimmerman, *Great Ideas in the History of Surgery*, Norman Publishing, 1993.

Venkatesanada, Swami, *The Concise Yoga Vasistha*, State University of New York Press, Albany, 1984.

Verma, Rajendra, *Vedic Cosmology*, New Age International, New Delhi, 1996.

Vinayasagar, M. 1965, *Vrtta Mauktika of Candrasekhara Bhatta*, Rajasthan Oriental Research Institute, Jodhpur, 1965.

Vogel, K., *Mohammed ibn Musa Alchwarizmi's Algorismus*, Das früheste Lehrbuch Rechnen mit indischen Ziffern, Aalen, 1963. [In German]

Voltaire, *The Works of Voltaire*, John Morley, William F. Fleming, and Oliver Herbrand Gordon Leigh [Editors], E.R. Du Mont, London, 1901.

Wade, Nicholas, Startling Scientists, Plant Fixes Its Flawed Gene, *The New York Times*, March 23, 2005.

Wagner, Donald B., Chinese steel making techniques with a note on Indian wootz steel in China, *Indian Journal of History of science*, 42 (3), 289

- 318, 2007.

Walsh, J. J., *The Popes and Science*, London, 1912.

Warmington, E. H., *The Commerce Between the Roman Empire and India*, Curzon Press, London, 1974. (first published 1928)

Wayne, Randy, Excitability in Plant Cells, *American Scientist*, vol. 81 , p. 140 - 151, March - April 1993.

Weber, Albrecht, *Uber die Metrik der Inder: zwei Abhandlungen*, Dümmler, Berlin, 1863.

Wei-Xing, Niu, An inquiry into the astronomical meaning of Rāhu and Ketu, *Chinese Astronomy and Astrophysics*, 19 (2), 259 - 266, 1995.

Weiss, Rick, When Plants Act Like An Animals, *National Wildlife*, p. 18 - 19, December/January 1995.

Weizman, Howard, More on India's Sacred Cattle, *Current Anthropology*, 15, 321 - 323, 1974.

West, M. L., *Early Greek Philosophy and the Orient*, Clarendon Press, Oxford, 1971.

White, Andrew D., *A History of the Warfare of Science with Theology in Christendom*, D. Appleton and Company, New York, 1897. (reprinted by Prometheus, 1993)

Widgery, Alban G., The Principles of Hindu Ethics, *International Journal of Ethics*, 40 (2), 232 - 245, 1930.

Willies, Lynn, in association with P. T. Craddock, L. J. Gurjar, and K. T. M. Hegde, Ancient Lead and Zinc Mining in Rajasthan, India, *World Archaeology*, 16 (2), 222 - 233, 1984.

Wilson, Jamal O., David Rosen, Brent A. Nelson and Jeannette Yen, The effects of biological examples in idea generation, *Design Studies*, 31 (2010)

169-186.

Wolf, A. M. and G. A. Colditz, Current estimates of the economic cost of obesity in the United States, *Obesity Research*, 6 (2), 97 - 106, 1998.

Worthington, Vivian, *The History of Yoga*, Routledge & Kegan Paul, 1982.

Wright, W., *A Short History of Syriac Literature*, Philo Press, Amsterdam, 1966 (first pub. 1894).

Xenophon, *Cyropaedia*, [Translator] W. Miller, Harvard University Press, Cambridge, 1960.

Yates, Frances Amelia, *Giordano Bruno and the Hermetic Tradition*, University of Chicago Press, Chicago, 1964.

Yoke, Ho Peng, *Li, Qi and Shu: An Introduction to Science and Civilization in China*, Dover Publications, inc., New York, 1985.

Zimmer, Henry, *Hindu Medicine*, The Johns Hopkins Press, Baltimore, 1948.

Zukav, Gary, *The Dancing Wu-Li Masters*, Fontana/Collins, London, 1979.

Zysk, Kenneth G., The Science of Respiration and the Doctrine of the Bodily Winds in Ancient India, *Journal of American Oriental Society*, 113 (2), 198 - 213, 1993.

Zysk, Kenneth, *Asceticism and healing in ancient India : medicine in the Buddhist monastery*, Oxford University Press, New York 1991.

# Index


Abū Ma'sher, 176
Adelard of Bath, 72, 75, 93, 96, 130, 175, 179, 357
*Agni-Purāṇa*, 45, 58, 59, 233
*Ahiṁsā*, 18, 269, 318
Al-Andalusī, Ṣā'id, 32
Al-Andalusī, Ṣā'id, 30, 72, 90, 176, 178, 349, 357, 358, 361
Al-Bīrūnī, 21, 32, 54, 68, 90, 140, 142, 150, 162, 172, 196, 211, 244, 348, 353, 358
Al-Battānī, 94, 130, 179
Al-Jāḥiz, 34, 72, 349, 357
Al-Khwārizmī, 32
Al-Khwārizmī, 17, 72, 75, 88, 174, 177, 179, 180, 357
Al-Kindī, 350
Al-Majrīṭī, 130, 133, 175, 179
Al-Mas'udī, 59, 357
Al-Uqlidīsī, 32
Al-Uqlidīsī, 68, 87, 357
*Almagest*, 131, 177, 179
*Āpastambā-Śulbasūtra*, 104, 129
Apollonius of Tyana, 62, 234, 264, 269, 270, 357, 380, 383
Apuleius, 387
Archimedes, 3, 62
Aristotle, 3, 11, 73, 157, 191, 204, 210, 221, 235, 359
Arjuna, 30, 36, 40, 41, 69, 119, 266
Arrian, 220, 382
*Arthaśāstra*, 40, 41, 51, 162, 189, 199, 216, 222, 240, 245
Āryabhaṭa I, 13, 24, 54, 59, 113, 145, 150, 189, 194
Aśoka, 50, 252, 318, 354
*Atharvaveda*, 64, 70, 103, 160, 189, 192, 215, 245, 250, 264, 286, 312, 323

Babylon, 4, 65, 87, 159, 203, 365
Bacon, Roger, 5, 72, 94, 137, 204, 232, 242, 349
*Bakhshālī*, 75
*Baudhayāna-Śulbasūtra*, 109, 119, 124, 187
*Bayt al-Ḥikma*, 10, 32, 174, 358
*Bhagavad-Gītā*, 30, 36, 171, 248, 266, 268, 358, 360, 367, 372
Bhārdvāja, 258
Bhīṣma, 272
*Bible*, 4, 169, 272
Biomimicry, 14, 16, 261, 281
Bonaparte, Napoleon, 9
Brahmgupta, 150, 174, 175, 177, 180, 360
Bṛhaspati, 147, 312

*Cakravyūha*, 119
*Candā-māmā*, 7

*Caraka-Saṁhitā*, 14, 42, 44, 244, 246, 247, 249, 263, 301, 302, 305–307, 309, 320, 322, 325, 327, 330, 337, 348, 351
*Chandāḥ-sūtra*, 74, 78, 82
*Chāndogya-Upaniṣad*, 24, 58, 102, 138, 150, 186, 187, 216, 220, 240, 260, 267, 289
Clement of Alexandria, 386
Columbus, 5, 95, 277, 356
Copernicus, 10, 13, 17, 149, 152, 174, 180, 210, 357, 360

Daśaratha, 30, 40
*Daśaharā*, 163
Democritus, 3, 4, 8, 17, 202, 209
*Dhātu*, 244, 304, 305, 351
*Dincarayā*, 19, 330
Dolomieu, 9
Duryodhana, 144

*Ebers Papyrus*, 9, 216
Egypt, 4, 8, 25, 31, 35, 46, 63, 91, 94, 203, 216, 230, 320, 364
Emerson, 27, 364, 372
Eusebius, 387

Fibbonacci, 17, 107
Fourier, 9
France, 356

Galileo, 46, 212, 376
Gaṅgā, 250, 252
*Gaṇita*, 102
*Gurukula*, 38

Hippocrates, 3, 8, 301
*Ḥisāb al-ghubār*, 104
*Holī*, 162
Hypatia, 46

Ibn al-Haytham (Alhazen), 94
Ibn Labbān, 72, 73, 357
Ibn Sīnā, 73, 94, 357
Indus-Sarasvatī, 25, 31, 34, 118, 159, 302, 317

Janaka, 41

Kaaba, 10
*Kāmasūtra*, 51
Kaṇāda, 1, 8, 14, 17, 59, 187, 197, 198, 201, 211
Kanaka, 32, 90, 176, 178
*Kathā-Upaniṣad*, 267, 282
*Kātyāyana-Śulbasūtra*, 104, 127, 128
Kaurava, 119, 144
Kauṭilaya, 41, 189, 199, 200, 216, 221, 223, 240, 273, 321, 351
Kepler, 174, 180, 212, 360
Ketu, 142, 143, 181, 362
*Khaṇḍa-Khādyaka*, 180, 360
Kṛṣṇa, 30, 36, 253, 268, 270, 356, 372

Lakṣmaṇa, 30, 313, 343
Lakṣmī, 18, 252, 253
Leucippus, 8

*Mahābhārata*, 15, 21, 30, 118, 119, 144, 205, 233, 251, 254, 258, 272, 360
*Manu-Saṁhitā*, 258, 270, 272, 305
Margenau, 8
*Mārkaṇdaya-purāṇa*, 187, 189
Megasthenes, 27, 220, 382

Nāgārjuna, 89, 219, 352, 354
*Nakṣatra*, 138, 165, 167, 181
Nālandā, 89, 150, 354, 377

Nārada, 24, 58, 94

*Pañca-karma*, 323, 332, 333
*Pañcatantra*, 233
Pāṇḍava, 118, 144
*Paramāṇu*, 189, 206, 207
Pārvatī, 253
Patañjali, 270, 281–283, 285, 358, 366, 375
Plastic surgery, 14, 19, 338, 344
Plato, 3, 4, 8, 73, 102, 192, 202, 264, 265, 270, 271, 348, 364
Plotinus, 73, 203, 264, 270, 357, 362, 380
*Prakṛti*, 249, 255, 272, 324
*Pṛthivī*, 197, 205, 249, 250, 323
Pythagoras, 3, 4, 8, 31, 33, 91, 126, 128, 203, 264, 268, 270, 357, 363, 365

Qin Dynasty, 11, 46

Rāhu, 142, 143, 145, 181, 182, 362
Rāma, 30, 40, 268, 313, 343
Rāmakṛṣṇa Parmhaṁsa, 36
Rāvaṇa, 70, 313, 343
*Rāmāyaṇa*, 15, 30, 212, 216, 233, 251, 313, 343
Redouté, 9
*Ṛgveda*, 31, 33, 35, 42, 45, 51, 64, 68, 78, 103, 138, 141, 144, 170, 173, 192, 215, 239, 245, 259, 282, 303, 311, 317, 338, 353

Saltpeter, 11, 232, 234, 236, 237
Sanatkumāra, 24
*Sapt-ṛṣi*, 7
Sarasvatī, 18, 270
*Śāstrārtha*, 16, 41, 42, 46, 70
Sebokht, 86
*Śilājīta*, 328
Sītā, 30, 252
Śiva, 20, 253, 268, 270
*Srimad-Bhāgvatam*, 171, 194, 195, 205, 207, 268
*Śulbasūtra*, 17, 109, 119, 127, 365
Sūrpaṇakhā, 343
*Suśruta-Saṁhitā*, 16, 215, 236, 245, 246, 301, 304, 307, 314, 337, 341, 346, 351
Sylvester, Pope, 96, 99, 357, 358, 360

Thales, 3, 8, 33, 202, 380, 384
Thoreau, 30, 269, 372
Tulsī Dās, 313

*Upaniṣad*, 15, 40, 244, 264, 266, 267, 303, 360, 363

*Vaiśeṣika-Sūtra*, 14, 17, 24, 59, 187, 192, 197, 201, 205, 206, 211, 271, 282
Varāhamihira, 32, 82, 133, 142, 150, 157, 172, 181, 182, 362
Vātsyāyana, 51
*Vāyu-Purāṇa*, 70, 205, 207
*Vedas*, 14, 24, 26, 31, 49, 65, 83, 150, 167, 169, 173, 189, 219, 233, 264, 322, 367
Viṣṇu, 144, 253, 373
*Viṣṇu-Purāṇa*, 149, 171, 213, 372
Viśvāmitra, 40, 307

Xenophon, 381

Yājñavalkya, 40, 41, 78
*Yajurveda*, 103, 138, 233, 240, 247
Yudhiṣṭhira, 30, 144, 248, 272

*Yuga*, 153, 165

*Zīj al-Sindhind*, 75, 112, 174, 175, 180

Made in the USA
Middletown, DE
18 February 2022

61509821R00258